TSINGHUA UNIVERSITY TEXTS

TSINGHUA UNIVERSITY TEXTS

Tsinghua University is one of the leading institutions in China. In Tsinghua University Texts, Tsinghua University Press and Springer will copublish texts originating from courses taught at Tsinghua University, either by lecturers of the University or lecturers from abroad. The Texts are intended for advanced undergraduate and graduate students but can also be used by researchers new to the field.

Tsinghua University Texts cover basic theories, techniques and methodologies, as well as advanced topics in current research areas. They focus on mathematics, statistics, physics, chemistry, geosciences, materials science, electrical engineering, computer sciences, life sciences, architecture, social sciences, and business & management. Other areas of contemporary interest will also be included.

Jiu Ding
Aihui Zhou

Statistical Properties of Deterministic Systems

With 4 figures

TSINGHUA UNIVERSITY PRESS

Springer

AUTHORS:

Jiu Ding
Ph.D., Professor
Department of Mathematics
The University of Southern Mississipi
Hattiesburg, MS 39406-0001, USA
Email: Jiu.Ding@usm.edu

Aihui Zhou
Ph.D., Professor
Academy of Mathematics and Systems Science
Chinese Academy of Sciences
Beijing 100080, China
Email: azhou@lsec.cc.ac.cn

ISBN 978-7-302-18296-1 **Tsinghua University Press, Beijing**
ISBN 978-3-540-85366-4 **Springer Berlin Heidelberg New York**
e ISBN 978-3-540-85367-1 **Springer Berlin Heidelberg New York**

Series ISSN 1993-0305 Tsinghua University Texts
Series e ISSN 1993-0313 Tsinghua University Texts

Library of Congress Control Number: 2008932537

Co-published by Tsinghua University Press, Beijing and Springer-Verlag GmbH Berlin Heidelberg

Springer is a part of Springer Science+Business Media
springer.com

Cover design: Frido Steinen-Broo, EStudio Calamar, Spain
Printed on acid-free paper

Dedicated to our respective families

Preface

Ergodic theory is a mathematical subject that studies the statistical properties of deterministic dynamical systems. It is a combination of several branches of pure mathematics, such as measure theory, functional analysis, topology, and geometry, and it also has applications in a variety of fields in science and engineering, as a branch of applied mathematics. In the past decades, the ergodic theory of chaotic dynamical systems has found more and more applications in mathematics, physics, engineering, biology and various other fields. For example, its theory and methods have played a major role in such emerging interdisciplinary subjects as computational molecular dynamics, drug designs, and third generation wireless communications in the past decade.

Many problems in science and engineering are often reduced to studying the asymptotic behavior of discrete dynamical systems. We know that in neural networks, condensed matter physics, turbulence in flows, large scale laser arrays, convection-diffusion equations, coupled mapping lattices in phase transition, and molecular dynamics, the asymptotic property of the complicated dynamical system often exhibits chaotic phenomena and is unpredictable. However, if we study chaotic dynamical systems from the statistical point of view, we find that chaos in the deterministic sense usually possesses some kind of regularity in the probabilistic sense. In this textbook, which is written for the upper level undergraduate students and graduate students, we study chaos from the statistical point of view. From this viewpoint, we mainly investigate the evolution process of density functions governed by the underlying deterministic dynamical system. For this purpose, we employ the concept of density functions in the study of the statistical properties of sequences of iterated measurable transformations. These statistical properties often depend on the existence and the properties of those probability measures which are absolutely continuous with respect to the Lebesgue measure and which are invariant under the transformation with respect to time. The existence of absolutely continuous invariant finite measures is equivalent to the existence of nontrivial fixed points of a class of stochastic operators (or Markov operators), called Frobenius-Perron operators by the great mathematician Stanislaw Ulam, who pioneered the exploration of nonlinear science, in his famous book "A Collection of Mathematical Problems" [120] in 1960.

In this book, we mainly study two kinds of problems. The first is the existence of nontrivial fixed points of Frobenius-Perron operators, and the other concerns the computation of such fixed points. They can be viewed as the functional analysis and the numerical analysis of Frobenius-Perron operators, respectively. For the first problem, many excellent books have been written, such as "Probabilistic Properties of Deterministic Systems" and its extended second edition "Chaos, Fractals, and Noise: Stochastic Aspects of Dynamics" by Lasota and Mackey [82], and "Law of Chaos: Invariant Measures and Dynamical Systems in One Dimension" by Boyarsky and Góra [14]. For the second problem, this book might be among the first ones in the form of a textbook on the computational ergodic theory of discrete dynamical systems. One feature that distinguishes this book from the others is that our textbook combines strict mathematical analysis and efficient computational methods as a unified whole. This is the authors' attempt to reduce the gap between pure mathematical theory and practical physical, engineering, and biological applications.

The first famous papers on the existence of nontrivial fixed points of Frobenius-Perron operators include the proof (see, e.g., Theorem 6.8.1 of [82]) of the existence of a unique smooth invariant measure for a second order continuously differentiable expanding transformation on a finite dimensional, compact, connected, smooth Riemann manifold by Krzyzewski and Szlenk [80] in 1969, and the pioneering work [83] on the existence of absolutely continuous invariant measures of piecewise second order differentiable and stretching interval mappings by Lasota and Yorke in 1973. The latter also answered a question posed by Ulam in his above mentioned book. In the same book, Ulam proposed a piecewise constant approximation method which became the first approach to the numerical analysis of Frobenius-Perron operators. A solution to Ulam's conjecture by Tien-Yien Li [86] in 1976 is a fundamental work in the new area of *computational ergodic theory.*

Our book has nine chapters. As an introduction, Chapter 1 leads the reader into a mathematical trip from order to chaos via the iteration of a one-parameter family of quadratic polynomials with the changing values of the parameter, from which the reader enters the new vision of "chaos from the statistical point of view." The fundamental mathematical knowledge used in the book – basic measure theory and functional analysis– constitutes the content of Chapter 2. In Chapter 3, we study the basic concepts and classic results in ergodic theory. The main linear operator studied in this book – the Frobenius-Perron operator – is introduced in Chapter 4, which also presents some general results that have not appeared in other books. Chapter 5 is exclusively devoted to the investigation of the existence problem of absolutely continuous invariant measures, and we shall prove several existence results for various classes of one-dimensional mappings and multi-dimensional transformations. The computational problem is studied in Chapter 6, in which two numerical methods are given for the approximation of Frobenius-Perron operators. One is the classic Ulam's piecewise constant method, and the other is its improvement with higher order approximation accuracy; that is, the piecewise linear Markov method which was mainly developed by the authors. In Chapter 7, we present Keller's result on the stability of Markov operators and its application to the convergence rate analysis of

Ulam's method under the L^1-norm and Murray's work for a more explicit upper bound of the error estimate. We also explore the convergence rate under the variation norm for the piecewise linear Markov method. Chapter 8 gives a simple mathematical description of the related concepts of entropy, in particular the Boltzmann entropy and its relationship with the iteration of Frobenius-Perron operators. Several modern applications of absolutely continuous invariant probability measures will be given in the last chapter.

This book can be used as a textbook for students of pure mathematics, applied mathematics, and computational mathematics as an introductory course on the ergodic theory of dynamical systems for the purpose of entering the related frontier of interdisciplinary areas. It can also be adopted as a textbook or a reference book for a specialized course for different areas of computational science, such as computational physics, computational chemistry, and computational biology. For students or researchers in engineering subjects such as electrical engineering, who want to study chaos and applied ergodic theory, this book can be used as a tool book. A good background of advanced calculus is sufficient to read and understand this book, except possibly for Section 2.4 on the modern definition of variation and Section 5.4 on the proof of the existence of multi-dimensional absolutely continuous invariant measures which may be omitted at the first reading. Some of the exercises at the end of each chapter complement the main text, so the reader should try to do as many as possible, or at least take a look and read appropriate references if possible. Each main topic of ergodic theory contains matter for huge books, but the purpose of this book is to introduce as many readers as possible with various backgrounds into fascinating new fields having great potential of ever increasing applications. Thus, our presentation is quite concise and elementary and as a result, some important but more specialized topics and results must be omitted, which can be found in other monographs.

Another feature of this textbook is that it contains much of our own joint research in the past fifteen years. In this sense it is like a monograph. Our joint research has been supported by the National Science Foundation of China, the National Basic Research Program of China, the Academy of Mathematics and Systems Science at the Chinese Academy of Sciences, the State Key Laboratory of Scientific and Engineering Computing at the Chinese Academy of Sciences, the Chinese Ministry of Education, the China Bridge Foundation at the University of Connecticut, and the Lucas Endowment for Faculty Excellence at the University of Southern Mississippi, among the others, for which we express our deep gratitude.

Jiu Ding would also like to thank his Ph.D. thesis advisor, University Distinguished Professor Tien-Yien Li of Michigan State University. It is Dr. Li's highly educative graduate course "Ergodic Theory on [0, 1]" for the academic year 1988-1989, based on the lecture notes [87] delivered at Kyoto University of Japan one year earlier, that introduced him into the new research field of computational ergodic theory and led him to write a related Ph.D. dissertation. Aihui Zhou is very grateful to his Ph.D. thesis advisor, Academician Qun Lin, of the Chinese Academy of Sciences, who with a great insight, encouraged him to enter this wide and exciting research area.

The first edition of this book was published in Chinese by the Tsinghua University Press in Beijing, China in January 2006 and reprinted in December in the same year. We thank editors Xiaoyan Liu, Lixia Tong, and Haiyan Wang and five former Ph.D. students of Aihui Zhou, Xiaoying Dai, Congming Jin, Fang Liu, Lihua Shen, and Ying Yang for their diligent editorial work and technical assistance, which made the fast publication of the Chinese edition possible. We thank Lixia Tong for her help during the preparation of this revised and expanded English edition of the book.

Jiu Ding and Aihui Zhou
Beijing, March 2008

Contents

Chapter 1
Introduction

JiuDing, Department of Mathematics, The University of Southern Mississippi in America.
Aihui Zhou, Academy of Mathematics and Systems Science, Chinese Academy of Sciences.

Abstract Using the famous logistic model $S_r(x) = rx(1-x)$ as an example, we give a brief survey of discrete dynamical systems for the purpose of leading the reader on a mathematical trip from order to chaos, and then we introduce basic ideas behind the statistical study of chaos, which is the main topic of the book.
Keywords Logistic model, period-doubling bifurcation, Li-Yorke chaos, Frobenius-Perron operator, absolutely continuous invariant measure.

In the modern statistical study of discrete deterministic dynamical systems and its applications to physical sciences, there are two important and mutually related problems. On the theoretical part, there is the problem of the *existence* of absolutely continuous invariant measures that give the statistical properties of the dynamics, such as the probability distribution of the orbits for almost all initial points and the speed of the decay of correlations. On the practical part, we encounter the problem of the *computation* of such invariant measures to any prescribed precision in order to numerically explore the chaotic behavior in many physical systems. In this textbook, we try to address these two problems. For this purpose, we need to study a class of positive linear operators, called Frobenius-Perron operators, that describe the density evolution governed by the underlying dynamical system. Density functions, which are the fixed points of Frobenius-Perron operators, define absolutely continuous invariant probability measures associated with the deterministic dynamical system, which can be numerically investigated via structure preserving computational methods that approximate such fixed density functions.

Before we begin to study the statistical properties of discrete dynamical systems, we first review the deterministic properties of one-dimensional mappings in this introductory chapter as a starting point. The well-known logistical model, which has played an important role in the history of the evolution of the concept of chaos in science and mathematics, will be studied in detail from the deterministic point of view. Then, we are naturally led to the statistical study of chaos by introducing the concept of Frobenius-Perron operators with an intuitive approach, which motivates the main topic of this book.

1.1 Discrete Deterministic Systems—from Order to Chaos

In their broad sense, dynamical systems provide rules under which phenomena (states) in the mathematical or physical world evolve with respect to time. Differential equations are widely used to model continuous time dynamical systems in many areas of science, such as classical mechanics, quantum mechanics, neural networks, mathematical biology, etc., as these equations describe mathematically the laws by which they are governed. Transformations on phase spaces not only determine a discrete time dynamical system[23], but also form the basis of investigating continuous time dynamical systems via such mathematical tools as the Poincare' map. Even simple nonlinear transformations may exhibit a quasi-stochastic or unpredictable behavior which is a key feature of the chaotic dynamics. Poincaré deduced this kind of chaotic motion for the three-body problem in celestial mechanics about fifty years before the advent of electronic computers in the 1940s, and about eighty years before Tien-Yien Li and James A. Yorke first coined the term "Chaos" in their seminar paper "Period Three Implies Chaos" [88] in 1975.

The discrete time evolution of a dynamical system in the N-dimensional Euclidean space $\mathbb{R}^N$ is usually given by a first order difference equation which is often written as a recurrence relation

$$\boldsymbol{x}_{n+1} = \boldsymbol{S}(\boldsymbol{x}_n), \quad n = 0, 1, \cdots,$$

where $\boldsymbol{S}$ is a transformation from a subset Ω of $\mathbb{R}^N$ into itself. For example, consider a population of organisms for which there is a constant supply of food and limited space, and no predators. In order to model the populations in successive generations, let x_n denote the population of the nth generation, and adjust the numbers so that the capacity of the environment is equal to 1, which means that $0 \leqslant x_n \leqslant 1$ for all n. One popular formula for the dynamics of the population is the so-called *logistic model*, after the differential equation studied by the Belgian mathematician Pierre F. Verhulst about 160 years ago [98]:

$$x_{n+1} = rx_n(1 - x_n), \quad n = 0, 1, \cdots,$$

where $r \in (0, 4]$ is a parameter. In the following, we study the deterministic properties of this logistical model to some extent when the parameter r varies from 0 to 4 and see how the dynamics will change from the regular behavior to the chaotic behavior as r increases toward 4.

First, we introduce some standard terms in discrete dynamical systems . Let X be a set and $S : X \to X$ be a transformation. A point $x \in X$ is called a *fixed point* of S if $S(x) = x$ and an *eventually fixed point* of S if there is a positive integer k such that $S^k(x)$ is a fixed point of S, where $S^k(x) = S(S(\cdots(S(x))\cdots))$ (i.e., S^k is the composition of S with itself $k - 1$ times) is the kth iterate of x. A point $x_0 \in X$ is called a *periodic point* of S with period $n \geqslant 1$ or a *period-n point* of S if $S^n(x_0) = x_0$ and if in addition, x_0, $S(x_0), S^2(x_0), \cdots, S^{n-1}(x_0)$

are distinct. A fixed point is a periodic point with period 1. An *eventually periodic point* is a point whose kth iterate is a periodic point for some $k > 0$. The *orbit* of an initial point x_0 is the sequence

$$x_0, S(x_0), S^2(x_0), \cdots, S^n(x_0), \cdots$$

of the iterates of x_0 under S. If x_0 is a period-n point, then the orbit

$$x_0, S(x_0), \cdots, S^{n-1}(x_0), \cdots$$

of x_0 is a *periodic orbit* which can be represented by $\{x_0, S(x_0), \cdots, S^{n-1}(x_0)\}$ called an *n-cycle* of S.

From the mean value theorem of calculus, a fixed point x of a differentiable mapping S of an interval is *attracting* or *repelling* if $|S'(x)| < 1$ or $|S'(x)| > 1$, respectively. Similarly, a period-n point x_0 of S is attracting or repelling when $|(S^n)'(x_0)| < 1$ or $|(S^n)'(x_0)| > 1$ respectively, and the corresponding n-cycle is *attracting* or *repelling*. Such information only gives the local dynamical properties of a fixed point or a periodic orbit, not the global ones which need more subtle arguments and more thorough analysis to obtain in general.

Now, we begin to study the iteration of the logistic model. Let

$$S_r(x) = rx(1-x), \quad \forall\, x \in [0,1],$$

where the parameter $r \in (0,4]$ so that S_r maps $[0,1]$ into itself. It is obvious that S_r has one fixed point 0 when $0 < r \leqslant 1$ and two fixed points 0 and $p_r \equiv 1-1/r$ when $r > 1$. Since $S_r'(0) = r$ and $S_r'(p_r) = 2-r$, one can see that the fixed point 0 is attracting for $r \leqslant 1$ and repelling for $r > 1$, and the fixed point p_r is attracting for $1 < r \leqslant 3$ and repelling for $r > 3$. In the remaining part of this section, we study the global properties of the fixed points and possible periodic points in more detail.

As will be shown below, the dynamics of S_r changes as the parameter r passes through each of the values 1, 2, 3, $1+\sqrt{6}, \cdots$, called the *bifurcation points* of the one-parameter family $\{S_r\}$ of the quadratic mappings, that is, the number and nature of the fixed points and/or the periodic points change when r passes through each of them. Hence, our discussion below will be split into four cases, from easy to more complicated ones. They are respectively $0 < r \leqslant 1$, $1 < r \leqslant 2$, $2 < r \leqslant 3$, and $3 < r \leqslant 4$. In the analysis, we often use the simple fact that the limit x^* of a convergent sequence $\{x_n\}$ of the iterates of a continuous mapping S must be a fixed point of S if x^* is in the domain of S.

Case 1. $0 < r \leqslant 1$ (see Figure 1.1).

Since $0 < S_r(x) = rx(1-x) < x$ for $0 < x < 1$, the iteration sequence $\{S_r^n(x)\}$ is positive and monotonically decreasing, and so it converges to the

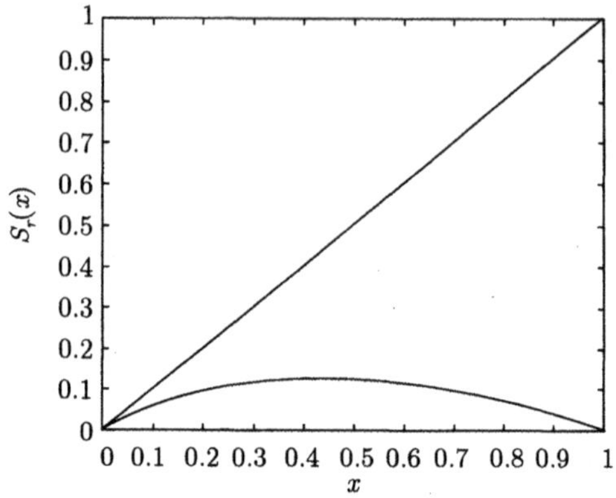

Figure 1.1 S_r at $r = 0.5$

unique fixed point 0 of S_r as n approaches infinity. It follows that the *basin of attraction* of 0, which is the set of all the initial points whose orbit converges to the fixed point 0 by definition, is the closed interval $[0, 1]$. So there are no periodic points except for the unique fixed point 0.

Case 2. $1 < r \leqslant 2$ (see Figure 1.2).

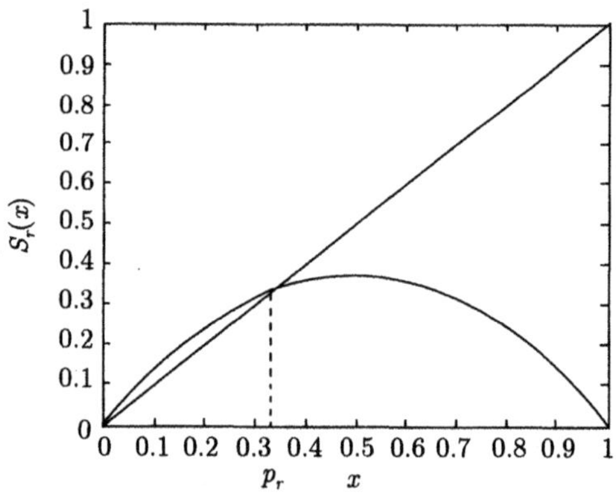

Figure 1.2 S_r at $r = 1.5$

Now, S_r has two fixed points, 0 and $p_r = 1 - 1/r$. We know that the fixed point 0 is repelling and the fixed point p_r is attracting. Let $0 < x < p_r$. Then, $1/r < 1 - x$, so $x < rx(1-x) = S_r(x)$. By induction we see that $x < S_r(x) < \cdots < S_r^n(x) < \cdots$. On the other hand, since S_r is strictly increasing on $[0, p_r]$,

$$S_r(x) < S_r(p_r) = p_r,$$

which implies that $S_r^n(x) < p_r$ for all n. Thus, the sequence $\{S_r^n(x)\}$ is strictly increasing, bounded above by p_r, and hence it converges to the fixed point p_r. Similarly, if $p_r < x \leqslant 1/2$, then $\{S_r^n(x)\}$ is a monotonically decreasing sequence bounded below by p_r, so it also converges to p_r. Finally, if $1/2 < x < 1$, then $0 < S_r(x) \leqslant 1/2$, so by the above argument, $\{S_r^n(x)\}$ converges to p_r. Therefore, when $1 < r \leqslant 2$, the basin of attraction of the fixed point p_r is the open interval $(0, 1)$, the basin of attraction of the fixed point 0 is the 2-point set $\{0, 1\}$, and there are no other periodic points besides the two fixed points.

Case 3. $2 < r \leqslant 3$. (see Figure 1.3).

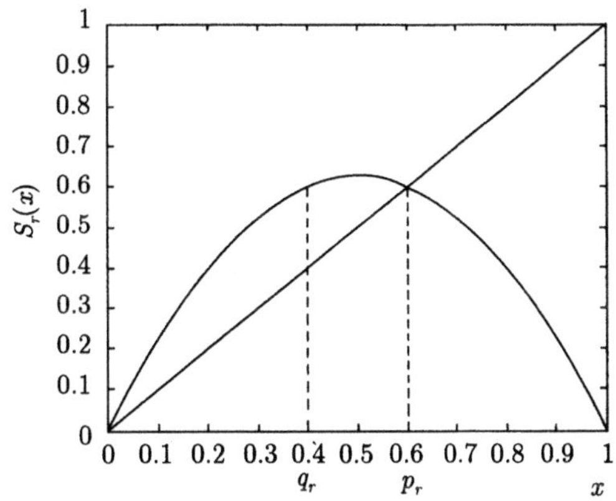

Figure 1.3 S_r at $r = 2.5$

When $r > 2$, the fixed point $p_r > 1/2$. Assume that $r < 3$ and let q_r be the unique number in $(0, 1/2)$, which is symmetric to p_r about $1/2$, such that $S_r(q_r) = S_r(p_r) = p_r$. Then, using the geometry of the graph of S_r and the fact that $q_r \leqslant S_r(r/4)$, one can show that (see Exercise 1.1):

(i) if $x \in (0, q_r)$, then x has an iterate $> q_r$;

(ii) if $q_r < x \leqslant p_r$, then $p_r \leqslant S_r(x) \leqslant r/4$;

(iii) if $p_r < x \leqslant r/4$, then $q_r \leqslant S_r(x) < p_r$;

(iv) if $r/4 < x < 1$, then $0 < S_r(x) < p_r$.

From (i)-(iv) it follows that if $0 < x < 1$, then x has an iterate in the interval $(q_r, p_r]$. Moreover, (ii) and (iii) imply that the iterates of x oscillate between the intervals $(q_r, p_r]$ and $[p_r, r/4]$. Thus,

(v) if x is in $(q_r, p_r]$, then so is the sequence $\{S_r^{2n}(x)\}$;

(vi) if x is in $[p_r, r/4]$, then so is the sequence $\{S_r^{2n}(x)\}$.

Since 0 and p_r are the fixed points of S_r, a simple calculation shows that

$$S_r^2(x) - x = rx(x - p_r)\left[-r^2x^2 + \left(r^2 + r\right)x - r - 1\right]. \tag{1.1}$$

The expression inside the brackets has no real roots when $2 < r < 3$. Therefore, if $2 < r < 3$, then the only fixed points of S_r^2 are 0 and p_r. Since $S_r^2(x) - x$ has no roots in (q_r, p_r), it has the same sign as $S_r^2(1/2) - 1/2$ which is positive. Consequently $x < S_r^2(x)$ for all $x \in (q_r, p_r)$, and by (v) the sequence $\{S_r^{2n}(x)\}$

is monotonically increasing, lies in $(q_r, p_r]$, and converges to the only positive fixed point p_r of S_r^2. Using the continuity of S_r, we find that

$$S_r^{2n+1}(x) = S_r(S_r^{2n}(x)) \to S_r(p_r) = p_r$$

as n increases without bound. Therefore, $S_r^n(x) \to p_r$ whenever $x \in (q_r, p_r]$. Since every x in $(0,1)$ has an iterate in $(q_r, p_r]$, we conclude that $S_r^n(x) \to p_r$ as n increases without bound, for all $x \in (0,1)$. In other words, the basin of attraction of p_r is $(0,1)$, so the basin of attraction of 0 is $\{0,1\}$. A consequence of this result is that there are no periodic points for S_r other than the fixed points. The same conclusion can be proven for $r = 3$ with a more careful analysis.

Case 4. $3 < r \leqslant 4$ (see Figure 1.4).

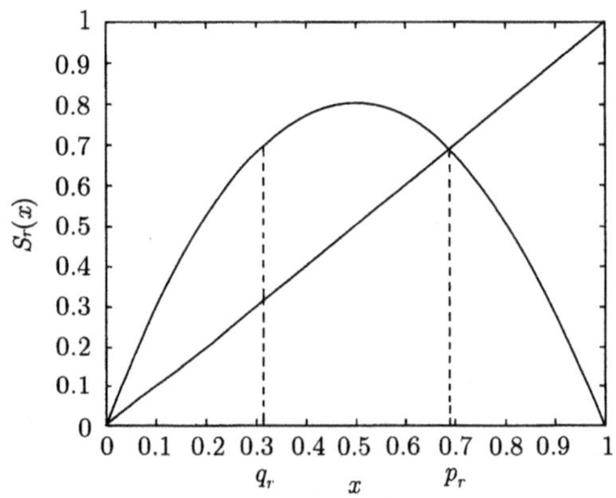

Figure 1.4 S_r at $r = 3.2$

We have learned that the dynamics of S_r is regular when $0 < r \leqslant 3$, and in particular the only periodic points are fixed points. When $3 < r \leqslant 4$, both 0 and $p_r = 1 - 1/r$ are repelling fixed points. Do the iterates of other points in $(0,1)$ converge, or oscillate, or have no pattern at all? Are there periodic points other than 0 and p_r? The analysis of the dynamics of S_r becomes more and more complicated as r increases from 3 to 4. We only study the case $3 < r < 1 + \sqrt{6}$ in detail and list the main results that follow.

For our purpose, we need to study the dynamics of S_r^2. When $r = 3$, the graph of S_r^2 is tangent to the diagonal $y = x$ at the point (p_r, p_r). From (1.1), the other two fixed points of S_r^2 besides 0 and p_r are the real roots of the quadratic equation

$$-r^2x^2 + (r^2 + r)x - r - 1 = 0,$$

which are

$$s_r = \frac{1}{2} + \frac{1}{2r} - \frac{1}{2r}\sqrt{(r-3)(r+1)} \text{ and } t_r = \frac{1}{2} + \frac{1}{2r} + \frac{1}{2r}\sqrt{(r-3)(r+1)}.$$

Since 0 and p_r are the only fixed points of S_r for $r > 1$, it is obvious that $\{s_r, t_r\}$ is a 2-cycle for $r > 3$. After a simple computation, we find that

$$(S_r^2)'(s_r) = S_r'(s_r)S_r'(t_r) = (r - 2rs_r)(r - 2rt_r) = -r^2 + 2r + 4.$$

Since $|-r^2 + 2r + 4| < 1$ if and only if $3 < r < 1 + \sqrt{6}$, the 2-cycle $\{s_r, t_r\}$ is attracting if $3 < r < 1 + \sqrt{6}$. It can further be shown that the basin of attraction of the 2-cycle $\{s_r, t_r\}$ consists of all $x \in (0, 1)$ except for the fixed point p_r and the points whose iterates are eventually p_r.

When $r > 1 + \sqrt{6}$, the 2-cycle $\{s_r, t_r\}$ becomes repelling. As we may expect, an attracting 4-cycle is born. Actually, there exists a sequence $\{r_n\}$ of the so-called *period-doubling* bifurcation values for the parameter r, with $r_0 = 3$ and $r_1 = 1 + \sqrt{6}$, such that

- if $r_0 < r \leqslant r_1$, then S_r has two repelling fixed points and one attracting 2-cycle;
- if $r_1 < r \leqslant r_2$, then S_r has two repelling fixed points, one repelling 2-cycle, and one attracting 2^2-cycle;
- if $r_2 < r \leqslant r_3$, then S_r has two repelling fixed points, one repelling 2-cycle, one repelling 2^2-cycle, and one attracting 2^3-cycle;

In general, for $n = 1, 2, \cdots$,

- if $r_{n-1} < r \leqslant r_n$, then S_r has two repelling fixed points, one repelling 2^k-cycle for $k = 1, 2, \cdots, n-1$, and one attracting 2^n-cycle.

It is well-known that $\lim_{n\to\infty} r_n = r_\infty = 3.561547\cdots$. This number r_∞ is called the *Feigenbaum number* for the quadratic family $\{S_r\}$. Moreover, the sequence $\{c_n\}$ of the ratios

$$c_n = \frac{r_n - r_{n-1}}{r_{n+1} - r_n}$$

converges to a number $c_\infty = 4.669202\cdots$, which is called the *universal constant* since for many other families of one-humped mappings, the bifurcations occur in such a regular fashion that the ratios of the distances between successive pairs of the bifurcation points approach the very same constant c_∞! This universal constant c_∞ is also referred to as the *Feigenbaum constant* because the physicist Michael Feigenbaum first found it and its universal property in 1978.

So far the dynamics of the quadratic family $\{S_r\}$ is still regular for $0 < r < r_\infty$ since every point $x \in (0, 1)$ is periodic, eventually periodic, or attracted to a fixed point or a periodic orbit. So, the eventual behavior of the orbits is *predictable*. When $r \geqslant r_\infty$, there could exhibit a complicated irregular or chaotic behavior for the dynamics of S_r. For example, if $3.829 \leqslant r \leqslant 3.840$, then S_r has period-3 points. The celebrated *Li-Yorke theorem* [88] says that if a continuous mapping S from an interval I into itself has a period-3 point, then it has a period-k point for any natural number k, and there is an uncountable set $\Lambda \subset I$, containing no periodic points, which satisfies the following conditions:

(i) For every pair of distinct numbers $x,\ y \in \Lambda$,

$$\limsup_{n\to\infty} |S^n(x) - S^n(y)| > 0 \quad \text{and} \quad \liminf_{n\to\infty} |S^n(x) - S^n(y)| = 0.$$

(ii) For every $x \in \Lambda$ and each periodic point $p \in I$,

$$\limsup_{n\to\infty} |S^n(x) - S^n(p)| > 0.$$

Thus, from the Li-Yorke theorem, the eventual behavior of the iterates of S_r with $3.829 \leqslant r \leqslant 3.840$ is *unpredictable.*

The case $r = 4$ is worth a special attention. It is well-known [7] that S_4 is *topologically conjugate* to the *tent function*

$$T(x) = \begin{cases} 2x, & \text{if } x \in \left[0, \dfrac{1}{2}\right], \\ 2(1-x), & \text{if } x \in \left[\dfrac{1}{2}, 1\right]. \end{cases} \tag{1.2}$$

That is, there is a homeomorphism $h : [0,1] \to [0,1]$ such that $S_4 \circ h = h \circ T$. Since T has a 3-cycle $\{2/7, 4/7, 6/7\}$, there is a period-3 orbit for S_4. By the Li-Yorke theorem, S_4 is chaotic. As a matter of fact, if we randomly pick an initial point $x_0 \in [0,1]$, then the *limit set* of the sequence $\{x_n\}$ with $x_n = S_4^n(x_0)$ is the whole interval $[0,1]$, that is, for each $x \in [0,1]$, there is a subsequence $\{x_{n_k}\}$ of $\{x_n\}$ such that $\lim_{k\to\infty} x_{n_k} = x$.

Chaotic dynamical systems are now very popular in science and engineering. Besides the original definition of Li-Yorke chaos in [88], there have been various definitions for "chaos" in the literature, and the most often used one is given by Devaney in [27]. Although there is no universal definition for chaos, the essential feature of chaos is *sensitive dependence on initial conditions* so that the eventual behavior of the dynamics is unpredictable. The theory and methods of chaotic dynamical systems have been of fundamental importance not only in mathematical sciences [22, 23, 27], but also in physical, engineering, biological, and even economic sciences [7, 18, 94, 98].

We have examined a family of discrete dynamical systems from the deterministic point of view and have observed the passage from order to chaos as the parameter value of the mappings changes. In the next section, we study chaos from another point of view, that is, from the probabilistic viewpoint.

1.2 Statistical Study of Chaos

Although a chaotic dynamical system exhibits unpredictability concerning the asymptotic behavior of the orbit starting from a generic point, it often behaves regularly as far as the statistical properties are concerned. In other words, a chaotic dynamical system in the deterministic sense may not be chaotic in the probabilistic sense.

In physical measurements, we often consider a probabilistic distribution of a physical quantity. Let $S : X \to X$ be a dynamical system on a phase space X of finite measure $\mu(X) < \infty$, and let A be a subset of X. Instead of observing the deterministic properties of individual orbits, let us consider the probabilistic properties by observing the *frequencies* of the first n terms of the orbit $\{S^n(x)\}$ of an initial point x that enter A for all natural numbers n. To calculate the frequency, let χ_A be the *characteristic function* of A, that is,

$$\chi_A(x) = \begin{cases} 1, & \text{if } x \in A, \\ 0, & \text{if } x \notin A. \end{cases} \tag{1.3}$$

Then, the frequency for a given n is exactly $n^{-1} \sum_{i=0}^{n-1} \chi_A(S^i(x))$. The *time average* or the *time mean*, which is the *asymptotic frequency* of all the terms of an orbit starting at $x \in X$ that enter A, is given by the limit

$$\lim_{n \to \infty} \frac{1}{n} \sum_{i=0}^{n-1} \chi_A(S^i(x))$$

if it exists, which measures how frequently the orbit stays in A. The classical ergodic theory deals with the existence of the time average, their metric properties, and their close relationships with other mathematical concepts and quantities, which originated from *Boltzmann's ergodic hypothesis* in statistical mechanics. In our context, this hypothesis concerns the following question: given a *measure preserving* transformation $S : X \to X$, i.e., $\mu(S^{-1}(A)) = \mu(A)$ for all measurable subsets A of X, and an integrable function $f : X \to \mathbb{R}$, find the conditions under which the limit

$$\lim_{n \to \infty} \frac{1}{n} \sum_{k=0}^{n-1} f(S^k(x)) \tag{1.4}$$

exists and is constant for $x \in X$ almost everywhere (a.e.).

In 1931, George D. Birkhoff proved that for any S and f the limit (1.4) exists for $x \in X$ almost everywhere, and furthermore, if S is *ergodic*, that is, $S^{-1}(A) = A$ implies that $A = \varnothing$ or X a.e., then the time average coincides with the *space average* or the *space mean*

$$\frac{\mu(A)}{\mu(X)} = \frac{1}{\mu(X)} \int_X \chi_A \mathrm{d}\mu$$

for $x \in X$ a.e. More specifically, the celebrated Birkhoff pointwise ergodic theorem [123] can be stated as follows (c.f. Theorem 3.3.1 in Chapter 3):

Theorem 1.2.1 (Birkhoff's pointwise ergodic theorem) *Let μ be a probability measure on X which is invariant under $S : X \to X$. Then, for any integrable function f defined on X and almost all $x \in X$, the time average*

$$\lim_{n\to\infty} \frac{1}{n} \sum_{i=0}^{n-1} f(S^i(x))$$

exists and is denoted as $\tilde{f}(x)$. Moreover,

$$\tilde{f}(S(x)) = \tilde{f}(x), \quad \forall\, x \in X \ \mu-\text{a.e.}$$

If in addition S is ergodic, then $\tilde{f}$ is the constant function $\int_X f d\mu$.

Now, another question arises naturally: given a transformation $S : X \to X$, what measure μ on X is invariant under S? If we do not impose more requirements for μ, the answer may be trivial or of no physical importance. For example, for the logistic model $S(x) = 4x(1-x)$, since 0 is a fixed point of S, it is easy to see that the *Dirac measure* δ_0 concentrated at 0 is invariant, where $\delta_0(A) = 1$ if $0 \in A$ and $\delta_0(A) = 0$ if $0 \notin A$. In general, any fixed point a of S gives rise to an invariant measure δ_a, the Dirac measure concentrated at a. Note that the Dirac measure δ_a with $a \in [0,1]$ is *not* absolutely continuous with respect to the Lebesgue measure of the unit interval. In other words, it cannot be represented as the integral of an integrable function on $[0,1]$.

The existence of an invariant measure for a continuous transformation on a compact metric space has been established by the following theorem [123], which will be proved in Section 3.4.

Theorem 1.2.2 (Krylov-Bogolioubov) *Let X be a compact metric space and let $S : X \to X$ be a continuous transformation. Then, there is an invariant probability measure μ under S.*

In many applications, we are more interested in the existence and computation of invariant probability measures which are *absolutely continuous* with respect to a given measure. In other words, we want to find invariant measures that can be expressed as integrals of *density functions* with respect to the given measure. In this textbook, we intend to study this problem. Here the concept of Frobenius-Perron operators, which gives the corresponding way the density functions change under the deterministic dynamical system, plays an important role. Considering the iteration of the Frobenius-Perron operator leads us to the following observation: *chaos in the deterministic sense may not be so in the probabilistic sense.*

Let (X, Σ, μ) be a σ-finite measure space, let $S : X \to X$ be a nonsingular transformation, i.e., $\mu(A) = 0$ implies $\mu(S^{-1}(A)) = 0$ for all $A \in \Sigma$, and let

$P : L^1(X) \to L^1(X)$ be the Frobenius-Perron operator associated with S which is defined *implicitly* by the relation

$$\int_A Pf \mathrm{d}\mu = \int_{S^{-1}(A)} f \mathrm{d}\mu, \quad \forall\, A \in \Sigma, \tag{1.5}$$

where $L^1(X)$ is the space of all integrable functions defined on X with respect to the measure μ (see Chapters 2 and 4 for their precise definitions). In Chapter 4, it will be proved that any fixed point f of P, which is also a density function, gives an absolutely continuous S-invariant probability measure μ_f on X defined by $\mu_f(A) = \int_A f \mathrm{d}\mu, \ \forall\, A \in \Sigma$.

The *existence* problem of fixed density functions of Frobenius-Perron operators is one of the main topics in modern ergodic theory. On the other hand, in physical sciences, one often needs to *compute* one or higher dimensional absolutely continuous invariant finite measures [7]. For example, in neural networks, condensed matter physics, turbulence in fluid flow, arrays of Josephson junctions, large-scale laser arrays, reaction-diffusion systems, etc., "coupled map lattices" often appear as models for phase transition, in which the evolution and convergence of density functions under the action of the Frobenius-Perron operator are examined. Understanding the statistical properties of these systems will become possible if we are able to calculate such global statistical quantities as invariant measures, entropy, Lyarpulov exponents, and moments. Thus, in many applied areas of physical sciences, not only the existence but also the computation of fixed density functions of Frobenius-Perron operators is essential for the investigation of the complicated dynamics.

However, the following two main difficulties make solving the above problems a challenge. First, the underlying space $L^1(X)$ is not reflexive in general, and second, the Frobenius-Perron operator P is usually not compact on $L^1(X)$. Thus, we can only apply some special techniques and the structure analysis to prove the existence and to develop convergent computational algorithms.

We use the following probabilistic argument to motivate the definition (1.5) of Frobenius-Perron operators, before we formally define this operator in Chapter 4. Consider again the dynamical system $S(x) = 4x(1-x)$. Instead of studying the eventual behavior of individual orbits, we investigate the asymptotic distribution of the iterates on $[0,1]$ under S. In other words, we examine the flow of density functions of these iterates' distributions if the density function of the initial distribution is known. Here, we give an intuitive description of this approach. Pick a large positive integer n and apply S to each of the n initial states

$$x_1^0,\ x_2^0,\ \cdots,\ x_n^0,$$

and then we have n new states

$$x_1^1 = S(x_1^0),\ x_2^1 = S(x_2^0),\ \cdots,\ x_n^1 = S(x_n^0).$$

The initial states can be represented by a function f_0 in the sense that the integral of f_0 over any interval I (not too small) is roughly the fraction of the number of the states in the interval, that is,

$$\int_I f_0(x)\mathrm{d}x \simeq \frac{1}{n}\sum_{i=1}^{n}\chi_I(x_i^0).$$

f_0 is called the *density function* of the initial states. Similarly, the density function f_1 for the states $x_1^1, x_2^1, \cdots, x_n^1$ satisfies

$$\int_I f_1(x)\mathrm{d}x \simeq \frac{1}{n}\sum_{i=1}^{n}\chi_I(x_i^1).$$

Our purpose is to find a relation between f_1 and f_0.

For the given $I \subset [0,1]$,

$$x_i^1 \in I \quad \text{if and only if} \quad x_i^0 \in S^{-1}(I).$$

Thus, from the equality $\chi_I(S(x)) = \chi_{S^{-1}(I)}(x)$, we have

$$\int_I f_1(x)\mathrm{d}x \simeq \frac{1}{n}\sum_{i=1}^{n}\chi_{S^{-1}(I)}(x_i^0),$$

which implies that

$$\int_I f_1(x)\mathrm{d}x = \int_{S^{-1}(I)} f_0(x)\mathrm{d}x.$$

If we write f_1 as Pf_0, then the above relationship between f_1 and f_0 is

$$\int_I Pf_0(x)\mathrm{d}x = \int_{S^{-1}(I)} f_0(x)\mathrm{d}x.$$

The operator P that maps the density function f of the initial states to the density function Pf of the next states is actually the Frobenius-Perron operator corresponding to the transformation S, as defined by (1.5).

Let $I = [0,x]$. Then, differentiating both sides of the equality

$$\int_0^x Pf(t)\mathrm{d}t = \int_{S^{-1}([0,x])} f(t)\mathrm{d}t$$

with respect to x gives

$$Pf(x) = \frac{\mathrm{d}}{\mathrm{d}x}\int_{S^{-1}([0,x])} f(t)\mathrm{d}t.$$

Since

$$S^{-1}([0,x]) = \left[0, \frac{1}{2} - \frac{1}{2}\sqrt{1-x}\right] \cup \left[\frac{1}{2} + \frac{1}{2}\sqrt{1-x}, 1\right],$$

after carrying out the indicated differentiation, we obtain

$$Pf(x) = \frac{1}{4\sqrt{1-x}}\left[f\left(\frac{1}{2} - \frac{1}{2}\sqrt{1-x}\right) + f\left(\frac{1}{2} + \frac{1}{2}\sqrt{1-x}\right)\right].$$

which is an explicit formula for the Frobenius-Perron operator corresponding to the quadratic mapping S. This formula tells us how S *transforms* a given density function f into a new density function Pf. In particular, if the initial density function $f(x) \equiv 1$, that is, if the initial distribution of the states is uniform, then the distribution of the new states under S is given by the density function

$$Pf(x) = \frac{1}{2\sqrt{1-x}}.$$

If we keep iterating, we can see that the density function sequence $\{P^n f(x)\}$ will approach the density function

$$f^*(x) = \frac{1}{\pi\sqrt{x(1-x)}}$$

as $n \to \infty$, which satisfies $Pf^* = f^*$. This fixed density function for the logistic model $S(x) = 4x(1-x)$ was found by Ulam and von Neumann [121] in 1947.

It turns out that the probability measure μ^* defined by

$$\mu^*(A) = \int_A f^*(x)\mathrm{d}x, \quad \forall \text{ measurable } A \subset [0,1],$$

which is absolutely continuous with respect to the Lebesgue measure on $[0,1]$, is invariant under the quadratic polynomial S. Thus, the chaotic dynamical system in the deterministic sense is *stable* in the probabilistic sense, that is, the probability distribution of the states of the iterates of S will approach eventually the stationary probability distribution given by f^*.

In this book, we shall mainly study Frobenius-Perron operators and the related concept of absolutely continuous invariant finite measures. There are two main issues that we would like to discuss: the existence of fixed density functions of Frobenius-Perron operators and their numerical computation. The main mathematical foundation for achieving our goals is integration theory and functional analysis, and a useful analytic tool is the concept of variation. So, in the next chapter we introduce the basis of measure theory and functional analysis as preliminaries for the subsequent chapters on the theoretical and numerical analysis of Frobenius-Perron operators.

Exercises

1.1 Prove the facts (i)-(vi) in the case of $2 < r \leqslant 3$ for the logistic model $S_r(x) = rx(1-x)$ in Section 1.1.
1.2 Find a homeomorphism $h : [0,1] \to [0,1]$ that makes S_4 topologically conjugate to the tent function T defined by (1.2).
1.3 Let $a \in X$ be a fixed point of $S : X \to X$. Show that the Dirac measure δ_a is S-invariant, and also show that if $\{a_1, a_2, \cdots, a_k\}$ is a k-cycle of S, then the measure $\delta = k^{-1} \sum_{i=1}^{k} \delta_{a_i}$ is S-invariant.

Chapter 2
Foundations of Measure Theory

JiuDing, Department of Mathematics, The University of Southern Mississippi in America.
Aihui Zhou, Academy of Mathematics and Systems Science, Chinese Academy of Sciences.

Abstract The fundamental mathematical knowledge used in the book is reviewed in this chapter, which includes basic measure theory and Lebesgue integration theory, L^1 spaces, the classic definition of variation for functions of one variable and the modern notion of variation for functions of several variables, compactness arguments for L^1 spaces, and quasi-compact operators on a Banach space which is a compactly-imbedded dense subspace of the L^1 space.
Keywords Radon-Nikodym theorem, L^p-space, BV-space, Helly's lemma, Cesáro convergence, quasi-compactness, Ionescu-Tulcea and Marinescu theorem.

The fundamental tools for studying the ergodic theory of chaotic dynamical systems are measure theory and functional analysis. In order to investigate theoretical and numerical aspects of Frobenius-Perron operators for the purpose of our book, we also need the important concept of functions of bounded variation. In this chapter, we introduce some useful concepts and basic results from real analysis and functional analysis.

Section 2.1 will give various definitions for measures and integration in the sense of Lebesgue. In Section 2.2, we shall present the most important integration theorems, Section 2.3 and Section 2.4 will be devoted to the study of functions of bounded variation in one variable and several variables, respectively. The concept of compactness and the class of quasi-compact linear operators will be introduced in Section 2.5. Our presentation is brief and only covers the materials that will be used in the sequel. For a more detailed study of measure theory, the reader can consult standard textbooks such as [109, 112] or the extensive monograph [57].

2.1 Measures and Integration

We start with the definition of a σ-algebra. Let X be an arbitrary non-empty set.

Definition 2.1.1 *A class Σ of subsets of X is called a* σ-algebra *on X if it satisfies*:

(i) *$A \in \Sigma$ implies that $A^c \in \Sigma$, where $A^c \equiv \{x \in X : x \notin A\}$ is the* complement *of A in X*;

(ii) $A_n \in \Sigma, n = 1, 2, \cdots$ *imply that the union* $\bigcup_{n=1}^{\infty} A_n \in \Sigma$;

(iii) $X \in \Sigma$.

Since the *empty set* $\varnothing = X^c$, by (i) and (iii) of the definition, $\varnothing \in \Sigma$. Furthermore, Σ is *closed* under the operation of countable intersection as per *de Morgen's formula*

$$\bigcap_{n=1}^{\infty} A_n = \left(\bigcup_{n=1}^{\infty} A_n^c \right)^c .$$

If Λ is a class of certain subsets of X, then the intersection of all the σ-algebras containing Λ is a σ-algebra and is called the σ-algebra *generated* by Λ.

Definition 2.1.2 *A real-valued (including ∞) nonnegative set function μ defined on a σ-algebra Σ is called a measure if $\mu(\varnothing) = 0$ and*

$$\mu \left(\bigcup_{n=1}^{\infty} A_n \right) = \sum_{n=1}^{\infty} \mu(A_n)$$

for any finite or infinite sequence $\{A_n\}$ of pairwise disjoint sets from Σ, that is $A_i \cap A_j = \varnothing$ for $i \neq j$. In other words, a measure μ is a countably additive *nonnegative set function.*

In the above definition, we use the convention that $a + \infty = \infty$ for any real number a and $\infty + \infty = \infty$.

Definition 2.1.3 *The ordered triple (X, Σ, μ) is called a* measure space *if there is given a σ-algebra Σ on a set X and a measure μ defined on Σ. If the measure μ is not specifically indicated, the ordered pair (X, Σ) is called a measurable space, and any $A \in \Sigma$ is called a Σ-measurable set, or simply a measurable set.*

Definition 2.1.4 *A real measure (or a complex measure) is a real-valued (or complex-valued) countably additive set function defined on a σ-algebra. If a real measure is nonnegative valued, it is called a positive measure.*

Remark 2.1.1 Clearly, a positive measure is also a measure. The value ∞ is admissible for a measure, but when we talk about a real measure μ, it is understood that $\mu(A)$ is a real number for every $A \in \Sigma$.

The *total variation* $|\mu|$ of a real measure μ on Σ is a measure defined on Σ by

$$|\mu|(A) = \sup \sum_{n=1}^{\infty} |\mu(A_n)|, \quad \forall\, A \in \Sigma,$$

the supremum being taken over all *measurable partitions* $\{A_n\}$ of A, i.e., $A_n \in \Sigma$, $\forall\, n$, $\bigcup\limits_{n=1}^{\infty} A_n = X$, and $A_n \cap A_k = \varnothing$ if $n \neq k$. For a real measure μ, we define its *positive part* μ^+ and its *negative part* μ^- by

$$\mu^+ = \frac{1}{2}(|\mu| + \mu), \quad \mu^- = \frac{1}{2}(|\mu| - \mu).$$

Then, both μ^+ and μ^- are measures on Σ and μ has the following *Jordan decomposition*:

$$\mu = \mu^+ - \mu^-, \quad |\mu| = \mu^+ + \mu^-.$$

Theorem 2.1.1 *If μ is a real measure on X, then $|\mu|(X) < \infty$. Hence, $|\mu|, \mu^+$, and μ^- are all positive measures.*

Remark 2.1.2 A simple example of a measure space is a so-called *counting measure space* (X, Σ, μ) in which X is any set, the σ-algebra Σ is the family of all subsets of X, and the measure $\mu(A)$ of A is the number (including ∞) of points in A. Another example is a finite *sample space* $X = \{x_1, x_2, \cdots, x_n\}$ in which the measure is defined by assigning to each single-element subset $\{x_i\}$ of X a nonnegative number p_i such that $\sum\limits_{i=1}^{n} p_i = 1$.

Remark 2.1.3 Let X be a locally compact Hausdorff space [57], let $\mathcal{B} \equiv \mathcal{B}(X)$ be the *Borel σ-algebra* which is the smallest σ-algebra containing all the open subsets of X, and let μ be a measure on Σ. Then, $(X, \mathcal{B}, \mu)$ is called a *Borel measure space*. In particular, if $X = \mathbb{R}^N (\mathbb{R}^1 \equiv \mathbb{R})$, the N-dimensional *Euclidean space* of all real column vectors $\boldsymbol{x} = (x_1, x_2, \cdots, x_N)^{\mathrm{T}}$ of N components under the usual addition and scalar multiplication with the standard *Euclidean inner product* $\langle \boldsymbol{x}, \boldsymbol{y} \rangle = \boldsymbol{x}^{\mathrm{T}} \boldsymbol{y} = \sum\limits_{i=1}^{N} x_i y_i$ for $\boldsymbol{x}, \boldsymbol{y} \in \mathbb{R}^N$, where $\boldsymbol{x}^{\mathrm{T}}$ is the transpose of x, then there exists a unique Borel measure μ on the Borel σ-algebra $\mathcal{B}(\mathbb{R}^N)$ such that

$$\mu\left(\prod_{i=1}^{N} [a_i, b_i]\right) = \prod_{i=1}^{N} (b_i - a_i),$$

where $\prod\limits_{i=1}^{N} [a_i, b_i]$ is the *Cartesian product* of the closed intervals $[a_1, b_1], \cdots, [a_N, b_N]$ and is called an *N-dimensional rectangle*. This measure μ will be denoted by m, which is called the *Lebesgue measure* on $\mathbb{R}^N$ (c.f. Remark 2.1.5 below). Whenever considering spaces $X = [0, 1]$ (or $\mathbb{R}$), $X = [0, 1]^N \equiv [0, 1] \times \cdots \times [0, 1]$ (or $\mathbb{R}^N$), or Borel subsets of these, we always assume that such sets are equipped with the standard Lebesgue measure unless indicated otherwise.

In this book, we are mainly interested in a more specific measure space defined as follows:

Definition 2.1.5 *A measure space (X, Σ, μ) is said to be* σ-finite *if X is a countable union of its subsets with finite measure,* i.e.,

$$X = \bigcup_{n=1}^{\infty} A_n, \quad A_n \in \Sigma, \quad \mu(A_n) < \infty, \quad n = 1, 2, \cdots .$$

Remark 2.1.4 $(\mathbb{R}^N, \mathcal{B}, m)$ is obviously σ-finite; A_n may be chosen as the N-dimensional ball of radius n centered at the origin for each positive integer n.

Definition 2.1.6 *A measure space (X, Σ, μ) is said to be* finite *if $\mu(X) < \infty$. In particular, if $\mu(X) = 1$, then the measure space is called a* probability space *or a* normalized measure space.

Definition 2.1.7 *A measure space (X, Σ, μ) is said to be* complete *if whenever $A \in \Sigma$ and $B \subset A$, $\mu(A) = 0$ implies $B \in \Sigma$.*

Remark 2.1.5 Every measure space can be extended uniquely to a complete measure space. Under the Lebesgue measure m, the completion of the Borel measure space $(\mathbb{R}^N, \mathcal{B}, m)$ is the *Lebesgue measure space* $(\mathbb{R}^N, \mathcal{L}, m)$, where $\mathcal{L}$ is the *Lebesgue σ-algebra* of Lebesgue measurable subsets of $\mathbb{R}^N$.

Remark 2.1.6 If a certain property involving the points of a measure space is true except for a set of measure zero, then we say that the property is true *almost everywhere* (abbreviated as a.e.). The notation μ-a.e. (or simply a.e. if μ is understood) is sometimes used if the property is true almost everywhere with respect to the measure μ.

We turn to the concept of measurable functions and the definition of their integration.

Definition 2.1.8 *Let (X, Σ, μ) be a measure space. A real-valued (or complex-valued) function $f : X \to \mathbb{R}$ (or $\mathbb{C}$) is said to be* measurable *if $f^{-1}(G) \in \Sigma$ for every open set $G \subset \mathbb{R}$ (or $\mathbb{C}$), where $f^{-1}(G) \equiv \{x \in X : f(x) \in G\}$ is the inverse image of G under f.*

Remark 2.1.7 More generally, a transformation $S : X \to Y$ from a *measurable space* (X, Σ) into a measurable space $(Y, \mathcal{A})$ is said to be *measurable* if $S^{-1}(A) \in \Sigma$ for each $A \in \mathcal{A}$. Thus, a measurable function f is a measurable transformation from (X, Σ) into $(\mathbb{R}, \mathcal{B})$. If X_1 and X_2 are topological spaces with their respective Borel σ-algebras $\mathcal{B}_1$ and $\mathcal{B}_2$, then a continuous transformation $S : X_1 \to X_2$ is a (Borel-) measurable transformation. In particular, a continuous function on a topological space is a measurable function.

Definition 2.1.9 *A measurable function $f : X \to [0, \infty)$ on a measurable space (X, Σ) is called a* simple function *if its range consists of finitely many* *points. In other words, f is a simple function if*

$$f = \sum_{i=1}^{n} \alpha_i \chi_{A_i}, \quad A_i \in \Sigma,$$

where χ_{A_i} is the characteristic function of A_i as defined by (1.3).

Definition 2.1.10 *Let (X, Σ, μ) be a measure space. If $f : X \to [0, \infty)$ is a simple function of the form*

$$f = \sum_{i=1}^{n} \alpha_i \chi_{A_i},$$

where α_i, $i = 1, 2, \cdots, n$ are the distinct values of f, and if $A \in \Sigma$, then the Lebesgue integral of f over A *is defined by*

$$\int_A f \mathrm{d}\mu = \sum_{i=1}^{n} \alpha_i \mu(A_i \cap A),$$

where the convention $0 \cdot \infty = 0$ is used.

Definition 2.1.11 *Let (X, Σ, μ) be a measure space, let $f : X \to \mathbb{R}$ be an arbitrary nonnegative measurable function, and let $A \in \Sigma$. Then, the* Lebesgue integral of f over A *is defined as*

$$\int_A f \mathrm{d}\mu = \sup \left\{ \int_A s \mathrm{d}\mu : 0 \leqslant s \leqslant f, \ s \text{ are simple functions} \right\}.$$

If $A = X$, then $\int_X f \mathrm{d}\mu$ is called the Lebesgue integral of f.

Given a real-valued function f on X, let

$$f^+(x) = \max\{f(x), 0\} \text{ and } f^-(x) = \max\{-f(x), 0\},$$

for $x \in X$. The functions f^+ and f^- are called the *positive part* and the *negative part* of f, respectively. We have the obvious equalities $f = f^+ - f^-$ and $|f| = f^+ + f^-$.

Definition 2.1.12 *Let (X, Σ, μ) be a measure space, let $f : X \to \mathbb{R}$ be a real-valued measurable function, and let $A \in \Sigma$. Then, the* Lebesgue integral of f over A *is defined by*

$$\int_A f \mathrm{d}\mu = \int_A f^+ \mathrm{d}\mu - \int_A f^- \mathrm{d}\mu,$$

if at least one of the two numbers $\int_A f^+ \mathrm{d}\mu$ and $\int_A f^- \mathrm{d}\mu$ is finite.

Definition 2.1.13 *Let (X, Σ, μ) be a measure space, let $f : X \to \mathbb{C}$ be a complex-valued measurable function, and let $A \in \Sigma$. If $f = u + \mathrm{i}v$ and $\int_A |f| \mathrm{d}\mu < \infty$, then the* Lebesgue integral of f over A *is defined by*

$$\int_A f \mathrm{d}\mu = \int_A u^+ \mathrm{d}\mu - \int_A u^- \mathrm{d}\mu + \mathrm{i}\left(\int_A v^+ \mathrm{d}\mu - \int_A v^- \mathrm{d}\mu\right).$$

If $\int_X |f| \mathrm{d}\mu < \infty$, then the function f is said to be Lebesgue integrable *(with respect to μ).*

Remark 2.1.8 A complex-valued measurable function is integrable if and only if its real part and imaginary part are both integrable. If f is a real-valued measurable function, then the equality

$$\int_X |f| \mathrm{d}\mu = \int_X f^+ \mu + \int_X f^- \mathrm{d}\mu$$

is always true no matter whether f is integrable or not.

Remark 2.1.9 Sometimes we write $\int_A f \mathrm{d}\mu$ as $\int_A f(x) \mathrm{d}\mu(x)$ to emphasize the independent variable x of the function f. In the case where A is an interval or an N-dimensional rectangle, we may employ more conventional Riemann integral notations for Lebesgue integrals, even if f may not be Riemann integrable. For example, we often write $\int_{[a,b]} f \mathrm{d}\mu$ as $\int_a^b f \mathrm{d}\mu$ and $\int_{[a,b]} f(x) \mathrm{d}\mu(x)$ as $\int_a^b f(x) \mathrm{d}\mu(x)$, and similarly for multi-dimensional integrals.

We give some basic properties of Lebesgue integration, the proof of which can be found in any textbook on real analysis.

Proposition 2.1.1 *Suppose that (X, Σ, μ) is a measure space and let f and g be measurable functions on X.*

(i) *If g is a nonnegative and integrable function, and $|f(x)| \leqslant g(x)$ for $x \in X$ μ-a.e., then f is integrable and*

$$\left|\int_X f \mathrm{d}\mu\right| \leqslant \int_X g \mathrm{d}\mu.$$

(ii) $\int_X |f| \mathrm{d}\mu = 0$ *if and only if* $f = 0$ μ-a.e.

(iii) *If f and g are integrable and a and b are numbers, then the function $af + bg$ is integrable and*

$$\int_X (af + bg) \mathrm{d}\mu = a \int_X f \mathrm{d}\mu + b \int_X g \mathrm{d}\mu.$$

(iv) *Suppose that f is integrable and $\{A_n\}$ is a sequence of disjoint measurable subsets of X. Then,*

$$\int_{\bigcup_{n=1}^{\infty} A_n} f \mathrm{d}\mu = \sum_{n=1}^{\infty} \int_{A_n} f \mathrm{d}\mu.$$

(v) *If f is integrable and $\int_A f \mathrm{d}\mu = 0$ for every $A \in \Sigma$, then $f(x) = 0$ for $x \in X$* μ-a.e.

Remark 2.1.10 Throughout this book, unless it is specifically stated to the contrary, a measure space will always be understood to be σ-finite and complete, and a function is always real-valued.

2.2 Basic Integration Theory

In this section, we list without proof the three fundamental convergence theorems and some other important theorems for Lebesgue integration that will be used throughout our book. Let (X, Σ, μ) be a measure space.

Theorem 2.2.1 (Lebesgue's dominated convergence theorem) *Suppose that $\{f_n\}$ is a sequence of measurable functions on X such that the limit*

$$\lim_{n\to\infty} f_n(x) = f(x)$$

exists for $x \in X$ a.e. *If there is an integrable function g such that*

$$|f_n(x)| \leqslant g(x) \quad x \in X \text{ a.e.}, \quad \forall\, n = 1, 2, \cdots,$$

then f and all f_n are integrable. Moreover,

$$\lim_{n\to\infty} \int_X |f_n - f| \mathrm{d}\mu = 0 \quad \text{and} \quad \lim_{n\to\infty} \int_X f_n \mathrm{d}\mu = \int_X f \mathrm{d}\mu.$$

Theorem 2.2.2 (Lebesgue's monotone convergence theorem) *Let $\{f_n\}$ be a sequence of real-valued measurable functions on X, and suppose that*
(i) $0 \leqslant f_1(x) \leqslant f_2(x) \leqslant \cdots$ *for $x \in X$* a.e.,
(ii) $\lim_{n\to\infty} f_n(x) = f(x)$ *for $x \in X$* a.e.
Then, f is measurable and

$$\lim_{n\to\infty} \int_X f_n \mathrm{d}\mu = \int_X f \mathrm{d}\mu.$$

Theorem 2.2.3 (Fatou's lemma) *If $f_n : X \to [0, \infty)$ is measurable for each $n = 1, 2, \cdots$, then*

$$\int_X \liminf_{n\to\infty} f_n \mathrm{d}\mu \leqslant \liminf_{n\to\infty} \int_X f_n \mathrm{d}\mu.$$

Measurable functions can be used to construct new measures on a measurable space, as the following theorem indicates.

Theorem 2.2.4 *Suppose that $f : X \to [0, \infty)$ is measurable and let the set function μ_f be defined by*

$$\mu_f(A) = \int_A f \mathrm{d}\mu, \ \ \forall\, A \in \Sigma.$$

Then, μ_f is a measure on Σ, and

$$\int_X g \mathrm{d}\mu_f = \int_X g f \mathrm{d}\mu$$

for every measurable function $g : X \to [0, \infty)$.

The measure μ_f satisfies the property that $\mu_f(A) = 0$ whenever $\mu(A) = 0$. Moreover, μ_f is a finite measure if and only if f is integrable. The following theorem gives a very important converse to the above conclusion, which is of fundamental importance for the definition of Frobenius-Perron operators that will be studied extensively in this book.

Definition 2.2.1 *Let μ be a measure on a σ-algebra Σ, and let ν be an arbitrary measure on Σ; ν may be positive, real, or complex. We say that ν is* absolutely continuous *with respect to μ, and write*

$$\nu \ll \mu,$$

if $\nu(A) = 0$ for every $A \in \Sigma$ such that $\mu(A) = 0$. If $\nu \ll \mu$ and $\mu \ll \nu$ both hold, we say that the measures μ and ν are equivalent, *written as $\mu \cong \nu$.*

Theorem 2.2.5 (Radon-Nikodym theorem) *Let (X, Σ, μ) be a σ-finite measure space and let ν be a real (or complex) measure which is absolutely continuous with respect to μ. Then, there exists a unique μ-integrable function $f : X \to \mathbb{R}$ (or $\mathbb{C}$) such that*

$$\nu(A) = \int_A f \mathrm{d}\mu, \ \ \forall\, A \in \Sigma. \tag{2.1}$$

Moreover, if $\nu \ll \mu$ and ν is σ-finite, then (2.1) is still valid with a nonnegative measurable function f which may not be μ-integrable.

The function f in (2.1) is called the *Radon-Nikodym derivative* of ν with respect to μ and is sometimes written as $f = \mathrm{d}\nu/\mathrm{d}\mu$.

We now introduce L^p-spaces which are very useful function spaces in ergodic theory.

Definition 2.2.2 *Let p be a real number such that $1 \leqslant p < \infty$. The family of all real-valued (or complex-valued) measurable functions $f : X \to \mathbb{R}$ (or $\mathbb{C}$) satisfying*

$$\int_X |f|^p \mathrm{d}\mu < \infty$$

is denoted by $L^p(X, \Sigma, \mu)$. The space $L^\infty(X, \Sigma, \mu)$ is defined as the family of all measurable functions which are bounded μ-a.e.

Two functions f_1, $f_2 \in L^p(X, \Sigma, \mu)$ are considered the same if $f_1(x) = f_2(x)$, $x \in X$ μ-a.e. Thus, $L^p(X, \Sigma, \mu)$ becomes a *vector space* under the usual function addition and scalar multiplication for any $p \in [1, \infty]$. We shall sometimes write L^p instead of $L^p(X, \Sigma, \mu)$ if the measure space is understood, $L^p(X)$ if Σ and μ are understood, $L^p(\mu)$ if X and Σ are understood, or $L^p(\Sigma)$ if X and μ are understood. We note that L^1 is exactly the space of all μ-integrable functions.

The number $\|f\|_p = \left(\int_X |f|^p \mathrm{d}\mu \right)^{1/p}$ is called the *L^p-norm* of $f \in L^p$ for $p < \infty$ and the number $\|g\|_\infty = \operatorname*{ess\,sup}_{x \in X} |g(x)|$ is referred to as the *L^∞-norm* of $g \in L^\infty$. $\|f\|_1$ will be written as $\|f\|$ or $\|f\|_\mu$ for the simplicity of notation since this norm is mainly used in our book. $\| \cdot \|_p$ does define a norm on L^p for each $p \in [1, \infty]$ since it satisfies the three axioms for a norm:

(i) $\|f\|_p = 0$ if and only if $f = 0$, or $f(x) = 0$ a.e.;
(ii) $\|af\|_p = |a| \|f\|_p$ for any $f \in L^p$ and any scalar a;
(iii) $\|f + g\|_p \leqslant \|f\|_p + \|g\|_p$ for all $f, g \in L^p$ (*triangle inequality*).

It is well-known that under the above defined L^p-norm, $L^p(X, \Sigma, \mu)$ is a *Banach space*, that is, a complete normed vector space. Moreover, under the natural ordering relation among real-valued functions, L^p becomes a *Banach lattice* [114]. Another useful property of the L^1 space is that

$$\|f\| = \|f - g\| + \|g\|, \tag{2.2}$$

for all real-valued functions f, $g \in L^1$ such that $f \geqslant g$.

The *dual space* or just the *dual* of a Banach space, by definition, is the space of all *bounded linear functionals* on it. The following theorem characterizes the dual of L^p.

Theorem 2.2.6 *Let $1 \leqslant p < \infty$. The dual of $L^p(X, \Sigma, \mu)$ is isomorphic to $L^{p'}(X, \Sigma, \mu)$, where $1/p + 1/p' = 1$ for $p > 1$ and $p' = \infty$ if $p = 1$.*

The dual relation between $f \in L^p$ and $g \in L^{p'}$ is given by

$$\langle f, g \rangle \equiv \int_X fg \mathrm{d}\mu,$$

which satisfies the *Cauchy-Hölder inequality*

$$|\langle f, g\rangle| \leqslant \|f\|_p \|g\|_{p'}, \quad \forall\, f \in L^p,\ g \in L^{p'}. \tag{2.3}$$

In ergodic theory, one often uses various notions of *convergence* for sequences of functions.

Definition 2.2.3 *A sequence of functions* $f_n \in L^p, 1 \leqslant p < \infty$ *is* weakly Cesáro convergent *to* $f \in L^p$ *if*

$$\lim_{n\to\infty} \frac{1}{n}\sum_{i=0}^{n-1}\langle f_i, g\rangle \;=\; \langle f, g\rangle, \quad \forall\, g \in L^{p'},$$

and is strongly Cesáro convergent *to* $f \in L^p$ *if*

$$\lim_{n\to\infty}\left\|\frac{1}{n}\sum_{i=0}^{n-1} f_i - f\right\|_p = 0.$$

Definition 2.2.4 *A sequence of functions* $f_n \in L^p, 1 \leqslant p < \infty$ *is* weakly convergent *to* $f \in L^p$ *if*

$$\lim_{n\to\infty}\langle f_n, g\rangle \;=\; \langle f, g\rangle, \quad \forall\, g \in L^{p'},$$

and is strongly convergent *to* $f \in L^p$ *if*

$$\lim_{n\to\infty}\|f_n - f\|_p = 0.$$

Remark 2.2.1 From the Cauchy-Hölder inequality (2.3), we see that strong convergence implies weak convergence. The converse is not true in general, as demonstrated by the classic example of $f_n(x) = \sin(nx)$ in $L^2(0,1)$ [109].

2.3 Functions of Bounded Variation in One Variable

The concept of variation plays an important role in the compactness argument for L^1 spaces, and thus will be a key concept for studying Frobenius-Perron operators that are defined on L^1 spaces. In this section, we first introduce the classic definition of variation for functions of one variable, which will be used for the statistical study of one-dimensional mappings. Then, in the next section, we study the modern notion of variation for functions of several variables in terms of Schwartz's distribution theory.

Definition 2.3.1 *Let* f *be a real-valued or complex-valued function defined on an interval* $[a, b]$. *The variation of* f *on* $[a, b]$ *is the nonnegative number (may be* ∞*)*

$$\bigvee_a^b f = \sup\left\{\sum_{i=1}^n |f(x_i) - f(x_{i-1})| : a = x_0 < x_1 < \cdots < x_n = b\right\}.$$

If $\bigvee_a^b f < \infty$, *we say that* f *is of bounded variation on* $[a, b]$.

Since two L^1 functions are viewed as the same in $L^1(a,b)$ if their values are identical for almost all $x \in [a,b]$ with respect to the Lebesgue measure, we define the variation of a function $f \in L^1(a,b)$ as follows.

Definition 2.3.2 *Let $f \in L^1(a,b)$. Then, its variation on $[a,b]$ is defined to be*

$$\bigvee_{[a,b]} f = \inf \left\{ \bigvee_a^b g : \; g(x) = f(x), \; \forall\, x \in [a,b] \text{ a.e.} \right\}.$$

If $\bigvee_{[a,b]} f < \infty$, then f is said to be of bounded variation on $[a,b]$.

Some well-known properties of variation are summarized in the following proposition. Their proofs are referred to, e.g., [109].

Proposition 2.3.1 (i) *If f is a monotonic function on $[a,b]$, then*

$$\bigvee_a^b f = |f(a) - f(b)|.$$

(ii) *If $f_1, f_2, \cdots, f_n$ are functions of bounded variation on $[a,b]$, then so is their sum, and*

$$\bigvee_a^b (f_1 + \cdots + f_n) \leqslant \bigvee_a^b f_1 + \cdots + \bigvee_a^b f_n.$$

(iii) *Let f be a function of bounded variation on $[a,b]$ and let $a = a_0 < a_1 < \cdots < a_n = b$. Then, f is of bounded variation on $[a_{i-1}, a_i]$ for $i = 1, \cdots, n$ and*

$$\bigvee_{a_0}^{a_1} f + \cdots + \bigvee_{a_{n-1}}^{a_n} f = \bigvee_a^b f.$$

(iv) *If $g : [s,t] \to [a,b]$ is monotonically increasing or decreasing on $[s,t]$ and f is of bounded variation on $[a,b]$, then the composition function $f \circ g$ is of bounded variation on $[s,t]$, and*

$$\bigvee_s^t f \circ g \leqslant \bigvee_a^b f.$$

(v) *If f is of bounded variation on $[a,b]$ and g is continuously differentiable on $[a,b]$, then the product function fg is of bounded variation on $[a,b]$, and*

$$\bigvee_a^b fg \leqslant \left(\sup_{x \in [a,b]} |g(x)| \right) \bigvee_a^b f + \int_a^b |f(x) g'(x)| \mathrm{d}x.$$

In the special case that $f(x) \equiv 1$, we actually have the equality

$$\bigvee_a^b g = \int_a^b |g'(x)|\mathrm{d}x. \tag{2.4}$$

Remark 2.3.1 The equality (2.4) is still true for absolutely continuous functions on $[a,b]$. Although a function g of bounded variation has a Lebesgue integrable derivative function, (2.4) may not hold in general (see [109]).

A less well-known but useful property for a function of bounded variation is the so-called *Yorke's inequality*, which will be needed in Section 5.2 for proving the existence of fixed density functions of Frobenius-Perron operators associated with piecewise second order continuously differentiable and stretching mappings of an interval.

Proposition 2.3.2 (Yorke's inequality) *Let f be a function defined on $[0,1]$ and let it be of bounded variation on $[a,b] \subset [0,1]$. Then, the product function $f\chi_{[a,b]}$ of f and the characteristic function $\chi_{[a,b]}$ is of bounded variation on $[0,1]$ and satisfies*

$$\bigvee_0^1 f\chi_{[a,b]} \leqslant 2\bigvee_a^b f + \frac{2}{b-a}\int_a^b |f(x)|\mathrm{d}x. \tag{2.5}$$

Proof Choose $c \in [a,b]$ such that

$$|f(c)| \leqslant \frac{1}{b-a}\int_a^b |f(x)|\mathrm{d}x.$$

Then,

$$\begin{aligned}
\bigvee_0^1 f\chi_{[a,b]} &\leqslant \bigvee_a^b f + |f(a)| + |f(b)| \\
&\leqslant \bigvee_a^b f + |f(a) - f(c)| + |f(c) - f(b)| + 2|f(c)| \\
&\leqslant 2\bigvee_a^b f + \frac{2}{b-a}\int_a^b |f(x)|\mathrm{d}x.
\end{aligned}$$

□

Let $BV \equiv BV(a,b)$ be the set of all the functions in $L^1(a,b)$ with bounded variation. Then, BV is clearly a vector subspace of L^1. The set of all $f \in BV$ with $\int_a^b f(x)\mathrm{d}x = 0$ is a vector subspace of BV and is denoted by $BV_0 \equiv BV_0(a,b)$.

For each $f \in BV$, we define the *BV-norm* of f as

$$\|f\|_{BV} = \|f\| + \bigvee_{[a,b]} f. \tag{2.6}$$

Then, the space BV equipped with the BV-norm $\|\cdot\|_{BV}$ is a Banach space (see [109] for a proof; see also Corollary 2.4.1 in the next section). A significant property of BV functions is the following classic result due to Helly [109].

Theorem 2.3.1 (Helly's lemma) *If a sequence $\{f_n\}$ of functions defined on $[a, b]$ is uniformly bounded in both integral and variation, that is, there is a constant K such that*

$$\|f_n\| \leqslant K, \quad \bigvee_a^b f_n \leqslant K, \quad \forall\, n = 1, 2, \cdots,$$

then there is a subsequence $\{f_{n_k}\}$ of $\{f_n\}$ such that

$$\lim_{k\to\infty} \|f_{n_k} - f\| = 0,$$

where f satisfies that $\|f\| \leqslant K$ and $\bigvee_a^b f \leqslant K$.

2.4 Functions of Bounded Variation in Several Variables

The modern definition of variation for functions of several variables is based on the concept of *distributions* or *generalized functions* in the theory of weakly differentiable functions [126], and has found applications in such fields as minimal surfaces [63] and the calculus of variation [105]. We employ this notion for the existence of absolutely continuous invariant measures for multi-dimensional transformations and the convergence analysis for their computation.

Let $\Omega \subset \mathbb{R}^N$ be a bounded open region and let k be a nonnegative integer. As standard notations in functions theory, $C^k(\Omega)$ is the space of all the functions which have continuous partial derivatives up to order k in Ω, and $C_0^k(\Omega)$ consists of those in $C^k(\Omega)$ with a compact support. $C^k(\overline{\Omega})$ consists of all $f \in C^k(\Omega)$ whose partial derivatives up to order k are bounded and uniformly continuous. $C^k(\Omega;\mathbb{R}^N)$, $C_0^k(\overline{\Omega};\mathbb{R}^N)$, and $C^k(\overline{\Omega};\mathbb{R}^N)$ denote the corresponding spaces of vector-valued functions, respectively. Elements in $C^k(\Omega)$ (or $C^k(\Omega;\mathbb{R}^N)$) are called *C^k-functions* (or *C^k-mappings*, or *C^k-transformations*). Let $\|\cdot\|_2 = \sqrt{\langle\cdot,\cdot\rangle}$ denote the usual *Euclidean vector 2-norm* on $\mathbb{R}^N$ induced by the Euclidean inner product, and let $\|\cdot\| \equiv \|\cdot\|_1$ be the *vector 1-norm* on $\mathbb{R}^N$. The space and norm notations for different function spaces, such as the *Sobolev spaces* $W^{k,p}(\Omega)$ and $W^{0,p}(\Omega) \equiv L^p(\Omega)$, are used in a standard way as in [63] or [126]. In particular, $\|\cdot\|_{k,p}$ is the *Sobolev norm* on $W^{k,p}(\Omega)$ and

$\|\cdot\|_{0,1} = \|\cdot\|$. The norm on $C_0^1(\Omega;\mathbb{R}^N)$, which will be used for the definition of variation below, is defined as $\|\boldsymbol{g}\|_\infty \equiv \|\boldsymbol{g}\|_{0,\infty} = \max\{\|\boldsymbol{g}(\boldsymbol{x})\|_2 : \forall\, \boldsymbol{x} \in \Omega\}$ for $\boldsymbol{g} \in C_0^1(\Omega;\mathbb{R}^N)$.

Definition 2.4.1 ([63]) *Let $f \in L^1(\Omega)$. The number (may be ∞),*

$$V(f;\Omega) = \sup\left\{\int_\Omega f \,\mathrm{div}\, \boldsymbol{g}\, \mathrm{d}m : \boldsymbol{g} \in C_0^1(\Omega;\mathbb{R}^N),\ \|\boldsymbol{g}(\boldsymbol{x})\|_2 \leqslant 1,\ \forall\, \boldsymbol{x} \in \Omega\right\}$$

is called the variation *of f in Ω. Here,* $\mathrm{div}\,\boldsymbol{g} = \sum_{i=1}^N \partial g_i/\partial x_i$ *is the* divergence *of $\boldsymbol{g}$.*

Remark 2.4.1 The *gradient* of $f \in L^1(\Omega)$ in the sense of distribution (or generalized functions) will be denoted by Df [63], and so we can write $V(f;\Omega)$ as $\int_\Omega \|Df\|$. The latter notation for the variation of f comes from the fact that if $f \in C^1(\Omega)$, then

$$\int_\Omega \|Df\| = \int_\Omega \|\mathbf{grad}\, f\| \mathrm{d}m,$$

where $\mathbf{grad}\, f = (\partial f/\partial x_1, \partial f/\partial x_2, \cdots, \partial f/\partial x_N)^{\mathrm{T}}$ is the gradient of f in the classic sense, which is deduced from integration by parts,

$$\int_\Omega f \,\mathrm{div}\, \boldsymbol{g}\, \mathrm{d}m = -\int_\Omega \sum_{i=1}^N \frac{\partial f}{\partial x_i} g_i\, \mathrm{d}m$$

for $\boldsymbol{g} \in C_0^1(\Omega;\mathbb{R}^N)$. More generally, if $f \in W^{1,1}(\Omega)$, then

$$\int_\Omega \|Df\| = \int_\Omega \|\mathbf{grad}\, f\| \mathrm{d}m,$$

where $\mathbf{grad}\, f = (f_1, f_2, \cdots, f_N)^{\mathrm{T}}$ and each f_i is the *weak derivative* of f in the sense of Sobolev (see [126]).

Remark 2.4.2 When $N = 1$, the new definition of variation and the classic one coincide [63]. That is,

$$\bigvee_{[a,b]} f = V(f;(a,b)).$$

Definition 2.4.2 *A function $f \in L^1(\Omega)$ is said to have* bounded variation *in Ω if $V(f;\Omega) < \infty$. We define $BV \equiv BV(\Omega)$ as the space of all functions in $L^1(\Omega)$ with bounded variation. $BV_0 \equiv BV_0(\Omega)$ is defined as the set of all $f \in BV(\Omega)$ with $\int_\Omega f \mathrm{d}m = 0$.*

Remark 2.4.3 The notation Df actually represents a vector-valued *Radon measure* $\boldsymbol{\omega} = (\omega_1, \omega_2, \cdots, \omega_N)$ for $f \in BV(\Omega)$, and in fact, $\int_\Omega \|Df\|$ is the total variation of $\boldsymbol{\omega}$ on Ω. More generally, $\int_A \|Df\|$ is the total variation of $\boldsymbol{\omega}$ on A for any measurable set $A \subset \Omega$ (see Remark 1.5 of [63] or [105]). Also, the notation $\int_A \langle Df, \boldsymbol{g}\rangle$ means the integral of the vector-valued function $\boldsymbol{g} = (g_1, g_2, \cdots, g_N)^{\mathrm{T}}$, where $g_i \in L^1(\omega_i)$ for each i, over $A \subset \Omega$ with respect to $\boldsymbol{\omega}$, i.e.,

$$\int_A \langle Df, \boldsymbol{g}\rangle \equiv \int_A \boldsymbol{g} \cdot \mathrm{d}\boldsymbol{\omega} = \sum_{i=1}^N \int_A g_i \, \mathrm{d}\omega_i.$$

Finally, we have that

$$\begin{aligned} V(f; \Omega) &= \sup \left\{ \int_\Omega \langle Df, \boldsymbol{g}\rangle : \boldsymbol{g} \in C_0^1(\Omega; \mathbb{R}^N), \ \|\boldsymbol{g}(\boldsymbol{x})\|_2 \leqslant 1, \ \forall\, \boldsymbol{x} \in \Omega \right\} \\ &= \sup \left\{ \int_\Omega \langle Df, \boldsymbol{g}\rangle : \boldsymbol{g} \in C^1(\Omega; \mathbb{R}^N), \ \|\boldsymbol{g}(\boldsymbol{x})\|_2 \leqslant 1, \ \forall\, \boldsymbol{x} \in \Omega \right\}. \end{aligned}$$

We present some important properties of variation in the remaining part of the section.

Theorem 2.4.1 *Let $\{f_n\}$ be a sequence of functions in $BV(\Omega)$ which converges in $L^1(\Omega)$ to a function f. Then,*

$$V(f; \Omega) \leqslant \liminf_{n\to\infty} V(f_n; \Omega). \tag{2.7}$$

Proof Let $\boldsymbol{g} \in C_0^1(\Omega; \mathbb{R}^N)$ be a vector-valued function such that $\|\boldsymbol{g}(\boldsymbol{x})\|_2 \leqslant 1$ for all $\boldsymbol{x} \in \Omega$. Then,

$$\int_\Omega f \,\mathrm{div}\, \boldsymbol{g} \,\mathrm{d}m = \lim_{n\to\infty} \int_\Omega f_n \,\mathrm{div}\, \boldsymbol{g} \,\mathrm{d}m \leqslant \liminf_{n\to\infty} V(f_n; \Omega).$$

Now (2.7) follows upon taking the supremum over all such $\boldsymbol{g}$. □

Corollary 2.4.1 *$BV(\Omega)$ is a Banach space under the BV*-norm

$$\|f\|_{BV} \equiv \|f\| + V(f; \Omega).$$

Proof The norm properties follow easily from the definitions of the L^1-norm and the variation of f, and so it remains to prove the completeness of the norm. Suppose that $\{f_n\}$ is a Cauchy sequence in $BV(\Omega)$. Then, it is obviously a Cauchy sequence in $L^1(\Omega)$, and hence, by the completeness of $L^1(\Omega)$, there is a function $f \in L^1(\Omega)$ such that $\lim_{n\to\infty} f_n = f$ in $L^1(\Omega)$. Since the number sequence

$\{V(f_n;\Omega)\}$ is uniformly bounded, by Theorem 2.4.1, $f \in BV(\Omega)$. Given any positive integer k, since $\lim_{n\to\infty}(f_n - f_k) = f - f_k$ under the L^1-norm, Theorem 2.4.1 again implies that

$$V(f - f_k;\Omega) \leqslant \liminf_{n\to\infty} V(f_n - f_k;\Omega).$$

From $\lim_{n,k\to\infty} V(f_n - f_k;\Omega) = 0$, we have $\lim_{k\to\infty} V(f - f_k;\Omega) = 0$. Therefore, $\lim_{n\to\infty} f_n = f$ in $BV(\Omega)$. □

Remark 2.4.4 It follows from Remark 2.4.1 that the Sobolev space $W^{1,1}(\Omega)$ is a closed subspace of $BV(\Omega)$ and $\|f\|_{1,1} = \|f\|_{BV}$ for any $f \in W^{1,1}(\Omega)$; see also Remark 2.4.5 below.

The proof of the following theorem is referred to [63]. Let $C^\infty(\Omega)$ denote the space of the functions which have all orders continuous partial derivatives in Ω.

Theorem 2.4.2 *Let $f \in BV(\Omega)$. Then, there exists a sequence $\{f_n\}$ in $C^\infty(\Omega)$ such that*

$$\lim_{n\to\infty} \|f_n - f\| = 0 \ \textit{and} \ \lim_{n\to\infty} V(f_n;\Omega) = V(f;\Omega).$$

Remark 2.4.5 As a contrast to Theorem 2.4.2, for any $f \in W^{1,1}(\Omega)$, there is a sequence $\{f_n\}$ in $C^\infty(\Omega)$ that satisfies

$$\lim_{n\to\infty} \|f_n - f\| = 0 \quad \text{and} \quad \lim_{n\to\infty} V(f_n - f;\Omega) = 0,$$

which implies that $\lim_{n\to\infty} \|f_n - f\|_{1,1} = 0$.

Now, we introduce the concept of the trace of a function in $BV(\Omega)$. In this book we assume that the bounded open region Ω and any subregion $\Omega_0 \subset \Omega$ involved are *admissible* (see [126] for its definition), for instance, a domain with a *piecewise Lipschitz continuous* boundary, so that for any $f \in BV(\Omega)$ the following *trace* function $\mathrm{tr}_\Omega f$ is well-defined on the boundary of Ω and the subsequent two results are valid (see [63] or [126] for their proofs).

Definition 2.4.3 *The trace function or just the trace of a function $f \in BV(\Omega)$ is defined as*

$$\mathrm{tr}_\Omega f(\boldsymbol{x}) = \lim_{r\to 0} \frac{1}{m(B(\boldsymbol{x},r)\cap\Omega)} \int_{B(\boldsymbol{x},r)\cap\Omega} f dm$$

for $\boldsymbol{x} \in \partial\Omega$ a.e. *with respect to the $(N-1)$-dimensional Hausdorff measure H for the boundary $\partial\Omega$ of Ω, where $B(\boldsymbol{x},r)$ is the N-dimensional ball centered at $\boldsymbol{x}$ with radius r.*

Theorem 2.4.3 *Let $\Omega_0 \subset \Omega$. If $f \in BV(\Omega)$, then for $\boldsymbol{g} \in C_0^1(\Omega; \mathbb{R}^N)$,*

$$\int_{\Omega_0} f \operatorname{div} \boldsymbol{g} \, \mathrm{d}m = -\int_{\Omega_0} \langle Df, \boldsymbol{g} \rangle + \int_{\partial\Omega_0} \mathrm{tr}_{\Omega_0} f \langle \boldsymbol{g}, \boldsymbol{n} \rangle \, \mathrm{d}H,$$

where $\mathbf{n}$ *is the unit outward normal vector to* $\partial\Omega_0$.

Theorem 2.4.4 *There exists a constant $\kappa(\Omega)$ such that*

$$\int_{\partial\Omega} |\mathrm{tr}_\Omega f| \, dH \leqslant \kappa(\Omega) \|f\|_{BV}, \quad \forall\, f \in BV(\Omega).$$

At the end of this section, we introduce a classical definition of variation due to Tonnelli [63] for functions defined on an N-dimensional rectangle $\Omega = \prod_{i=1}^{N} [a_i, b_i]$.

Definition 2.4.4 *Denote by $\boldsymbol{P}_i$ the coordinate projection of $\mathbb{R}^N$ onto $\mathbb{R}^{N-1}$ given by $\boldsymbol{P}_i(x_1, x_2, \cdots, x_N) = (x_1, \cdots, x_{i-1}, x_{i+1}, \cdots, x_N)$. Then, the* Tonnelli variation *of a function $f : \Omega \to \mathbb{R}$ is defined as*

$$\bigvee(f; \Omega) = \max_{1 \leqslant i \leqslant N} \int_{\boldsymbol{P}_i(\Omega)} \bigvee_{[a_i, b_i]} f \, \mathrm{d}m_{N-1},$$

where $\bigvee_{[a_i,b_i]} f$ is the variation of the single-variable function

$$f_{\boldsymbol{P}_i(x_1, x_2, \cdots, x_N)}(x_i) \equiv f(x_1, \cdots, x_i, \cdots, x_N)$$

on $[a_i, b_i]$ and m_{N-1} denotes the Lebesgue measure on $\mathbb{R}^{N-1}$. If $\bigvee(f; \Omega) < \infty$, then we say that f is of bounded variation in the sense of Tonnelli.

From Theorem 5.3.5 in Ziemer [126] and its proof, we have

Proposition 2.4.1 *For the two definitions $V(f; \Omega)$ and $\bigvee(f; \Omega)$ of the variation of a function f defined on an N-dimensional rectangle $\Omega = \prod_{i=1}^{N} [a_i, b_i]$,*

$$\bigvee(f; \Omega) \leqslant V(f; \Omega) \leqslant N \bigvee(f; \Omega).$$

2.5 Compactness and Quasi-compactness

The concept of compactness is a key one in studying the convergence problem of a sequence of the iterates related to Frobenius-Perron operators in the sequel. In the first part of this section, we shall study the strong compactness and the weak compactness for L^1 spaces, while the concept of quasi-compactness for a bounded linear operator on a Banach space which is a subspace of an L^1 space will be studied in the second part of the section.

2.5.1 Strong and Weak Compactness

Let (X, Σ, σ) be a measure space and let $\mathcal{F}$ be a subset of $L^1(X)$.

Definition 2.5.1 *The set $\mathcal{F}$ is said to be (strongly)* precompact *if for every sequence $\{f_n\}$ in $\mathcal{F}$, there is a subsequence $\{f_{n_i}\}$ of $\{f_n\}$ such that*

$$\lim_{i\to\infty} \|f_{n_i} - \bar{f}\| = 0$$

for some $\bar{f} \in L^1$. If in addition $\mathcal{F}$ is closed in L^1, then $\mathcal{F}$ is said to be compact.

The Euclidean space $\mathbb{R}^N$ can be viewed as $L^1(X)$, where $X = \{1, 2, \cdots, N\}$ with the counting measure. It is well-known that a subset A of $\mathbb{R}^N$ is precompact if and only if A is bounded. Thus, any closed and bounded subset of $\mathbb{R}^N$ is compact.

Definition 2.5.2 *The set $\mathcal{F}$ is said to be* weakly precompact *if every sequence $\{f_n\}$ in $\mathcal{F}$ contains a subsequence $\{f_{n_i}\}$ which converges weakly to some $\bar{f} \in L^1(X)$, i.e.,*

$$\lim_{i\to\infty} \langle f_{n_i}, g\rangle = \langle \bar{f}, g\rangle, \ \forall\, g \in L^\infty(X).$$

If in addition $\mathcal{F}$ is closed in the weak topology of $L^1(X)$, then $\mathcal{F}$ is said to be weakly compact.

Remark 2.5.1 Since strong convergence implies weak convergence, it is obvious that if $\mathcal{F}$ is precompact, then it is also weakly precompact. Any subset of a strongly or weakly precompact set is itself strongly or weakly precompact.

The proof of the following criteria for the compactness can be seen in, e.g., [57].

Proposition 2.5.1 *Let $g \in L^1$. Then, the set*

$$\mathcal{F} = \{f \in L^1 : \ |f(x)| \leqslant |g(x)|, \ \forall\, x \in X \text{ a.e.}\}$$

is weakly compact. In particular, if $\mu(X) < \infty$, then any bounded subset of $L^\infty(X)$ is weakly precompact in $L^1(X)$.

Proposition 2.5.2 *Suppose that $\mu(X) < \infty$ and $p > 1$. Then any bounded subset of $L^p(X)$ is weakly precompact in $L^1(X)$.*

A very useful technique in the compactness argument for L^1 spaces is the following generalization of the classic Helly's lemma.

Theorem 2.5.1 *Let Ω be a bounded region in $\mathbb{R}^N$ such that its boundary $\partial\Omega$ is Lipschitz continuous. Then, bounded closed subsets of $BV(\Omega)$ are compact in $L^1(\Omega)$.*

Proof Suppose that $\{f_n\}$ is a sequence in $BV(\Omega)$ such that $\|f_n\|_{BV} \leqslant M$ uniformly, where M is a constant. For each n, by Theorem 2.4.2, we can choose $g_n \in C^\infty(\Omega)$ such that

$$\int_\Omega |g_n - f_n| \, \mathrm{d}m < \frac{1}{n} \quad \text{and} \quad \|g_n\|_{BV} \leqslant M + 2, \;\; n = 1, 2, \cdots .$$

By the Rellich theorem [1], the sequence $\{g_n\}$ is precompact in $L^1(\Omega)$, so it contains a subsequence which converges in $L^1(\Omega)$ to a function f. By Theorem 2.4.1, $f \in BV(\Omega)$ and $\|f\|_{BV} \leqslant M$, and f is the limit of a subsequence extracted from the original sequence $\{f_n\}$. □

We end this subsection by stating a characterization for the weak compactness of subsets of $L^1(X)$ with $\mu(X) < \infty$. Its proof is referred to [57] (Theorem IV.8.9).

Theorem 2.5.2 *Let (X, Σ, μ) be a finite measure space. Then, $\mathcal{F} \subset L^1(X)$ is* weakly precompact *if and only if*

(i) *$\mathcal{F}$ is bounded in $L^1(X)$, and*

(ii) *the integrals $\int_A f \mathrm{d}\mu$ are* uniformly countably additive *in the sense that for any $\epsilon > 0$, there is $\delta > 0$ such that*

$$\left| \int_A f \mathrm{d}m \right| < \epsilon$$

for all $A \subset X$ with $\mu(A) < \delta$ and all $f \in \mathcal{F}$.

2.5.2 Quasi-Compactness

Let $T : V \to V$ be a bounded linear operator on a Banach space $(V, \|\cdot\|_V)$.

Definition 2.5.3 *T is said to be compact if it maps bounded sets to precompact sets.*

The following is a classic result (see, e.g., [57]).

Proposition 2.5.3 *Suppose that for any sequence $\{f_n\}$ in V such that $\{f_n\}$ converges weakly to some $f \in V$, the sequence $\{Tf_n\}$ converges strongly to f. Then, $T : V \to V$ is* compact.

Compact operators constitute a class of important bounded linear operators. However, the Frobenius-Perron operator, which will be defined formally in Chapter 4, is not compact on L^1 in general. The concept of quasi-compactness plays a more important role in the analysis of Frobenius-Perron operators on different function subspaces of L^1.

Definition 2.5.4 ([79]) *T is said to be* quasi-compact *if there exists a positive integer r and a compact operator K such that*

$$\|T^r - K\|_V < 1. \tag{2.8}$$

In the book, we often let $(V, \|\cdot\|_V) = (BV(\Omega), \|\cdot\|_{BV})$, where Ω is a bounded region of $\mathbb{R}^N$, and we shall study quasi-compact Frobenius-Perron operators restricted to the dense subspace $BV(\Omega)$ of $L^1(\Omega)$. From Corollary 2.4.1 we know that $BV(\Omega)$ is itself a Banach space under the variation norm $\|\cdot\|_{BV}$, and Theorems 2.4.1 and 2.5.1 imply that closed bounded subsets of $(BV(\Omega), \|\cdot\|_{BV})$ are compact in $L^1(\Omega)$.

There are a number of equivalent definitions for quasi-compact linear operators. One of them can be stated as the following theorem (see [14, 57]).

Theorem 2.5.3 *$T : V \to V$ is quasi-compact if and only if there are bounded linear operators $\{Q_\lambda : \lambda \in \Lambda\}$ and R on V such that*

$$\begin{aligned}
&T^n = \sum_{\lambda\in\Lambda} \lambda^n \Phi_\lambda + R^n, \quad \forall\, n = 1, 2, \cdots,\\
&\Phi_\lambda \Phi_{\lambda'} = 0 \quad if \quad \lambda \neq \lambda',\\
&\Phi_\lambda^2 = \Phi_\lambda, \quad \forall\, \lambda \in \Lambda,\\
&\Phi_\lambda R = R\Phi_\lambda = 0, \quad \forall\, \lambda \in \Lambda,\\
&\Phi_\lambda V = D(\lambda), \quad \forall\, \lambda \in \Lambda,\\
&r(R) < 1,
\end{aligned}$$

where Λ is the set of the eigenvalues of T with modulus 1, $D(\lambda) = \{f \in V : Tf = \lambda f\}$ is the eigenspace *of T associated with eigenvalue λ, and $r(R) = \lim_{n\to\infty} \|R^n\|_V^{1/n}$ is the* spectral radius *of R.*

The following classical Ionescu-Tulcea and Marinescu theorem [14] gives a very useful sufficient condition for the quasi-compactness of a bounded linear operator on a dense subspace of L^1.

Theorem 2.5.4 (Ionescu-Tulcea and Marinescu theorem) *Let Ω be a bounded region of $\mathbb{R}^N$, and let $(V, \|\cdot\|_V)$ be a Banach space such that V is a dense vector subspace of $L^1(\Omega)$. Let $T : V \to V$ be a bounded linear operator with respect to both the norm $\|\cdot\|_V$ and the norm $\|\cdot\|$. Assume that*

(i) *if $f_n \in V$ for $n = 1, 2, \cdots$, $f \in L^1(\Omega)$, $\lim_{n\to\infty} \|f_n - f\| = 0$, and $\|f_n\|_V \leqslant M$ for all n, then $f \in V$ and $\|f\|_V \leqslant M$, where M is a constant;*

(ii) $\sup_{n\geqslant 0} \{\|T^n f\|/\|f\| : f \in V, f \neq 0\} < \infty$;

(iii) *there exist $k \geqslant 1$, $0 < \alpha < 1$, and $\beta < \infty$ such that*

$$\|T^k f\|_V \leqslant \alpha\|f\|_V + \beta\|f\|, \quad \forall\, f \in V; \tag{2.9}$$

(iv) *if* V_0 *is a bounded subset of* $(V, \|\cdot\|_V)$, *then* $T^k V_0$ *is precompact in* $L^1(\Omega)$. *Then,* Λ *has only a finite number of elements,* $D(\lambda)$ *is finite dimensional for each* $\lambda \in \Lambda$, *and* $T : (V, \|\cdot\|_V) \to (V, \|\cdot\|_V)$ *is quasi-compact.*

Exercises

2.1 Let $f(x) = \sin x$. Show that $f \in BV[0, 2\pi]$ and calculate $V(x) \equiv \bigvee_0^x f$ for $x > 0$.

2.2 Let

$$f(x) = \begin{cases} x^2 \sin \frac{1}{x}, & 0 < x \leqslant 2\pi, \\ 0, & x = 0. \end{cases}$$

Show that $f \in BV(0, 2\pi)$.

2.3 Let

$$f(x) = \begin{cases} x \sin \frac{1}{x}, & 0 < x \leqslant 2\pi, \\ 0, & x = 0. \end{cases}$$

Show that f is not of bounded variation on $[0, 2\pi]$.

2.4 Show that if a function f is of bounded variation on $[a, b]$, then it is bounded on $[a, b]$.

2.5 Show that the product of two functions of bounded variation on $[a, b]$ is also of bounded variation on $[a, b]$.

2.6 Let f, g be functions of bounded variation on $[a, b]$ such that $g(x) \geqslant c > 0$ for $x \in [a, b]$. Show that the quotient function f/g is also of bounded variation on $[a, b]$.

2.7 Show that Lipschitz continuous functions on $[a, b]$ are of bounded variation on $[a, b]$.

2.8 Let (X, Σ, μ) be a finite measure space and let $1 \leqslant p_1 \leqslant p_2 \leqslant \infty$. Show that every strongly precompact subset of L^{p_2} is also strongly precompact in L^{p_1}. Is the same true for weak precompactness?

2.9 Consider the following four families of functions:

(i) $f_a(x) = ae^{-ax}$, $a \geqslant 1$;
(ii) $f_a(x) = ae^{-ax}$, $0 \leqslant a \leqslant 1$;
(iii) $f_a(x) = e^{-x} \sin ax$, $a \geqslant 1$;
(iv) $f_a(x) = e^{-x} \sin ax, 0 \leqslant a \leqslant 1$.

Which of these families is weakly and/or strongly precompact in $L^1(0, \infty)$?

Chapter 3
Rudiments of Ergodic Theory

JiuDing, Department of Mathematics, The University of Southern Mississippi in America.
Aihui Zhou, Academy of Mathematics and Systems Science, Chinese Academy of Sciences.

Abstract We give a short introduction to ergodic theory and its applications to topological dynamical systems. First we study the general properties of measure preserving transformations. Then we introduce the concepts of ergodicity, mixing, and exactness that describe different levels of chaotic behavior of the deterministic dynamics. The classic Birkhoff pointwise ergodic theorem and von Neumann mean ergodic theorem are stated, and some characteristics of ergodicity, mixing and exactness in terms of function sequences convergence are also presented.
Keywords Measure preserving transformation, ergodicity, mixing, exactness, ergodic theorems, topological dynamical systems, Frobenius-Perron operators on measures, unique ergodicity.

There are two fundamental questions in ergodic theory and its applications. One is the investigation of useful properties of measure preserving transformations on measure spaces, ergodic invariants, and the spectral analysis of related linear operators. The other is about the existence of important invariant measures for various classes of measurable transformations, and their spectral properties and statistical consequences. We study the first question in this chapter. We shall introduce some classic ergodic theorems, and apply these convergence theorems to characterizing ergodicity, mixing, and exactness of the given dynamical systems. These properties represent different levels of the degree of chaotic behaviors of deterministic systems.

In the first three sections, we study the general properties of measure preserving transformations on measure spaces, while in Section 3.4 we focus on the ergodic theory of continuous transformations defined on compact metric spaces. The concept of Frobenius-Perron operators will be introduced in the next chapter to study the existence of absolutely continuous invariant finite measures, that is an answer to the second question above.

3.1 Measure Preserving Transformations

Let (X, Σ, μ) be a measure space. As usual, it is assumed to be σ-finite.

Definition 3.1.1 *A transformation $S : X \to X$ is called* measurable *if* $S^{-1}(A)$

$\in \Sigma$ *for each* $A \in \Sigma$.

Definition 3.1.2 *A measurable transformation* $S : X \to X$ *is said to be* measure preserving *or to* preserve μ, *or alternatively, the measure* μ *is said to be* S-invariant *or* invariant under S, *if*

$$\mu(S^{-1}(A)) = \mu(A), \ \ \forall\, A \in \Sigma. \tag{3.1}$$

Remark 3.1.1 It is practically difficult to check (3.1) for every measurable set in the above definition. In practice, if (3.1) is valid for all measurable sets in a subclass π which is closed under the intersection operation of its members and which generates Σ, i.e., Σ is the smallest σ-algebra that contains π, then μ is invariant under S [8]. The subclass π with the above property is called a π-*system.*

Another sufficient and necessary condition for the S-invariance of μ is given by the following result, which is also a consequence of the dual relation of Frobenius-Perron operators and Koopman operators under the additional assumption of nonsingularity for S in the next chapter.

Proposition 3.1.1 *Let* $S : X \to X$ *be a measurable transformation of a finite measure space* (X, Σ, μ). *Then, the measure* μ *is* S-invariant *if and only if*

$$\int_X g(x)\mathrm{d}\mu(x) = \int_X g(S(x))\mathrm{d}\mu(x), \ \ \forall\, g \in L^\infty. \tag{3.2}$$

Proof If (3.2) is satisfied, then by letting $g = \chi_A$ for any $A \in \Sigma$ and noting that $\chi_A \circ S = \chi_{S^{-1}(A)}$, we have

$$\mu(A) = \int_X \chi_A \mathrm{d}\mu = \int_X \chi_A \circ S \mathrm{d}\mu = \int_X \chi_{S^{-1}(A)}\mathrm{d}\mu = \mu(S^{-1}(A)),$$

so S preserves μ.

Conversely, if μ is invariant under S, then $\int_X \chi_A \mathrm{d}\mu = \int_X \chi_A \circ S \mathrm{d}\mu$ for all $A \in \Sigma$. Thus, (3.2) holds for all simple functions, and so a limit process shows that (3.2) is true for all $g \in L^\infty$. □

In the following, we give several examples of measure preserving transformations.

Example 3.1.1 The *identity transformation* $I : X \to X$ of any measure space (X, Σ, μ) defined by $I(x) = x$, $\forall\, x \in X$ is measure preserving.

Example 3.1.2 The tent function $T : [0, 1] \to [0, 1]$ defined by (1.2) is measure preserving with respect to the Lebesgue measure m on the interval $[0, 1]$.

Example 3.1.3 Let k be a positive integer greater than or equal to 2. Then, the *k-adic transformation*

$$S(x) = kx \text{ (mod 1)}, \ \forall \in [0,1]$$

is m-invariant.

Proof We first note that the family of all closed intervals $[a,b] \subset [0,1]$ is a π-system which generates the Borel σ-algebra of $[0,1]$. The inverse image $S^{-1}([a,b])$ of $[a,b]$ is the disjoint union of k subintervals of length $(b-a)/k$, so $m(S^{-1}([a,b])) = m([a,b])$. Thus, m is S-invariant from Remark 3.1.1. □

The same argument applies to the next several examples.

Example 3.1.4 The *translation mapping*

$$S(x) = x + a \text{ (mod 1)}, \ \forall\, x \in [0,1],$$

where a is a real number, preserves the Lebesbue measure m on $[0,1]$.

Example 3.1.5 Let $\partial\mathbb{D}$ be the unit circle of the complex plane $\mathbb{C}$. Given $a \in \partial\mathbb{D}$, then the *rotation mapping* $R : \partial\mathbb{D} \to \partial\mathbb{D}$ defined by

$$R(z) = az, \ \forall\, z \in \partial\mathbb{D}$$

preserves the Harr measure on $\partial\mathbb{D}$.

Proof Note that the Harr measure is exactly the same as the normalized Lebesgue measure on the circle, and the invariance follows from the facts that the family of all closed arcs of $\partial\mathbb{D}$ is a π-system and is a generator for the Borel σ-algebra of $\partial\mathbb{D}$ and that the rotation mapping preserves the length of any arc. □

Remark 3.1.2 By the same token, any translation mapping $S(x) = ax$ on a *compact group* G preserves the Harr measure on G, where a is a fixed element of G. More generally, any continuous isomorphism of G is Harr measure invariant.

Our last example concerns the logistic model that we encountered in Chapter 1.

Example 3.1.6 The quadratic mapping

$$S(x) = 4x(1-x), \ \forall\, x \in [0,1]$$

preserves the absolutely continuous probability measure μ of the interval $[0,1]$ given by

$$\mu(A) = \int_A \frac{1}{\pi\sqrt{x(1-x)}}\mathrm{d}x, \ \forall\, A \in \mathcal{B}.$$

Proof. We use Proposition 3.1.1 since verification of (3.1) is tedious even for $A = [a, b]$. A direct computation shows that

$$\int_0^1 g(x)\mathrm{d}\mu(x) = \int_0^1 g(x)\frac{1}{\pi\sqrt{x(1-x)}}\mathrm{d}x = \frac{2}{\pi}\int_0^{\frac{\pi}{2}} g(\sin^2\theta)\mathrm{d}\theta$$

and

$$\begin{aligned}\int_0^1 g(S(x))\mathrm{d}\mu(x) &= \int_0^1 g(4x(1-x))\frac{1}{\pi\sqrt{x(1-x)}}\mathrm{d}x\\ &= \frac{2}{\pi}\int_0^{\frac{\pi}{2}} g(\sin^2(2\theta))\mathrm{d}\theta = \frac{1}{\pi}\int_0^{\pi} g(\sin^2 t)\mathrm{d}t = \frac{2}{\pi}\int_0^{\frac{\pi}{2}} g(\sin^2 t)\mathrm{d}t,\end{aligned}$$

so the result follows. □

3.2 Ergodicity, Mixing and Exactness

Chaotic discrete dynamical systems are closely related to the concept of *transitivity* of orbits, which is related to the property of *indecomposability* of the transformation. In this section, we shall introduce the concepts of ergodicity, mixing, and exactness that give different levels of the chaotic behavior.

3.2.1 Ergodicity

If for a measure preserving transformation $S : (X, \Sigma, \mu) \to (X, \Sigma, \mu)$, there is a nontrivial set $A \in \Sigma$ different from X such that $S^{-1}(A) = A$, then $S^{-1}(A^c) = A^c$, and so the dynamics of S on X can be decomposed into two parts: $S|_A : A \to A$ and $S|_{A^c} : A^c \to A^c$. If this cannot happen, then we have the property of *ergodicity.* A set A is said to be *S-invariant* or an *invariant set* of S if it satisfies

$$S^{-1}(A) = A.$$

Definition 3.2.1 *Let (X, Σ, μ) be a measure space. A measurable transformation $S : X \to X$ is said to be* ergodic *if every invariant set $A \in \Sigma$ of S is such that either $\mu(A) = 0$ or $\mu(A^c) = 0$. In other words, S is ergodic if and only if its invariant sets or their complements are equal to the empty set almost everywhere. Such sets are called* trivial subsets *of X.*

Determining whether a given transformation is ergodic or not is difficult in general from the definition. However, the following theorem may be of use sometimes. First, we need the definition of nonsingular transformations.

Definition 3.2.2 *A measurable transformation S from a measure space (X, Σ, μ) into itself is said to be* nonsingular *if $\mu(A) = 0$ implies $\mu(S^{-1}(A)) = 0$.*

Theorem 3.2.1 *Let* (X, Σ, μ) *be a measure space and let* $S : X \to X$ *be a nonsingular transformation. Then,* S *is ergodic if and only if, for every bounded measurable function* $f : X \to \mathbb{R}$,

$$f(S(x)) = f(x), \quad \forall\, x \in X \ \mu\text{-a.e.} \tag{3.3}$$

implies that $f(x)$ *is constant almost everywhere.*

Proof Suppose that S is ergodic. If (3.3) is satisfied by a nonconstant bounded function f, then there is some real number a such that the sets $A = S^{-1}((-\infty, a])$ and $B = S^{-1}((a, \infty))$ have positive measures. Since

$$\begin{aligned} S^{-1}(A) &= \{x : S(x) \in A\} = \{x : f(S(x)) \leqslant a\} \\ &= \{x : f(x) \leqslant a\} = A \end{aligned}$$

and a similar equality holds for B, we have a decomposition of X into nontrivial invariant sets A and B of S, which contradicts the assumption that S is ergodic.

Conversely, assume that S is not ergodic. By Definition 3.2.1, there is a nontrivial invariant set $A \in \Sigma$ of S. Then, the function χ_A is nonconstant. Moreover, since $S^{-1}(A) = A$,

$$\chi_A(S(x)) = \chi_{S^{-1}(A)}(x) = \chi_A(x), \quad \forall\, x \in X \ \mu\text{-a.e.},$$

and so (3.3) is satisfied by the nonconstant function χ_A. □

Any function f satisfying (3.3) is called an *invariant function.*

Example 3.2.1 The rotation mapping of Example 3.1.5 is not ergodic if $a = e^{i\theta}$, where $\theta/2\pi$ is rational. For example, if $\theta = \pi/3$, then the union of the arcs with end points $e^{i(k-1)\pi/6}$ and $e^{ik\pi/6}$ for $k = 1, 2, \cdots, 6$ is a nontrivial invariant set of the rotation. When $\theta/2\pi$ is irrational, the rotation mapping becomes ergodic, which can be proved by using Theorem 4.3.3 (i) which will be presented in Section 4.3 (see Example 4.4.1 in [82]).

Theorem 3.2.2 below provides another equivalent condition of ergodicity in terms of the limit of a sequence of *Cesáro averages* for measure preserving transformations.

Theorem 3.2.2 *Let* (X, Σ, μ) *be a probability measure space and let* $S : X \to X$ *be measure preserving. Then,* S *is ergodic if and only if for all* $A, B \in \Sigma$,

$$\lim_{n\to\infty} \frac{1}{n} \sum_{i=0}^{n-1} \mu(S^{-i}(A) \cap B) = \mu(A)\mu(B). \tag{3.4}$$

Proof For the sufficiency part, suppose that S is not ergodic with respect to μ. Then, there are $A,\ B \in \Sigma$ that are nontrivial disjoint invariant sets of S. Since $S^{-n}(A) \cap B = \varnothing$ for each n, the left-hand side of (3.4) is zero, while the right-hand side $\mu(A)\mu(B) > 0$. The necessity of (3.4) for ergodicity is an easy consequence of the Birkhoff pointwise ergodic theorem in the next section. □

3.2.2 Mixing and Exactness

The characterization (3.4) of ergodicity motivates the following definitions of mixing and weakly mixing, which give stronger notions of irregularity for the deterministic dynamics.

Definition 3.2.3 *Let (X, Σ, μ) be a probability measure space and let $S : X \to X$ be a measure preserving transformation. S is said to be* mixing *if for all $A, B \in \Sigma$,*

$$\lim_{n\to\infty} \mu(S^{-n}(A) \cap B) = \mu(A)\mu(B). \tag{3.5}$$

Definition 3.2.4 *Let (X, Σ, μ) be a probability measure space and let $S : X \to X$ be a measure preserving transformation. S is said to be* weakly mixing *if for all $A, B \in \Sigma$,*

$$\lim_{n\to\infty} \frac{1}{n}\sum_{i=0}^{n-1} \left|\mu(S^{-i}(A) \cap B) - \mu(A)\mu(B)\right| = 0.$$

By using the simple fact that

$$\lim_{n\to\infty} a_n = L \text{ implies } \lim_{n\to\infty} \frac{1}{n}\sum_{i=0}^{n-1} a_i = L \tag{3.6}$$

for any sequence $\{a_n\}$ of numbers, we see that if S is mixing, then it is weakly mixing which obviously implies ergodicity of S by Theorem 3.2.2. Mixing can be interpreted as that the fraction of the points starting in B that ended up in A after n iterations as n increases without bound is just given by the product of the measures of A and B and is independent of where they are in X.

Example 3.2.2 The *dyadic transformation* $S(x) = 2x \pmod 1$ preserves the Lebesgue measure m. Let $A = [0, a]$. Then, we see easily that $S^{-n}(A)$ consists of the 2^n subintervals

$$\left[\frac{i}{2^n}, \frac{i+a}{2^n}\right], \quad i = 0, 1, \cdots, 2^n - 1.$$

It follows that $m(S^{-n}(A) \cap B) \to m(A)m(B)$ as $n \to \infty$. More generally, the k-adic transformation is mixing.

For the dyadic transformation, since $S^n(x) = 2^n x \pmod 1$, we deduce that $S^n([0, a]) = [0, 1]$ eventually for any $a > 0$. This implies that $m(S^n(A)) \to 1$ for any nontrivial measurable set $A \subset [0, 1]$ as $n \to \infty$. This property of "exactness" is made precise by the following definition due to Rochlin [112].

Definition 3.2.5 *Let (X, Σ, μ) be a probability measure space and let $S : X \to X$ be a measure preserving transformation such that $S\Sigma \subset \Sigma$, where $S\Sigma \equiv \{S(A) : A \in \Sigma\}$. S is said to be* exact *if for all $A \in \Sigma$ such that $\mu(A) > 0$,*

$$\lim_{n\to\infty} \mu(S^n(A)) = 1. \tag{3.7}$$

It can be shown that exactness implies mixing; see Theorem 4.3.3 in the next chapter. Thus, the four notions of exactness, mixing, weakly mixing, and ergodicity give a hierarchy of chaotic behaviors. The following theorem characterizes exactness in terms of a specially defined sub-σ-algebra.

Theorem 3.2.3 *Let (X, Σ, μ) be a probability measure space and let $S : X \to X$ be a measure preserving transformation such that $S\Sigma \subset \Sigma$. Then, S is exact if and only if the sub-σ-algebra*

$$\Sigma^{\mathrm{T}} \equiv \bigcap_{n=0}^{\infty} S^{-n}\Sigma$$

consists of only the sets of μ-measure 0 or 1.

Proof Suppose that there is $A \in \Sigma^{\mathrm{T}}$ with $0 < \mu(A) < 1$. Then, $A = S^{-n}A_n$, where $A_n \in \Sigma$ for each n. We have $\mu(A) = \mu(A_n)$ since S preserves μ. The inclusion $S^n(A) = S^n(S^{-n}A_n) \subset A_n$ implies that $\mu(S^n(A)) \leqslant \mu(A) < 1$ for all n. Hence, S is not exact.

Now, assume that Σ^{T} consists of the sets of μ-measure 0 or 1. Suppose that there is $A \in \Sigma$ such that $\mu(A) > 0$ and (3.7) is not true. Then, without loss of generality, we may assume that $\mu(S^n(A)) \leqslant r$ for some $r < 1$ and all n. Since the set sequence $\{S^{-n}(S^n(A))\}$ is monotonically increasing with respect to the inclusion relation, the set $B = \bigcup_{n=0}^{\infty} S^{-n}(S^n(A)) \in \Sigma^{\mathrm{T}}$. Since $B \supset A$, we have $\mu(B) \geqslant \mu(A) > 0$, so $\mu(B) = 1$ by the assumption. On the other hand,

$$\mu(B) = \lim_{n\to\infty} \mu(S^{-n}(S^n(A))) = \lim_{n\to\infty} \mu(S^n(A)) \leqslant r < 1,$$

hence we are led to a contradiction. □

Remark 3.2.1 Invertible transformations cannot be exact. In fact, if S is an invertible measure preserving transformation on a probability measure space, then for $A \in \Sigma$ with $\mu(A) < 1$,

$$\mu(S(A)) = \mu(S^{-1}(S(A))) = \mu(A).$$

This implies by induction that $\mu(S^n(A)) = \mu(A)$ for any n, which violates (3.7).

Remark 3.2.2 Exactness means that the images of any nontrivial set $A \in \Sigma$ will spread and completely fill the space X eventually. Mixing means that the images of any set $B \in \Sigma$ under the iteration of S become independent of any fixed set $A \in \Sigma$ asymptotically. Weak mixing means that B becomes independent of A if we neglect a finite number of times. Ergodicity means that B becomes independent of A on the average.

Remark 3.2.3 It is sufficient to check the convergence properties on a π-system generating Σ in Theorem 3.2.2, Definitions 3.2.3—3.2.5 for ergodicity, mixing, weak mixing, and exactness.

3.3 Ergodic Theorems

Invariant measures describe the asymptotic statistical distribution of individual orbits of the iterates under the measure preserving transformation for almost all initial points with respect to the measure. Let $S : (X, \Sigma, \mu) \to (X, \Sigma, \mu)$ be a measure preserving transformation. A typical question about the statistical properties of the dynamics is on the *frequency* of the iterates $S^n(x)$ of x that fall into a given set $A \in \Sigma$.

The nth iterate $S^n(x)$ belongs to A if and only if $\chi_A(S^n(x)) = 1$. Therefore, the number of the elements among the first n iterates

$$\{x, S(x), S^2(x), \cdots, S^{n-1}(x)\}$$

that visits the set A is exactly the same as $\sum_{i=0}^{n-1} \chi_A(S^i(x))$, and so the relative frequency of such elements entering A is equal to the fraction $n^{-1} \sum_{i=0}^{n-1} \chi_A(S^i(x))$. The asymptotic behavior of these frequencies is indicated by the following Birkhoff pointwise ergodic theorem published in 1931.

Theorem 3.3.1 (Birkhoff's pointwise ergodic theorem) *Let (X, Σ, μ) be a measure space and let $S : (X, \Sigma, \mu) \to (X, \Sigma, \mu)$ be a measure preserving transformation. Then, for each $f \in L^1(\mu)$, there exists a function $\tilde{f} \in L^1(\mu)$ such that $\tilde{f}(S(x)) = \tilde{f}(x)$, $x \in X$ μ-a.e. and*

$$\lim_{n\to\infty} \frac{1}{n} \sum_{i=0}^{n-1} f(S^i(x)) = \tilde{f}(x), \quad \forall\, x \in X \ \mu\text{-a.e.}$$

Furthermore, if $\mu(X) < \infty$, then $\int_X \tilde{f} d\mu = \int_X f d\mu.$

While we do not intend to prove this classic result, we point out that the last conclusion of the above theorem follows from Theorem 2.2.1 and Proposition 3.1.1.

By Theorem 3.2.1, if S is ergodic, then $f \circ S = f$ implies that f is constant almost everywhere. Thus, immediately we have

Corollary 3.3.1 *If S is ergodic, then $\tilde{f}$ is a constant function, and if in addition $\mu(X) < \infty$, then*

$$\tilde{f} = \frac{1}{\mu(X)} \int_X f \mathrm{d}\mu.$$

Consequently, if $\mu(X) = 1$ and S is ergodic, then

$$\lim_{n\to\infty} \frac{1}{n} \sum_{i=0}^{n-1} \chi_A(S^i(x)) = \mu(A), \quad \forall\, x \in X \ \mu\text{-a.e.}, \tag{3.8}$$

which means that the orbit of μ almost every point of X enters the set A with an asymptotic relative frequency $\mu(A)$.

Remark 3.3.1 (3.8) gives immediately the necessity of (3.4) for the ergodicity of a measure preserving transformation on a finite measure space.

Using functional analysis for unitary operators, von Neumann proved the following mean ergodic theorem for $p = 2$ in 1931, the proof of which is referred to [79].

Theorem 3.3.2 (von Neumann's mean ergodic theorem) *Let (X, Σ, μ) be a finite measure space, let $S : (X, \Sigma, \mu) \to (X, \Sigma, \mu)$ be a measure preserving transformation, and let $1 \leqslant p < \infty$. Then, for each $f \in L^p(\mu)$, there exists a function $\tilde{f} \in L^p(\mu)$ such that $\tilde{f}(S(x)) = \tilde{f}(x)$, $\forall\, x \in X$ μ-a.e., and*

$$\lim_{n\to\infty} \left\| \frac{1}{n} \sum_{i=0}^{n-1} f \circ S^i - \tilde{f} \right\|_p = 0.$$

The following ergodic theorem expresses the concepts of ergodicity, weak mixing, mixing, and exactness in functional form. This is useful for checking whether concrete transformations have the mixing properties. It will also be used in the proof of Theorem 3.4.9.

Theorem 3.3.3 *Let (X, Σ, μ) be a probability measure space and let $S : (X, \Sigma, \mu) \to (X, \Sigma, \mu)$ be a measure preserving transformation.*

(i) *The following are equivalent:*

(1) *S is ergodic.*

(2) *For all f, $g \in L^2(\mu)$,*

$$\lim_{n\to\infty} \frac{1}{n} \sum_{i=0}^{n-1} \int_X (f \circ S^i) g \mathrm{d}\mu = \int_X f \mathrm{d}\mu \int_X g \mathrm{d}\mu.$$

(3) *For all $f \in L^2(\mu)$,*

$$\lim_{n\to\infty} \frac{1}{n} \sum_{i=0}^{n-1} \int_X (f \circ S^i) f \mathrm{d}\mu = \left(\int_X f \mathrm{d}\mu \right)^2.$$

(ii) *The following are equivalent:*
(1) S *is weakly mixing.*
(2) *For all* $f,\ g \in L^2(\mu)$,

$$\lim_{n\to\infty}\frac{1}{n}\sum_{i=0}^{n-1}\left|\int_X (f\circ S^i)g\mathrm{d}\mu - \int_X f\mathrm{d}\mu\int_X g\mathrm{d}\mu\right| = 0. \tag{3.9}$$

(3) *For all* $f \in L^2(\mu)$,

$$\lim_{n\to\infty}\frac{1}{n}\sum_{i=0}^{n-1}\left|\int_X (f\circ S^i)f\mathrm{d}\mu - \left(\int_X f\mathrm{d}\mu\right)^2\right| = 0.$$

(4) *For all* $f \in L^2(\mu)$,

$$\lim_{n\to\infty}\frac{1}{n}\sum_{i=0}^{n-1}\left[\int_X (f\circ S^i)f\mathrm{d}\mu - \left(\int_X f\mathrm{d}\mu\right)^2\right]^2 = 0.$$

(iii) *The following are equivalent:*
(1) S *is mixing.*
(2) *For all* $f,\ g \in L^2(\mu)$,

$$\lim_{n\to\infty}\int_X (f\circ S^n)g\mathrm{d}\mu = \int_X f\mathrm{d}\mu\int_X g\mathrm{d}\mu. \tag{3.10}$$

(3) *For all* $f \in L^2(\mu)$,

$$\lim_{n\to\infty}\int_X (f\circ S^n)f\mathrm{d}\mu = \left(\int_X f\mathrm{d}\mu\right)^2. \tag{3.11}$$

(iv) *Under the additional assumption that* $S\Sigma \subset \Sigma$, S *is exact if for all* $f \in L^2(\mu)$,

$$\lim_{n\to\infty}\int_X \left(f\circ S^n - \int_X f\mathrm{d}\mu\right)^2 \mathrm{d}\mu = 0. \tag{3.12}$$

Proof (i), (ii), (iii) can be proved using similar methods. We just prove (iii) to illustrate the ideas before proving (iv). Readers can finish the proof of (i) and (ii) with a slight modification.

The proof of (iii):

(2) $\Rightarrow$ (1). This follows by putting $f = \chi_A$, $g = \chi_B$ in (3.10) for $A,\ B \in \Sigma$ and Definition 3.2.3 for a mixing transformation.

(1) $\Rightarrow$ (3). Definition 3.2.3 means that (3.10) is valid for $f = \chi_A$, $g = \chi_B$ with $A,\ B \in \Sigma$. Using linear combinations of characteristic functions, we see

that (3.10) is true for all simple functions f and g. In particular, (3.11) is satisfied by all simple functions h.

Now, suppose that $f \in L^2(\mu)$, and let $\epsilon > 0$ be given. Choose a simple function h such that $\|f-h\|_2 < \epsilon$, and choose a positive integer $n(\epsilon)$ such that $n \geqslant n(\epsilon)$ implies

$$\left| \int_X (h \circ S^n) h \mathrm{d}\mu - \left(\int_X h \mathrm{d}\mu \right)^2 \right| < \epsilon.$$

Then, for $n \geqslant n(\epsilon)$,

$$\begin{aligned}
&\left| \int_X (f \circ S^n) f \mathrm{d}\mu - \left(\int_X f \mathrm{d}\mu \right)^2 \right| \\
&\leqslant \left| \int_X (f \circ S^n) f \mathrm{d}\mu - \int_X (h \circ S^n) f \mathrm{d}\mu \right| \\
&\quad + \left| \int_X (h \circ S^n) f \mathrm{d}\mu - \int_X (h \circ S^n) h \mathrm{d}\mu \right| \\
&\quad + \left| \int_X (h \circ S^n) h \mathrm{d}\mu - \left(\int_X h \mathrm{d}\mu \right)^2 \right| \\
&\quad + \left| \left(\int_X h \mathrm{d}\mu \right)^2 - \int_X f \mathrm{d}\mu \int_X h \mathrm{d}\mu \right| \\
&\quad + \left| \int_X f \mathrm{d}\mu \int_X h \mathrm{d}\mu - \left(\int_X f \mathrm{d}\mu \right)^2 \right| \\
&\leqslant \left| \int_X [(f-h) \circ S^n] f \mathrm{d}\mu \right| + \left| \int_X (h \circ S^n)(f-h) \mathrm{d}\mu \right| + \epsilon \\
&\quad + \left| \int_X h \mathrm{d}\mu \right| \left| \int_X (h-f) \mathrm{d}\mu \right| + \left| \int_X f \mathrm{d}\mu \right| \left| \int_X (h-f) \mathrm{d}\mu \right| \\
&\leqslant \|f-h\|_2 \|f\|_2 + \|f-h\|_2 \|h\|_2 + \epsilon \\
&\quad + \|h\|_2 \|f-h\|_2 + \|f\|_2 \|h-f\|_2 \\
&\leqslant \epsilon \|f\|_2 + \epsilon(\|f\|_2 + \epsilon) + \epsilon + (\|f\|_2 + \epsilon)\epsilon + \epsilon \|f\|_2,
\end{aligned}$$

where the third inequality is from the Cauchy-Schwarz inequality. Therefore, (3.11) holds for all $f \in L^2(\mu)$.

(3) $\Rightarrow$ (2). Let $f \in L^2(\mu)$ and let W denote the smallest closed vector subspace of $L^2(\mu)$ that contains f and all the constant functions and satisfies the condition that $g \circ S \in W$ for all $g \in W$. The set

$$V \equiv \left\{ g \in L^2(\mu) : \lim_{n \to \infty} \int_X (f \circ S^n) g \mathrm{d}\mu = \int_X f \mathrm{d}\mu \int_X g \mathrm{d}\mu \right\}$$

is a closed vector subspace of $L^2(\mu)$ that contains f and all the constant functions. It is clear that $g \in V$ implies $g \circ S \in V$. So $V \supset W$. On the other hand, if $g \in W^{\perp}$, the *orthogonal complement* of W in $L^2(\mu)$, then $\int_X g d\mu = 0$ and $\int_X (f \circ S^n) g d\mu = 0$ for all $n \geqslant 0$. Therefore, $W^{\perp} \subset V$, which implies that $V = L^2(\mu)$.

The proof of (iv):

Suppose that (3.12) is true. Let $A \in \Sigma$ be such that $\mu(A) > 0$ and define $f_A(x) = \chi_A(x)/\mu(A)$ for $x \in X$. Then, $\lim_{n\to\infty} \|f_A \circ S^n - 1\|_2 = 0$. Since μ is S-invariant, by Proposition 3.1.1, $\int_X \chi_{S^n(A)} \circ S^n d\mu = \int_X \chi_{S^n(A)} d\mu$, so

$$
\begin{aligned}
\mu(S^n(A)) &= \int_X \chi_{S^n(A)} d\mu \\
&= \int_X f_A(\chi_{S^n(A)} \circ S^n) d\mu - \int_X (f_A - 1)\chi_{S^n(A)} \circ S^n d\mu \\
&= \int_{S^{-n}(S^n(A))} f_A d\mu - \int_{S^{-n}(S^n(A))} (f_A - 1) d\mu \\
&= 1 - \int_{S^n(A)} (f_A - 1) \circ S^n d\mu = 1 - \int_{S^n(A)} (f_A \circ S^n - 1) d\mu \\
&\geqslant 1 - \int_X |f_A \circ S^n - 1| d\mu \geqslant 1 - \|f_A \circ S^n - 1\|_2 \to 1
\end{aligned}
$$

as $n \to \infty$. Thus, S is exact. □

Remark 3.3.2 A subset A of a Banach space B is said to be a *fundamental set* of B if all linear combinations of elements of A are *dense* in B. In the above theorem, it is enough to require the convergence conditions for f and g belonging to fundamental sets of $L^2(\mu)$.

3.4 Topological Dynamical Systems

In the previous sections, we have studied the measure-theoretic properties of measure preserving transformations of general measure spaces. The subject of *topological dynamical systems* studies the asymptotic properties of continuous transformations of topological spaces. In this section, we present some basic concepts of topological dynamics and several general results concerning invariant measures for continuous transformations on compact metric spaces.

Let $(X, d(\cdot,\cdot))$ be a compact metric space and let $C(X)$ be the Banach space of all continuous functions defined on X with the *max-norm* $\|f\|_\infty = \max_{x\in X} |f(x)|$ for $f \in C(X)$. When $X = [a, b]$, we write $C[a, b]$ instead of $C([a, b])$. Given a continuous transformation $S: X \to X$, we denote by $U \equiv U_S : C(X) \to C(X)$ the *composition operator* defined by $Uf(x) = f(S(x))$. In the next chapter, when we study Frobenius-Perron operators, this operator U will be called the Koopman operator which is defined on $L^\infty(X)$ with a given measure μ on X. The composition operator U is clearly linear and *multiplicative*, i.e., $U(f \cdot g) = (Uf)(Ug)$. It can also be used in the statement of the ergodic theorems in the previous section.

Example 3.4.1 Let $Y = \{0, 1, \cdots, k-1\}$ with the *discrete topology.* Let $X = \prod_{n=-\infty}^{\infty} Y$ with the *product topology* which is equivalent to the topology induced by the metric on X given by

$$d(x, y) = \sum_{n=-\infty}^{\infty} \frac{|x_n - y_n|}{2^{|n|}},$$

where $x = \{x_n\}_{n=-\infty}^{\infty}$, $y = \{y_n\}_{n=-\infty}^{\infty} \in X$. Then, the *two-sided shift* $S: X \to X$, defined by $S(x) = y$ with $y_n = x_{n+1}$ for all n, is a *homeomorphism* of X, that is, S is one-to-one and onto, and both $S: X \to X$ and $S^{-1}: X \to X$ are continuous. If we let $X = \prod_{n=0}^{\infty} Y$ with the product topology, then the *one-sided shift* $S(\{x_0, x_1, x_2, \cdots\}) = \{x_1, x_2, x_3, \cdots\}$ is a continuous transformation on X. The inverse image under S of any point consists of k points.

Example 3.4.2 Suppose that $S: X \to X$ is the two-sided shift as above. Let $\boldsymbol{A} = (a_{ij})_{i,j=0}^{k-1}$ be a (0, 1) matrix of order $k \times k$, i.e., $a_{ij} \in \{0, 1\}$ for all $i, j = 0, 1, \cdots, k-1$. Denote $X_{\boldsymbol{A}} = \{\{x_n\}_{n=-\infty}^{\infty} \in X : a_{x_n x_{n+1}} = 1, \ \forall\, n\}$. In other words, $X_{\boldsymbol{A}}$ consists of all the bi-sequences $\{x_n\}_{n=-\infty}^{\infty}$ whose neighboring pairs are "allowed" by the matrix $\boldsymbol{A}$. Then, $X_{\boldsymbol{A}}$ is a closed subset of X and $S(X_{\boldsymbol{A}}) = X_{\boldsymbol{A}}$, so that $S: X_{\boldsymbol{A}} \to X_{\boldsymbol{A}}$ is a homeomorphism of $X_{\boldsymbol{A}}$, which is called the *sub-shift of finite type* (or *two-sided topological Markov chain*) determined by the matrix $\boldsymbol{A}$. In particular, if $a_{ij} \equiv 1$, then $X_{\boldsymbol{A}} = X$, and if $\boldsymbol{A} = \boldsymbol{I}$, the identity matrix, then $X_{\boldsymbol{A}}$ consists of the k points $\{0\}_{-\infty}^{\infty}, \{1\}_{-\infty}^{\infty}, \cdots, \{k-1\}_{-\infty}^{\infty}$. One can define *one-sided topological Markov chains* similarly. Sub-shifts of finite type are important models in some diffeomorphisms with significant applications in statistical physics [13].

In the following, we first study the topological properties of continuous transformations, and then we explore the ergodic properties of such transformations.

Definition 3.4.1 *A homeomorphism $S: X \to X$ of a compact metric space X is said to be* minimal *if for any $x \in X$, the bi-sequence $\{S^n(x)\}_{n=-\infty}^{\infty}$, which is also called the* orbit *of x, is dense in X.*

Theorem 3.4.1 *Let $S: X \to X$ be a homeomorphism of a compact metric space X. Then, the following are equivalent:*

(i) *S is minimal.*

(ii) *If F is a closed subset of X such that $S(F) = F$, then $F = \varnothing$ or $F = X$.*

(iii) $\bigcup_{n=-\infty}^{\infty} S^n(G) = X$ *for any nonempty open subset G of X.*

Proof (i) $\Rightarrow$ (ii). Suppose that S is minimal and let F be a nonempty closed subset of X such that $S(F) = F$. Let $x \in F$. Since the orbit of x is contained in F and is dense in X, it is clear that $F = X$.

(ii) $\Rightarrow$ (iii). If G is a nonempty open subset of X, then the set $F = \left(\bigcup_{-\infty}^{\infty} S^n(G)\right)^c$ is closed and $S(F) = F$. Since $F \neq X$, we have $F = \varnothing$.

(iii) $\Rightarrow$ (i). Given $x \in X$. Let G be any nonempty open subset of X. By (iii), $x \in S^n(G)$ for some integer n. So, $S^{-n}(x) \in G$, and thus the orbit of x is dense in X. □

Theorem 3.4.1 (ii) means that a minimal transformation of a compact metric space is "indecomposable" like an ergodic transformation of a measure space. The following theorem shows that it has no invariant nonconstant continuous function, similar to Theorem 3.2.1 for ergodic transformations.

Theorem 3.4.2 *Suppose that $S : X \to X$ is a minimal homeomorphism of a compact metric space X. If $U_S f = f$ for some $f \in C(X)$, then f is a constant function.*

Proof The condition $f \circ S = f$ implies that $f \circ S^n = f$ for all integers n. Since S is minimal, we know that f is constant on the dense orbit of any $x \in X$. The fact that f is continuous ensures that it must be a constant function. □

Now we introduce *topological transitivity*, a concept which is a bit weaker than minimality.

Definition 3.4.2 *A continuous transformation $S : X \to X$ on a compact metric space X is called* one-sided topologically transitive *if there exists some $x \in X$ such that the* orbit *$\{S^n(x)\}_{n=0}^{\infty}$ is dense in X. A homeomorphism $S : X \to X$ is said to be* topologically transitive *if there is some $x \in X$ such that the orbit $\{S^n(x)\}_{n=-\infty}^{\infty}$ is dense in X.*

Theorem 3.4.3 *Let $S : X \to X$ be a homeomorphism of a compact metric space X. Then, the following are equivalent:*

(i) *S is topologically transitive.*

(ii) *If F is a closed subset of X such that $S(F) = F$, then either $F = X$ or F is nowhere dense.*

(iii) *If G_1 and G_2 are nonempty open subsets of X, then there exists some integer n such that $S^n(G_1) \cap G_2 \neq \varnothing$.*

(iv) *The set of all points $x \in X$ whose orbit $\{S^n(x)\}_{n=-\infty}^{\infty}$ is dense in X is a dense intersection of a countable collection of open sets.*

Proof (i) $\Rightarrow$ (ii). Suppose that $\{S^n(x_0)\}_{n=-\infty}^{\infty}$ is dense in X and let $F \subset X$ be closed with $S(F) = F$. If F has no interior, then there is a nonempty open set $G \subset F$. Thus, there is an integer n with $S^n(x_0) \in G$. It follows from $G \subset F$ that the orbit of x_0 is contained in F. Since the orbit of x_0 is dense in X, we must have $F = X$.

(ii) $\Rightarrow$ (iii). Suppose that G_1 and G_2 are nonempty open subsets of X. Then, $\bigcup_{n=-\infty}^{\infty} S^n(G_1)$ is open and S-invariant, so it is dense by condition (ii). Thus, $\bigcup_{n=-\infty}^{\infty} S^n(G_1) \cap G_2 \neq \varnothing$, which implies that $S^n(G_1) \cap G_2 \neq \varnothing$ for some integer n.

(iii) $\Rightarrow$ (iv). Let $G_1, G_2, \cdots, G_n, \cdots$ be a countable base for X. Then, $\bigcap_{n=1}^{\infty} \bigcup_{i=-\infty}^{\infty} S^i(G_n)$ is the set of all points $x \in X$ whose orbit is dense in X. Since $\bigcup_{i=-\infty}^{\infty} S^i(G_n)$ is dense for each n by condition (iii), we have (iv).

(iv) $\Rightarrow$ (i). This is obvious. □

Similarly, we can prove the following theorem for one-sided topological transitivity.

Theorem 3.4.4 *Let $S : X \to X$ be a continuous transformation of a compact metric space X with $S(X) = X$. Then, the following are equivalent:*

(i) *S is one-sided topologically transitive.*

(ii) *If F is a closed subset of X such that $S(F) \subset F$, then either $F = X$ or F is nowhere dense.*

(iii) *If G_1 and G_2 are nonempty open subsets of X, then there exists an integer $n \geqslant 1$ such that $S^{-n}(G_1) \cap G_2 \neq \varnothing$.*

(iv) *The set of all points $x \in X$ whose orbit $\{S^n(x)\}_{n=0}^{\infty}$ is dense in X is a dense intersection of a countable collection of open sets.*

The following theorem indicates that ergodicity implies topological transitivity in some sense.

Theorem 3.4.5 *Let $S : X \to X$ be a homeomorphism of a compact metric space X and let μ be a probability measure on the Borel σ-algebra of X such that $\mu(G) > 0$ for every nonempty open subset G of X. If S is an ergodic measure preserving transformation with respect to μ, then the orbit $\{S^n(x)\}_{n=-\infty}^{\infty}$ of x is dense in X for $x \in X$ μ-a.e. Consequently, S is topologically transitive.*

Proof As in the proof of Theorem 3.4.3, let $G_1, G_2, \cdots, G_n, \cdots$ be a countable base for X. Then, $\bigcap_{n=1}^{\infty} \bigcup_{i=-\infty}^{\infty} S^i(G_n)$ is the set of all points $x \in X$ whose orbit is dense in X. For each n the open set $\bigcup_{i=-\infty}^{\infty} S^i(G_n)$ is S-invariant, so by ergodicity it has the measure 0 or 1. But, since this set contains G_n and since $\mu(G_n) > 0$ by assumption, we have $\mu\left(\bigcup_{i=-\infty}^{\infty} S^i(G_n)\right) = 1$ for every n. Therefore,

$$\mu\left(\bigcap_{n=1}^{\infty}\bigcup_{i=-\infty}^{\infty} S^i(G_n)\right) = 1. \qquad \square$$

We turn to the study of invariant measures for continuous transformations on compact metric spaces. Let $\mathcal{B}(X)$ be the σ-algebra of Borel subsets of a compact metric space X generated by the family of all the open subsets of X (also by the family of all the closed subsets of X). We denote by $M(X)$ the collection of all probability measures defined on the measurable space $(X, \mathcal{B}(X))$. Any $\mu \in M(X)$ is called a *Borel probability measure* on X. The following Riesz's representation theorem [57] basically gives an isomorphic relation between $M(X)$ and the collection of all normalized positive linear functionals on $C(X)$.

Theorem 3.4.6 (Riesz's representation theorem) *Let X be a compact metric space and let w be a continuous linear functional on $C(X)$ such that $w(f) \geqslant 0$ for $f \geqslant 0$ and $w(1) = 1$. Then, there exists a unique $\mu \in M(X)$ such that $w(f) = \int_X f\mathrm{d}\mu$ for all $f \in C(X)$.*

A direct corollary of Theorem 3.4.6 is

Corollary 3.4.1 *If μ and ν are two Borel probability measures on X such that $\int_X f\mathrm{d}\mu = \int_X f\mathrm{d}\nu$ for all $f \in C(X)$, then $\mu = \nu$.*

From Theorem 3.4.6, $M(X)$ can be identified with a convex subset of the unit closed ball in the dual space $C(X)^*$ of all continuous linear functionals on $C(X)$. The induced topology of $M(X)$ is called the *weak*-topology* of $M(X)$ inherited from $C(X)^*$, which is the smallest topology of $M(X)$ that makes each of the functionals $f(\mu) = \int_X f\mathrm{d}\mu$, $\forall\, \mu \in M(X)$ continuous for all $f \in C(X)$.

Thus, a sequence $\{\mu_n\}$ of measures in $M(X)$ converges to a measure $\mu \in M(X)$ under the weak*-topology if and only if

$$\lim_{n\to\infty}\int_X f\mathrm{d}\mu_n = \int_X f\mathrm{d}\mu$$

for each $f \in C(X)$. Moreover, this topology is induced by the metric

$$d(\mu,\nu) = \sum_{n=1}^{\infty}\frac{\left|\int_X f_n\mathrm{d}\mu - \int_X f_n\mathrm{d}\nu\right|}{2^n\|f_n\|_\infty},$$

where $\{f_n\}_{n=1}^{\infty}$ is a dense subset of $C(X)$ [123].

By functional analysis [57], the unit ball of $C(X)^*$ is compact in the weak*-topology. Therefore, we have the following important result which plays a key role in the proof of the existence theorem of invariant measures in this section.

Theorem 3.4.7 *If X is a compact metric space, then $M(X)$ is compact under the weak*-topology.*

We shall show that for any continuous transformation S of a compact metric space X, there exists an S-invariant probability measure. For this purpose, we define an operator $P \equiv P_S : M(X) \to M(X)$ by $P\mu = \mu \circ S^{-1}$, i.e.,

$$(P\mu)(A) = \mu(S^{-1}(A)), \ \ \forall\, A \in \mathcal{B}(X), \ \ \forall\, \mu \in M(X).$$

The correspondence from μ to $P\mu$ is called the *Frobenius-Perron operator on measures* (in contrast with Frobenius-Perron operators on functions defined in the next chapter), which is well-defined since S is continuous implies that S is measurable with respect to $\mathcal{B}(X)$. Furthermore, it is easy to show that $P : M(X) \to M(X)$ is continuous in the weak*-topology.

Lemma 3.4.1 *For any $f \in C(X)$,*

$$\int_X f \mathrm{d}(P_S\mu) = \int_X f \circ S \mathrm{d}\mu. \tag{3.13}$$

Proof By the definition of P_S we have $\int_X \chi_A \mathrm{d}(P_S\mu) = \int_X \chi_A \circ S \mathrm{d}\mu$ for any $A \in \mathcal{B}(X)$. Thus, (3.13) is valid for simple functions. By Lebesgue's monotone convergence theorem, (3.13) holds when f is a nonnegative measurable function via choosing a monotonically increasing sequence of simple functions converging to f. It follows that (3.13) is true for any $f \in C(X)$. □

Corollary 3.4.2 *$\mu \in M(X)$ is S-invariant if and only if $\int_X f \mathrm{d}\mu = \int_X f \circ S \mathrm{d}\mu$ for all $f \in C(X)$.*

We are ready to prove the following existence theorem due to Krylov and Bogolioubov .

Theorem 3.4.8 (Krylov-Bogolioubov theorem) *If $S : X \to X$ is a continuous transformation of a compact metric space X, then there is an S-invariant measure $\mu \in M(X)$.*

Proof Let $P : M(X) \to M(X)$ be the corresponding Frobenius-Perron operator on measures. Choose $\mu \in M(X)$ and define a sequence $\{\mu_n\}$ of measures in $M(X)$ by

$$\mu_n = \frac{1}{n} \sum_{i=0}^{n-1} P^i \mu.$$

By the compactness of $M(X)$ in the weak*-topology, there is a subsequence $\{\mu_{n_j}\}$ of $\{\mu_n\}$ such that $\lim_{j\to\infty}\mu_{n_j}=\mu^*\in M(X)$. For any $f\in C(X)$,

$$\begin{aligned}
&\left|\int_X f\circ S\mathrm{d}\mu^* - \int_X f\mathrm{d}\mu^*\right| \\
&= \lim_{j\to\infty}\left|\int_X f\circ S\mathrm{d}\mu_{n_j} - \int_X f\mathrm{d}\mu_{n_j}\right| \\
&= \lim_{j\to\infty}\left|\frac{1}{n_j}\int_X\sum_{i=0}^{n_j-1}\left(f\circ S^{i+1}-f\circ S^i\right)\mathrm{d}\mu\right| \\
&= \lim_{j\to\infty}\left|\frac{1}{n_j}\int_X\left(f\circ S^{n_j}-f\right)\mathrm{d}\mu\right| \\
&\leqslant \lim_{j\to\infty}\frac{2\|f\|_\infty}{n_j}=0.
\end{aligned}$$

By Corollary 3.4.2, μ^* is S-invariant. □

Let $\mu\in M(X)$ be an invariant measure of $S:X\to X$. We say that μ is *ergodic*, *weakly mixing*, or *mixing* if the measure preserving transformation S of the measure space $(X,\mathcal{B}(X),\mu)$ has the corresponding property, respectively.

Theorem 3.3.3 says that μ is ergodic if and only if $\forall\ f,\ g\in L^2(\mu)$,

$$\lim_{n\to\infty}\frac{1}{n}\sum_{i=0}^{n-1}\int_X(f\circ S^i)g\mathrm{d}\mu=\int_X f\mathrm{d}\mu\int_X g\mathrm{d}\mu,$$

and that μ is weakly mixing or mixing if and only if (3.9) or (3.10) is valid, respectively. The following Theorem 3.4.9 characterizes such properties in the context of topological dynamics. For its proof, a modification of Theorem 3.3.3 is needed here.

Lemma 3.4.2 *Let $S:X\to X$ be a continuous transformation of a compact metric space X and let $\mu\in M(X)$ be an invariant measure of S. Then,*

(i) *μ is ergodic if and only if for all $f\in C(X)$ and $g\in L^1(\mu)$,*

$$\lim_{n\to\infty}\frac{1}{n}\sum_{i=0}^{n-1}\int_X(f\circ S^i)g\mathrm{d}\mu=\int_X f\mathrm{d}\mu\int_X g\mathrm{d}\mu.$$

(ii) *μ is weakly mixing if and only if there exists a set $\mathcal{J}$ of positive integers of* density zero, i.e.,

$$\lim_{n\to\infty}\frac{\text{cardinality}(\mathcal{J}\cap\{0,1,\cdots,n-1\})}{n}=0,$$

such that for all $f\in C(X)$ and $g\in L^1(\mu)$,

$$\lim_{\mathcal{J}\not\ni n\to\infty}\int_X(f\circ S^n)g\mathrm{d}\mu=\int_X f\mathrm{d}\mu\int_X g\mathrm{d}\mu.$$

(iii) *μ is mixing if and only if for all $f \in C(X)$ and $g \in L^1(\mu)$,*

$$\lim_{n\to\infty} \int_X (f\circ S^n) g d\mu = \int_X f d\mu \int_X g d\mu.$$

Proof (i) Suppose that the convergence condition holds. Let $F, G \in L^2(\mu)$. Then, $G \in L^1(\mu)$, so

$$\lim_{n\to\infty} \frac{1}{n}\sum_{i=0}^{n-1} \int_X (f\circ S^i) G d\mu = \int_X f d\mu \int_X G d\mu, \quad \forall\, f \in C(X).$$

Now, approximate F in $L^2(\mu)$ by continuous functions to get

$$\lim_{n=\infty} \frac{1}{n}\sum_{i=0}^{n-1} \int_X (F\circ S^i) G d\mu = \int_X F d\mu \int_X G d\mu,$$

which, by means of Theorem 3.3.3 (i), implies that μ is ergodic.

Conversely, suppose that μ is ergodic. Let $f \in C(X)$. Then $f \in L^2(\mu)$. So, if $h \in L^2(\mu)$, we have

$$\lim_{n\to\infty} \frac{1}{n}\sum_{i=0}^{n-1} \int_X (f\circ S^i) h d\mu = \int_X f d\mu \int_X h d\mu.$$

If $g \in L^1(\mu)$, then by approximating g with $h \in L^2(\mu)$ under the L^1-norm, we obtain

$$\lim_{n\to\infty} \frac{1}{n}\sum_{i=0}^{n-1} \int_X (f\circ S^i) g d\mu = \int_X f d\mu \int_X g d\mu.$$

The proofs of (ii) and (iii) are similar by using Theorem 3.3.3 (ii) and (iii), respectively. And in the proof of (ii) we also need the following proposition about sequences of real numbers whose proof is referred to [123] (Theorem 1.20). □

Proposition 3.4.1 *If $\{a_n\}$ is a bounded sequence of real numbers, then the following are equivalent:*

(i) $\lim\limits_{n\to\infty} n^{-1} \sum\limits_{i=0}^{n-1} |a_i| = 0.$

(ii) $\lim\limits_{n\to\infty} n^{-1} \sum\limits_{i=0}^{n-1} |a_i|^2 = 0.$

(iii) *There exists a subset $\mathcal{J}$ of positive integers of density zero such that* $\lim\limits_{\mathcal{J}\ni n\to\infty} a_n = 0.$

Theorem 3.4.9 *Let $S : X \to X$ be a continuous transformation of a compact metric space X, let P be the corresponding Frobenius-Perron operator on measures, and let $\mu \in M(X)$ be an S-invariant measure. Then,*

(i) *μ is ergodic if and only if for any $\nu \in M(X)$ and $\nu \ll \mu$,*

$$\lim_{n\to\infty} \frac{1}{n} \sum_{i=0}^{n-1} P^i \nu = \mu. \tag{3.14}$$

(ii) *μ is weakly mixing if and only if there exists a set $\mathcal{J}$ of positive integers of density zero such that*

$$\lim_{\mathcal{J} \not\ni n \to \infty} P^n \nu = \mu$$

for any $\nu \in M(X)$ and $\nu \ll \mu$.

(iii) *μ is mixing if and only if for any $\nu \in M(X)$ and $\nu \ll \mu$, we have*

$$\lim_{n\to\infty} P^n \nu = \mu.$$

Proof (i) Suppose that μ is ergodic. Let $\nu \in M(X)$ and $\nu \ll \mu$. Then, $g = \mathrm{d}\nu/\mathrm{d}\mu \in L^1(\mu)$ and is a density function. If $f \in C(X)$, then Lemma 3.4.2 (i) implies that

$$\begin{aligned}\int_X f \mathrm{d}\left(\frac{1}{n}\sum_{i=0}^{n-1} P^i \nu\right) &= \frac{1}{n}\sum_{i=0}^{n-1} \int_X f \circ S^i \mathrm{d}\nu \\ &= \frac{1}{n}\sum_{i=0}^{n-1} \int_X (f \circ S^i) g \mathrm{d}\mu \to \int_X f \mathrm{d}\mu \int_X g \mathrm{d}\mu \\ &= \int_X f \mathrm{d}\mu \int_X 1 \mathrm{d}\nu = \int_X f \mathrm{d}\mu.\end{aligned}$$

Therefore, (3.14) is true.

Conversely, suppose that the convergence condition (3.14) holds. Let $g \in L^1(\mu)$ and $g \geqslant 0$. Define $\nu \in M(X)$ by

$$\nu(A) = \frac{\int_A g \mathrm{d}\mu}{\int_X g \mathrm{d}\mu}, \quad \forall\, A \in \mathcal{B}(X).$$

Then, for $f \in C(X)$, (3.14) implies that

$$\lim_{n\to\infty} \frac{1}{n}\sum_{i=0}^{n-1} \int_X (f \circ S^i) g \mathrm{d}\mu = \int_X f \mathrm{d}\mu \int_X g \mathrm{d}\mu. \tag{3.15}$$

Let $g \in L^1(\mu)$. Then, $g = g^+ - g^-$ and $g^+, g^- \geqslant 0$. Applying the above to g^+ and g^-, we see that (3.15) is satisfied by all $f \in C(X)$ and $g \in L^1(\mu)$. Hence, μ is ergodic thanks to Lemma 3.4.2 (i).

The proofs of (ii) and (iii) are similar and use the corresponding parts of Lemma 3.4.2. □

The following theorem strengthens Birkhoff's pointwise ergodic theorem in the case of topological dynamical systems.

Theorem 3.4.10 *Let $S : X \to X$ be a continuous transformation of a compact metric space X and let $\mu \in M(X)$ be S-invriant and ergodic. Then, there exists $Y \in \mathcal{B}(X)$ with $\mu(Y) = 1$ such that*

$$\lim_{n\to\infty} \frac{1}{n} \sum_{i=0}^{n-1} f(S^i(x)) = \int_X f d\mu, \ \ \forall\, x \in Y, \ \ \forall\, f \in C(X). \tag{3.16}$$

Proof Let $\{f_k\}_{k=1}^{\infty}$ be a countable dense subset of $C(X)$. By Birkhoff's pointwise ergodic theorem, for each k there exists $X_k \in \mathcal{B}(X)$ such that $\mu(X_k) = 1$ and

$$\lim_{n\to\infty} \frac{1}{n} \sum_{i=0}^{n-1} f_k(S^i(x)) = \int_X f_k d\mu, \ \ \forall\, x \in X_k.$$

Denote $Y = \bigcap_{k=1}^{\infty} X_k$. Then, $\mu(Y) = 1$ and

$$\lim_{n\to\infty} \frac{1}{n} \sum_{i=0}^{n-1} f_k(S^i(x)) = \int_X f_k d\mu, \ \ \forall\, x \in Y, \ \ \forall\, k \geqslant 1.$$

Thus, (3.16) follows from approximating a given $f \in C(X)$ by members of $\{f_k\}_{k=1}^{\infty}$. □

Theorem 3.4.10 implies another characterization of ergodicity.

Theorem 3.4.11 *Let $S : X \to X$ be a continuous transformation of a compact metric space X and let $\mu \in M(X)$ be S-invariant. Then, μ is ergodic if and only if*

$$\lim_{n\to\infty} \frac{1}{n} \sum_{i=0}^{n-1} \delta_{S^i(x)} = \mu, \ \ \forall\, x \in X \ \mu\text{-a.e.}$$

Proof The necessity part is from Theorem 3.4.10 immediately. For the sufficiency part, suppose that there is $Y \in \mathcal{B}(X)$ with $\mu(Y) = 1$ such that

$$\lim_{n\to\infty} \frac{1}{n} \sum_{i=0}^{n-1} \delta_{S^i(x)} = \mu, \ \ \forall\, x \in Y.$$

Then, for all $x \in Y$, $f \in C(X)$, and $g \in L^1(\mu)$, we have

$$\lim_{n\to\infty} \frac{1}{n} \sum_{i=0}^{n-1} f(S^i(x))g(x) = \int_X f d\mu \cdot g(x).$$

Lebesgue's dominated convergence theorem now yields

$$\lim_{n\to\infty} \frac{1}{n} \sum_{i=0}^{n-1} \int_X f(S^i(x))g(x)\mathrm{d}\mu(x) = \int_X f\mathrm{d}\mu \int_X g\mathrm{d}\mu,$$

which implies that μ is ergodic. □

To end this section, we study the situation when there is only one invariant measure for the given S.

Definition 3.4.3 *A continuous transformation $S : X \to X$ of a compact metric space X is called* uniquely ergodic *if there is only one S-invariant Borel probability measure.*

Unique ergodicity is closely related to minimality as the following result shows.

Theorem 3.4.12 *Let $S : X \to X$ be a homeomorphism of a compact metric space X. Suppose that S is uniquely ergodic with the invariant probability measure μ. Then, S is minimal if and only if $\mu(G) > 0$ for all nonempty open subsets G of X.*

Proof Suppose that S is minimal. If G is nonempty and open, then $X = \bigcap_{n=-\infty}^{\infty} S^n(G)$ by Theorem 3.4.1 (iii). So $\mu(G) > 0$.

Conversely, suppose that $\mu(G) > 0$ for all nonempty open sets G. If S is not minimal, then by Theorem 3.4.1 (ii), there is a nonempty closed set F such that $S(F) = F$ and $F \neq X$. By Theorem 3.4.8, the homeomorphism $S : F \to F$ has an invariant Borel probability measure μ_F. Then, the measure ν on X defined by $\nu(A) = \mu_F(F \cap A)$ for all Borel subsets A of X is S-invariant. Since F^c is nonempty and open, $\mu(F^c) > 0$ by the assumption. Thus, $\nu \neq \mu$ since $\nu(F^c) = 0$. This contradicts the unique ergodicity of S. □

If S is uniquely ergodic, then the conclusion of Theorem 3.4.10 can be strengthened.

Theorem 3.4.13 *Let $S : X \to X$ be a continuous transformation of a compact metric space X. Then, the following are equivalent:*

(i) *S is uniquely ergodic.*

(ii) *For each $f \in C(X)$, the sequence $\left\{ n^{-1} \sum_{i=0}^{n-1} f(S^i(x)) \right\}$ converges uniformly to a constant for all $x \in X$.*

(iii) *For each $f \in C(X)$, the sequence $\left\{ n^{-1} \sum_{i=0}^{n-1} f(S^i(x)) \right\}$ converges pointwisely to a constant for all $x \in X$.*

(iv) *There exists an invariant probability measure* μ *such that for all* $f \in C(X)$ *and all* $x \in X$,

$$\lim_{n\to\infty} \frac{1}{n}\sum_{i=0}^{n-1} f(S^i(x)) = \int_X f\mathrm{d}\mu.$$

Proof (i) $\Rightarrow$ (ii). Let μ be the unique invariant Borel probability measure of S. If (ii) is false, then there exists a function $g \in C(X)$ and a number $\epsilon > 0$ such that for any positive integer n_0 there is $n > n_0$ and $x_n \in X$ with

$$\left| \frac{1}{n}\sum_{i=0}^{n-1} g(S^i(x_n)) - \int_X g\mathrm{d}\mu \right| \geqslant \epsilon. \tag{3.17}$$

In other words, $\left|\int_X g\mathrm{d}\mu_n - \int_X g\mathrm{d}\mu\right| \geqslant \epsilon$, where

$$\mu_n = \frac{1}{n}\sum_{i=0}^{n-1} \delta_{S^i(x_n)} = \frac{1}{n}\sum_{i=0}^{n-1} P_S^i \delta_{x_n}.$$

Since the sequence $\{\mu_n\}$ is precompact in the weak*-topology, it has a subsequence $\{\mu_{n_i}\}$ that converges to some $\nu \in M(X)$. Moreover, the proof of Theorem 3.4.8 implies that ν is an invariant probability measure. Taking limit $n_i \to \infty$ in (3.17) gives $\left|\int_X g\mathrm{d}\nu - \int_X g\mathrm{d}\mu\right| \geqslant \epsilon$, so $\nu \neq \mu$. This contradicts the unique ergodicity of S.

(ii) $\Rightarrow$ (iii). It is obvious.

(iii) $\Rightarrow$ (iv). The given condition defines a linear functional

$$w(f) = \lim_{n\to\infty} \frac{1}{n}\sum_{i=0}^{n-1} f(S^i(x)), \ \forall\, f \in C(X).$$

This functional is continuous since for all n,

$$\left| \frac{1}{n}\sum_{i=0}^{n-1} f(S^i(x)) \right| \leqslant \|f\|_\infty.$$

It is clear that $w(1) = 1$ and $w(f) \geqslant 0$ if $f \geqslant 0$. Thus, by the Riesz representation theorem there exists a Borel probability measure μ such that $w(f) = \int_X f\mathrm{d}\mu$, $\forall\, f \in C(X)$. Since $w(f \circ S) = w(f)$ by a direct computation, $\int_X f \circ S\mathrm{d}\mu = \int_X f\mathrm{d}\mu$ for all $f \in C(X)$, which implies that μ is S-invariant by Corollary 3.4.2.

(iv) $\Rightarrow$ (i). Suppose that ν is an S-invariant Borel probability measure. Since

$$\lim_{n\to\infty}\frac{1}{n}\sum_{i=0}^{n-1} f(S^i(x)) = \tilde{f}, \quad \forall\, x\in X,$$

where $\tilde{f} = \int_X f\mathrm{d}\mu$. Integrating the above equality with respect to ν and using Lebesgue's dominated convergence theorem give

$$\int_X f\mathrm{d}\nu = \int_X \tilde{f}\mathrm{d}\nu = \tilde{f} = \int_X f\mathrm{d}\mu.$$

Hence, $\nu = \mu$ by Corollary 3.4.1. Therefore, S is uniquely ergodic. □

Exercises

3.1 Show that

(i) $S^{-1}(A)\cap S^{-1}(B) = S^{-1}(A\cap B)$.

(ii) $S^{-1}(A)\cup S^{-1}(B) = S^{-1}(A\cup B)$.

(iii) $S(S^{-1}(A))\subset A$ and $S^{-1}(S(A))\supset A$.

Give examples with strict inclusions. Prove: if S is one-to-one, then the first inclusion is reduced to equality, and if S is onto, then equality holds in the second inclusion.

3.2 Show by examples that $S^{-1}(S(A))\neq A$ in general for a measurable transformation S.

3.3 Let (X,Σ,μ) be a measure space, let $S: X\to X$ be a measurable transformation, and let $A\in\Sigma$. Prove that:

(i) If either $A\cap(S^{-1}(A))^c$ or $S^{-1}(A)\cap A^c$ has measure 0, then A is S-invariant.

(ii) If $\mu(X)<\infty$ and $\mu(A)\leqslant\mu(S^{-1}(A))$ for all $A\in\Sigma$, then μ is invariant with respect to S. Is the assumption $\mu(X)<\infty$ essential?

(iii) A is S-invariant if and only if $\chi_A\circ S=\chi_A$.

(iv) If S is nonsingular and if either $\mu(A) = 0$ or $\mu(A^c) = 0$, then A is S-invariant.

3.4 Let $S:[0,1]\to[0,1]$ be the *Gauss map* defined by

$$S(x) = \begin{cases} 0, & \text{if } x = 0,\\ \left\{\dfrac{1}{x}\right\}, & \text{if } x\neq 0,\end{cases}$$

where $\{t\}$ is the fractional part of t. Show that the probability measure μ defined by

$$\mu(A) = \frac{1}{\ln 2}\int_A \frac{1}{1+x}\mathrm{d}x$$

is S-invariant.

3.5 Let $S(x) = 2x$, $\forall\ x \in \mathbb{R}$. Show that S is not measure preserving with respect to the Lebesgue measure on $\mathbb{R}$. If we let $\boldsymbol{S}(x,y) = (2x, y/2)$, then $\boldsymbol{S}$ is measure preserving with respect to the Lebesgue measure on $\mathbb{R}^2$.

3.6 Let $\boldsymbol{S} : [0,1]^2 \to [0,1]^2$ be the *baker transformation* defined by

$$\boldsymbol{S}(x,y) = \begin{cases} \left(2x, \dfrac{1}{2}y\right), & x \in \left[0, \dfrac{1}{2}\right), \\ \left(2x-1, \dfrac{1}{2}y + \dfrac{1}{2}\right), & x \in \left[\dfrac{1}{2}, 1\right]. \end{cases}$$

Show that $\boldsymbol{S}$ preserves the Lebesgue measure m on $[0,1]^2$.

3.7 Let $S : X \to X$ be a measurable transformation on a measure space (X, Σ, μ). Show that the set function defined by $\nu(A) = \mu(S^{-1}(A))$, $\forall\ A \in \Sigma$ defines a measure on X.

3.8 Let $S : X \to X$ be a measurable transformation on a measure space (X, Σ, μ). Show that if μ is S^n-invariant for some positive integer n, then the measure $\eta = n^{-1} \sum\limits_{i=0}^{n-1} \mu \circ S^{-i}$ is S-invariant.

3.9 Let $S : X \to X$ and $A \subset X$. Define

$$B = \lim_{n\to\infty} S^{-n}(A) \equiv \bigcap_{i=1}^{\infty} \left(\bigcup_{n=i}^{\infty} S^{-n}(A) \right).$$

Prove that

(i) $B = \{x \in X : x \in S^{-n}(A) \text{ for infinitely many } n\}$.

(ii) $S^{-1}(B) = B$.

(iii) If S is measure preserving, and if A is measurable and S-invariant, then $\mu(B) = \mu(A)$.

3.10 Let $S : [0,1] \to [0,1]$ be the translation mapping defined in Example 3.1.4. Show that S is ergodic with respect to the Lebesgue measure m if and only if a is irrational.

3.11 Let S be an ergodic transformation on a measure space (X, Σ, μ). Let $A_0 \in \Sigma$ and define $A_n = S^{-n}(A_0)$ for $n = 1, 2, \cdots$. Denote $B_n = A_n^c$. Show that the set $\bigcap\limits_{n=0}^{\infty} B_n$ is S-invariant.

3.12 Show that the tent function T of Example 3.1.2 is ergodic with respect to the Lebesgue measure m on $[0,1]$. Moreover, let $\boldsymbol{S} : [0,1]^2 \to [0,1]^2$ be defined by $\boldsymbol{S}(x,y) = (T(x), T(y))$. Show that $\boldsymbol{S}$ is ergodic.

3.13 Consider the measure space (X, Σ, μ), where X is the set of all integers, Σ is the family of all subsets of X, and μ is the counting measure. Let $S(x) = x + k$, $x \in X$, where k is an integer. For what values of k is the transformation S ergodic?

3.14 Show that the baker transformation of Exercise 3.6 is ergodic.
3.15 Let (X, Σ, μ) be a probability measure space and let $S : X \to X$ be measure preserving. Suppose that Σ_0 is an *algebra* (i.e., Σ_0 is closed under finite union of its elements) that generates Σ. Show that if $\lim_{n\to\infty} \mu(S^{-n}(A) \cap B) = \mu(A)\mu(B)$ for all $A,\ B \in \Sigma_0$, then the same is true for all $A,\ B \in \Sigma$.
3.16 Show that $S(x) = 2x \pmod 1$ is mixing on $[0, 1]$.
3.17 Let $S : X \to X$ be a measurable transformation on a measurable space (X, Σ). Suppose that μ and ν are two S-invariant probability measures such that S is ergodic with respect to both μ and ν. Show that there exist sets $A,\ B \in \Sigma$ such that $A \cap B = \varnothing$ and $\mu(A) = \nu(B) = 1$.
3.18 Let $S : X \to X$ be a measurable transformation on a measurable space (X, Σ). Suppose that μ is the unique measure on X with respect to which S is measure preserving. Prove that S is ergodic with respect to μ.
3.19 Let S be a measure preserving transformation on a probability measure space (X, Σ, μ). For any $A \in \Sigma$ and $x \in X$, define

$$\chi_A^*(x) = \lim_{n\to\infty} \frac{1}{n} \sum_{i=0}^{n-1} \chi_A(S^i(x))$$

if the limit exists. Show that S is ergodic if and only if $\chi_A^*(x) = \mu(A)$ for almost every $x \in X$.
3.20 Let S be an ergodic measure preserving transformation on a probability measure space (X, Σ, μ). Show that the set

$$B = \bigcap_{i=0}^{\infty} \left(S^{-i}(A)\right)^c$$

is an invariant set under S.
3.21 Discuss Birkhoff's pointwise ergodic theorem applied to a finite space $X = \{a_1, a_2, \cdots, a_k\}$ with counting measure μ.
3.22 Let $Y = \{0, 1, \cdots, k-1\}$ and let μ be a measure on Y such that $\mu(\{i\}) = p_i, i = 0, 1, \cdots, k-1$, and $\sum_{i=0}^{k-1} p_i = 1$. Define $X = \prod_{n=-\infty}^{\infty} Y$ installed with the product measure $\hat{\mu}$. Show that the two-sided $(p_0, \cdots, p_{k-1})$-shift $S : X \to X$, defined by $S\left(\{x_n\}_{n=-\infty}^{\infty}\right) = \{y_n\}_{n=-\infty}^{\infty}$ with $y_n = x_{n+1}$ for all integers n, preserves $\hat{\mu}$.
3.23 Let Y be as in the previous problem, and let $X = \prod_{n=0}^{\infty} Y$ with the product measure $\hat{\mu}$. Let the one-sided $(p_0, \cdots, p_{k-1})$-shift $S : X \to X$ be defined by $S(\{x_0, x_1, \cdots\}) = \{x_1, x_2, \cdots\}$. Show that $\hat{\mu}$ is S-invariant.
3.24 Prove that the two-sided shift S in Problem 3.23 is ergodic with respect to $\hat{\mu}$.
3.25 Show that the Frobenius-Perron operator on measures $P : M(X) \to M(X)$ is continuous under the weak*-topology.

Chapter 4
Frobenius-Perron Operators

JiuDing, Department of Mathematics, The University of Southern Mississippi in America.
Aihui Zhou, Academy of Mathematics and Systems Science, Chinese Academy of Sciences.

Abstract In this chapter we first introduce Markov operators and study their general properties. Then we define Frobenius-Perron operators, a special class of Markov operators that will be mainly studied in the book. We also present a decomposition theorem for the Frobenius-Perron operator and its spectral analysis. We employ the Frobenius-Perron operator and its dual operator, the Koopman operator, to study ergodicity, mixing, and exactness of chaotic transformations.
Keywords Markov operator, stationary density, nonsingular transformation, Frobenius-Perron operator, Koopman operator, Decomposition theorem, spectrum.

From the Birkhoff pointwise ergodic theorem in the previous chapter, we know that absolutely continuous invariant probability measures determine the statistical properties of the underlying deterministic system for almost all initial points. Now a natural question arises: given a measurable transformation on a measure space, can we determine one or all absolutely continuous probability measures preserved by the given transformation so that we can apply ergodic theorems to the statistical study of the deterministic dynamical system? Trying to answer this question leads to the introduction of the concept of Frobenius-Perron operators. Exploring the existence of fixed density functions of this linear operator and developing efficient numerical methods for computing them are the two main topics of this book.

Since Frobenius-Perron operators constitute a special class of Markov operators, we first study general properties of Markov operators in Section 4.1. Section 4.2 is exclusively devoted to the definition and basic properties of Frobenius-Perron operators. The dual operator of the Frobenius-Perron operator, the Koopman operator named after Koopman was first presented it in 1931 [78], is introduced in Section 4.3, and with this operator notion we reformulate Birkhoff's pointwise ergodic theorem and von Neumann's mean ergodic theorem. We use Frobenius-Perron operators and Koopman operators to further study ergodicity, mixing, and exactness of chaotic transformations in Section 4.4. A decomposition theorem for general Frobenius-Perron operators and its application to the spectral analysis is presented in the final section.

4.1 Markov Operators

Let (X, Σ, μ) be a measure space which is σ-finite by assumption. As before, if not indicating explicitly, we always assume that all the function spaces are real. Since density functions will be used frequently in the sequel, we need a notation for all of them. Denote

$$\mathcal{D} \equiv \mathcal{D}(X, \Sigma, \mu) = \{f \in L^1(X, \Sigma, \mu) : \ f \geqslant 0, \ \|f\| = 1\}.$$

Any function $f \in \mathcal{D}$ will be called a *density*. If a probability measure μ_f is defined by $\mu_f(A) = \int_A f \mathrm{d}\mu$, $\forall\, A \in \Sigma$ for some $f \in \mathcal{D}$, then we say that f is the *density* of μ_f with respect to μ.

Definition 4.1.1 *A linear operator $P : L^1(\mu) \to L^1(\mu)$ is called a* Markov operator *if $P\mathcal{D} \subset \mathcal{D}$.*

Thus, P is a Markov operator if and only if P satisfies the condition that $Pf \geqslant 0$ and $\|Pf\| = \|f\|$ for all $f \geqslant 0$, $f \in L^1$.

Remark 4.1.1 Since f and Pf are L^1 functions, $f \geqslant 0$ and $Pf \geqslant 0$ are understood to hold almost everywhere with respect to μ. In the following, equalities and inequalities among L^p functions are in the above sense without adding the phrase "almost everywhere" when it is clear.

Since a Markov operator is a *positive operator*, that is, it maps nonnegative functions to nonnegative functions, so it is *monotone* in the sense that

$$Pf(x) \geqslant Pg(x), \ \ \forall\, x \in X \ \ \text{if} \ \ f(x) \geqslant g(x), \ \ \forall\, x \in X.$$

Some more properties of P defined on the *real* L^1 space are contained in the following proposition.

Proposition 4.1.1 *Let $P : L^1 \to L^1$ be a Markov operator. Then, for every $f \in L^1$,*

(i) $(Pf)^+ \leqslant Pf^+$;
(ii) $(Pf)^- \leqslant Pf^-$;
(iii) $|Pf| \leqslant P|f|$;
(iv) $\|Pf\| \leqslant \|f\|$.

Proof (i) and (ii) are obtained from

$$\begin{aligned}(Pf)^+(x) &= (Pf^+ - Pf^-)^+(x) = \max\{0, (Pf^+ - Pf^-)(x)\} \\ &\leqslant \max\{0, Pf^+(x)\} = Pf^+(x);\end{aligned}$$

$$\begin{aligned}(Pf)^-(x) &= (Pf^+ - Pf^-)^-(x) = \max\{0, -(Pf^+ - Pf^-)(x)\} \\ &= \max\{0, Pf^-(x) - Pf^+(x)\} \leqslant \max\{0, Pf^-(x)\} \\ &= Pf^-(x).\end{aligned}$$

(iii) follows from (i) and (ii), namely,

$$\begin{aligned}|Pf| &= (Pf)^+ + (Pf)^- \leqslant Pf^+ + Pf^- \\ &= P(f^+ + f^-) = P|f|.\end{aligned}$$

Finally, using (iii), we have

$$\|Pf\| = \int_X |Pf| \mathrm{d}\mu \leqslant \int_X P|f| \mathrm{d}\mu = \int_X |f| \mathrm{d}\mu = \|f\|,$$

which is (iv). □

Property (iv) above means that P is a (weak) *contraction* and it is still true when P is defined on a complex L^1 space (Exercise 4.31). In Section 4.4, we will give conditions under which P is actually an *isometry* (i.e., $\|Pf\| = \|f\|$ for all $f \in L^1$) on L^1 for Frobenius-Perron operators. For a given $f \in L^1$, to obtain an equivalent condition for the actual equality $\|Pf\| = \|f\|$, we need the concept of the support of a function.

Definition 4.1.2 *Let f be a measurable function with domain X. The* support *of f is defined to be*

$$\operatorname{supp} f = \{x \in X : f(x) \neq 0\},$$

which holds in the sense of modulo zero, *that is, $A = B$* modulo zero *means that the measure of the* symmetric difference $A \triangle B \equiv (A \cap B^c) \cup (B \cap A^c)$ *is zero.*

Proposition 4.1.2 *Let $f \in L^1$ and let $P : L^1 \to L^1$ be a Markov operator. Then, $\|Pf\| = \|f\|$ if and only if Pf^+ and Pf^- have disjoint supports.*

Proof Using the fact that for two nonnegative real numbers a and b, $|a-b| < a+b$ if and only if both $a > 0$ and $b > 0$, and the obvious inequality

$$|Pf^+(x) - Pf^-(x)| \leqslant Pf^+(x) + Pf^-(x), \quad \forall\, x \in X,$$

we have that

$$\int_X |Pf^+ - Pf^-| \mathrm{d}\mu = \int_X Pf^+ \mathrm{d}\mu + \int_X Pf^- \mathrm{d}\mu$$

if and only if there is no set $A \in \Sigma$ of positive measure such that $Pf^+(x) > 0$ and $Pf^-(x) > 0$ for $x \in A$, in other words, Pf^+ and Pf^- have disjoint supports. Since $f = f^+ - f^-$ and $|f| = f^+ + f^-$, the left-hand side integral is simply $\|Pf\|$, and the right-hand side is $\|Pf^+\| + \|Pf^-\| = \|f^+\| + \|f^-\| = \|f\|$, so the proposition is proved. □

The above proposition has the following useful consequence.

Proposition 4.1.3 *Let $f \in L^1$ and let $P : L^1 \to L^1$ be a Markov operator. Then, $Pf = f$ if and only if $Pf^+ = f^+$ and $Pf^- = f^-$.*

Proof We only need to prove the necessity part of the proposition. Suppose that $Pf = f$. Then,

$$f^+ = (Pf)^+ \leqslant Pf^+ \text{ and } f^- = (Pf)^- \leqslant Pf^-.$$

Hence,

$$\begin{aligned} &\int_X (Pf^+ - f^+)\mathrm{d}\mu + \int_X (Pf^- - f^-)\mathrm{d}\mu \\ &= \int_X (Pf^+ + Pf^-)\mathrm{d}\mu - \int_X (f^+ + f^-)\mathrm{d}\mu \\ &= \int_X P|f|\mathrm{d}\mu - \int_X |f|\mathrm{d}\mu = \|P|f|\| - \||f|\| \leqslant 0. \end{aligned}$$

Since both the integrands $Pf^+ - f^+$ and $Pf^- - f^-$ are nonnegative, it follows that $Pf^+ = f^+$ and $Pf^- = f^-$. □

Definition 4.1.3 *Let (X, Σ, μ) be a measure space and let P be a Markov operator on $L^1(X)$. Any* fixed density function *f of P, that is, $f \in \mathcal{D}$ and $Pf = f$, is called a* stationary density *of P.*

The concept of stationary densities of Markov operators is extremely important in ergodic theory, stochastic analysis, partial differential equations, and many applied fields [7, 14, 82]. We give four examples of Markov operators to end this section. More examples will be given in the exercise set at the end of this chapter.

Example 4.1.1 (*solutions of evolution equations*) A continuous dynamical system is often described by a differential equation defined on a region or manifold. The solution of an evolution equation can sometimes be expressed in terms of a *semigroup of Markov operators* $\{P_t\}_{t \geqslant 0}$ depending on the continuous time t, that is, P_t is a Markov operator for each fixed t and $P_t P_s = P_{t+s}$ for all $t, s \geqslant 0$. A simple example is the *heat equation*

$$\frac{\partial u}{\partial t} = \frac{\sigma^2}{2} \frac{\partial^2 u}{\partial x^2}, \quad t > 0, \ x \in \mathbb{R}$$

with the initial value condition

$$u(0, x) = f(x), \quad x \in \mathbb{R},$$

for which the solution $u(t, x)$ can be written as

$$u(t, x) = P_t f(x) = \int_{-\infty}^{\infty} K(t, x, y) f(y)\mathrm{d}y, \quad P_0 f(x) = f(x),$$

where the *stochastic kernel* is

$$K(t,x,y)=\frac{1}{\sqrt{2\pi\sigma^2 t}}\exp\left(-\frac{(x-y)^2}{2\sigma^2 t}\right).$$

Example 4.1.2 (*dynamical geometry*) We consider a dynamical system problem in Euclidean geometry. Given an inscribed N-sided polygon on the unit circle $\partial\mathbb{D}$ of the complex plane $\mathbb{C}$ with vertices $\boldsymbol{z}_1, \boldsymbol{z}_2, \cdots, \boldsymbol{z}_N \in \partial\mathbb{D}$. This *cyclic polygon* is determined by the corresponding N-dimensional vector $(\theta_1, \theta_2, \cdots, \theta_N)$ of the central angles subtended by each side. If by some rule we choose a point on each side of the polygon and join the center of the circle to each chosen point and extend them to meet the circle at the points $\boldsymbol{w}_1, \boldsymbol{w}_2, \cdots, \boldsymbol{w}_N$, then we obtain a new cyclic polygon, determined by its N central angles $(\phi_1, \phi_2, \cdots, \phi_N)$. Repeating this process we have a sequence of cyclic N-sided polygons. What interests us is the asymptotic behavior of the sequence. Since the sum of the N central angles of a polygon is always the same constant 2π, the rule to get the polygon $(\phi_1, \phi_2, \cdots, \phi_N)$ from the polygon $(\theta_1, \theta_2, \cdots, \theta_N)$ is given by a Markov operator on $\mathbb{R}^N$ for linear processes. For example, the rule given by the following process

$$\begin{bmatrix}\phi_1\\ \phi_2\\ \vdots\\ \phi_N\end{bmatrix}=\begin{bmatrix}1-\lambda_1 & \lambda_2 & 0 & \cdots & 0\\ 0 & 1-\lambda_2 & \lambda_3 & \cdots & 0\\ \vdots & \vdots & \vdots & & \vdots\\ 0 & 0 & \cdots & 1-\lambda_{N-1} & \lambda_n\\ \lambda_1 & 0 & \cdots & 0 & 1-\lambda_N\end{bmatrix}\begin{bmatrix}\theta_1\\ \theta_2\\ \vdots\\ \theta_N\end{bmatrix},$$

where λ_i's are real numbers between 0 and 1, corresponds to a *column stochastic matrix* which represents a Markov operator on $\mathbb{R}^N$. In this case, we could expect a regular eventual behavior for the dynamics. In particular, if $N = 3$ and $\lambda_1 = \lambda_2 = \lambda_3 = 1/2$, we have the following process: given any scalene triangle T_0, which is always cyclic since any triangle can be inscribed by a circle, and construct the inscribed circle. The points of tangency form the second triangle T_1. Then, construct the inscribed circle for T_1. The points of tangency on the three sides of T_1 form the third triangle T_2. Continuing this process, one gets a sequence $\{T_n\}$ of triangles. Although such triangles will shrink to a point eventually, the shape of T_n will approach that of an equilateral triangle. However, the sequence of *pedal triangles*, which is obtained from the process of constructing the three altitudes of the three sides of the current triangle, is chaotic [85]. We note that the corresponding operator is no longer a Markov operator, but a nonlinear operator that can be represented by four 3×3 matrices with all the column sums 1 (such matrices are said to be *column quasi-stochastic*). See [34, 65] for more details about the dynamic geometry of polygons.

Example 4.1.3 (*fractal geometry*) In the measure theory of fractal geometry, the *Gibbs measure* [4, 13] is perhaps the most important measure. Let

$$X = \prod_{n=0}^{\infty} \{0, 1, \cdots, k-1\}$$

be the one-sided symbolic space with the product topology of the discrete topology of $\{0, 1, \cdots, k-1\}$, as given in Example 3.4.1. Let a function $\phi \in C(X)$ satisfy some conditions. Then, the existence of a Gibbs measure on the compact space X is closely related to the maximum eigenvalue problem of the *Ruelle operator* $L_\phi : C(X) \to C(X)$ defined by

$$L_\phi f(x) = \sum_{y \in S^{-1}(\{x\})} \exp(\phi(x)) f(y), \quad \forall\, f \in C(X),$$

where $S : X \to X$ is the one-sided shift , that is,

$$S(\{x_0, x_1, x_2, \cdots\}) = \{x_1, x_2, x_3, \cdots\}.$$

Clearly, L_ϕ is a positive operator.

Let L_ϕ^* be the dual operator of L_ϕ, which is defined on the space of all Borel measures on X. The famous Ruelle theorem [113] asserts that there is a constant $\zeta > 0$, a positive function $f^* \in C(X)$, and an S-invariant probability measure μ such that

$$L_\phi f^* = \zeta f^*, \quad L_\phi^* \mu = \zeta \mu, \quad \int_X f^* \mathrm{d}\mu = 1.$$

Denote $P = \zeta^{-1} L_\phi$ and extend its domain from $C(X)$ to $L^1(\nu)$ by a continuous extension. Then, for all $f \in L^1(\mu)$,

$$\begin{aligned} \int_X P f \mathrm{d}\mu &= \zeta^{-1} \int_X L_\phi f \mathrm{d}\mu = \zeta^{-1} \int_X f \mathrm{d}(L_\phi^* \mu) \\ &= \zeta^{-1} \zeta \int_X f \mathrm{d}\mu = \int_X f \mathrm{d}\mu. \end{aligned}$$

Hence, $P : L^1(\mu) \to L^1(\mu)$ is a Markov operator and f^* is a stationary density of P.

Example 4.1.4 (*wavelets construction*) The so-called *scaling equation* in the theory of wavelets is the function equation of the form:

$$\phi(x) = \sum_{n \in \mathcal{I}} a_n \phi(2x - n), \quad \forall\, x \in \mathbb{R},$$

where ϕ is a scaling function called the father wavelet for the multi-resolution analysis in the wavelets theory, the index set $\mathcal{I}$ is either $\{1, 2, \cdots, k\}$, $\{1, 2, 3, \cdots\}$, or $\{0, 1, 2, \cdots\}$, and all a_n's are real numbers such that $\sum_{n\in\mathcal{I}} a_n = 2$. If we define a linear operator P by

$$Pf(x) = \sum_{n\in\mathcal{I}} a_n f(2x - n), \quad \forall\, x \in \mathbb{R},$$

then the solution ϕ of the above scaling equation is just a fixed point of P. If $a_n \geqslant 0$ for all $n \in \mathcal{I}$, then from

$$\begin{aligned}\int_{-\infty}^{\infty} Pf(x)\mathrm{d}x &= \sum_{n\in\mathcal{I}} a_n \int_{-\infty}^{\infty} f(2x - n)\mathrm{d}x \\ &= \frac{1}{2}\sum_{n\in\mathcal{I}} a_n \int_{-\infty}^{\infty} f(x)\mathrm{d}x = \int_{-\infty}^{\infty} f(x)\mathrm{d}x,\end{aligned}$$

the operator $P : L^1(\mathbb{R}) \to L^1(\mathbb{R})$ is a Markov operator.

In the next section, we shall introduce a subclass of Markov operators, the class of Frobenius-Perron operators which are induced by measurable transformations of measure spaces and which constitute the main topic of this book. For more extensive studies of Markov operators, see monographs [58, 82].

4.2 Frobenius-Perron Operators

Let (X, Σ, μ) be a measure space. For a given measurable transformation $S : X \to X$, the corresponding Frobenius-Perron operator to be defined below gives the evolution of probability density functions governed by the deterministic dynamical system. First we recall the definition of nonsingular transformations.

Definition 4.2.1 *A measurable transformation $S : X \to X$ on a measure space (X, Σ, μ) is* nonsingular *if $\mu(S^{-1}(A)) = 0$ for all $A \in \Sigma$ such that $\mu(A) = 0$.*

Note that every measure preserving transformation is necessarily nonsingular with respect to the invariant measure. For a nonsingular transformation $S : (X, \Sigma, \mu) \to (X, \Sigma, \mu)$, the problem that we are concerned with is the one of finding an invariant finite measure which is *absolutely continuous* with respect to μ. To motivate the definition of the Frobenius-Perron operator, we use the idea of the Frobenius-Perron operator on measures introduced in Section 3.4 that maps a measure ν to the measure $\nu \circ S^{-1}$ which is defined as

$$(\nu \circ S^{-1})(A) = \nu(S^{-1}(A)), \quad \forall\, A \in \Sigma.$$

It is clear that every fixed point of this operator is an invariant measure of S. Since we are more interested in absolutely continuous invariant measures which are important measures in many applications, we restrict the domain of the operator to the set of all probability measures on X which are absolutely continuous with respect to μ. By the Radon-Nikodym theorem, such a set is equivalent to $\mathcal{D}$, the set of all densities of $L^1(X, \Sigma, \mu)$. This observation leads to the following definition of Frobenius-Perron operators on functions.

Let (X, Σ, μ) be a measure space and let $S : X \to X$ be a nonsingular transformation. For a given function $f \in L^1$, define a real measure

$$\mu_f(A) = \int_{S^{-1}(A)} f d\mu, \quad \forall\, A \in \Sigma.$$

Since S is nonsingular, $\mu(A) = 0$ implies $\mu_f(A) = 0$. Thus, Theorem 2.2.5 implies that there exists a unique function $\hat{f} \in L^1$, denoted as Pf, such that

$$\mu_f(A) = \int_A \hat{f} d\mu, \quad \forall\, A \in \Sigma.$$

Definition 4.2.2 *The operator $P : L^1(\mu) \to L^1(\mu)$ defined by*

$$\int_A Pf d\mu = \int_{S^{-1}(A)} f d\mu, \quad \forall\, A \in \Sigma, \ \ \forall\, f \in L^1(\mu) \tag{4.1}$$

is called the Frobenius-Perron operator *associated with S.*

Sometimes we write P_S for P to emphasize the dependence of the operator P on the transformation S. It is straightforward to show that P has the following properties:

(i) P is linear, that is, for all $a, b \in \mathbb{R}$ and $f_1, f_2 \in L^1$,

$$P(af_1 + bf_2) = aPf_1 + bPf_2;$$

(ii) P is a positive operator, that is, $Pf \geqslant 0$ if $f \geqslant 0$;

(iii) $\langle Pf, 1\rangle = \langle f, 1\rangle$, that is, $\int_X Pf d\mu = \int_X f d\mu$; and

(iv) $P_{S_1 \circ S_2} = P_{S_1} P_{S_2}$ for nonsingular transformations S_1 and S_2 from X into itself. In particular, $P_{S^n} = (P_S)^n$.

The properties (i)−(iii) above show that Frobenius-Perron operators are also Markov operators. The importance of introducing the concept of Frobenius-Perron operators is evidenced from the following theorem.

Theorem 4.2.1 *Let P be the Frobenius-Perron operator associated with a nonsingular transformation $S : X \to X$ and let $f \in L^1$ be nonnegative. The finite measure μ_f defined by*

$$\mu_f(A) = \int_A f d\mu, \quad \forall\, A \in \Sigma$$

is invariant under S if and only if f is a fixed point of P.

Proof Assume that μ_f is S-invariant, that is,

$$\mu_f(A) = \mu_f(S^{-1}(A)), \ \forall A \in \Sigma.$$

Then, by the definition of μ_f,

$$\int_A f \mathrm{d}\mu = \int_{S^{-1}(A)} f \mathrm{d}\mu, \ \forall A \in \Sigma.$$

So, from the definition (4.1) of P, we have

$$\int_A f \mathrm{d}\mu = \int_A P f \mathrm{d}\mu, \ \forall A \in \Sigma.$$

Hence, $Pf = f$ by Proposition 2.1.1 (v). The converse is also true from the above. □

Remark 4.2.1 Note from the above theorem that the original measure μ is invariant under S if and only if $P1 = 1$, where 1 is the constant 1 function.

The following proposition provides a relationship between the supports of f and Pf.

Proposition 4.2.1 *Let $A \in \Sigma$ and $f \geqslant 0$. Then, $f(x) = 0$ for all $x \in S^{-1}(A)$ if and only if $Pf(x) = 0$ for all $x \in A$, and in particular,*

$$S^{-1}(\text{supp } Pf) \supset \text{supp } f.$$

Proof By the definition (4.1) of the Frobenius-Perron operator P,

$$\int_X \chi_A P f \mathrm{d}\mu = \int_X \chi_{S^{-1}(A)} f \mathrm{d}\mu.$$

Suppose that $f \in L^1(X)$ is nonnegative. Then, $Pf = 0$ on A implies that $f = 0$ on $S^{-1}(A)$ and vice versa. Since

$$\int_{\text{supp } f} f \mathrm{d}\mu = \int_{\text{supp } Pf} P f \mathrm{d}\mu = \int_{S^{-1}(\text{supp } Pf)} f \mathrm{d}\mu,$$

we have $S^{-1}(\text{supp } Pf) \supset \text{supp } f$. □

We give examples of Frobenius-Perron operators associated with several kinds of transformations. First we consider one dimensional mappings. Suppose that $X = [a, b]$ and μ is the Lebesgue measure m. From Definition 4.2.2 of the Frobenius-Perron operator, for all $x \in [a, b]$

$$\int_a^x P f \mathrm{d}m = \int_{S^{-1}([a,x])} f \mathrm{d}m.$$

Then, taking derivatives to both sides above with respect to x and using the fundamental theorem of calculus give an *explicit* expression of Frobenius-Perron operators associated with interval mappings:

$$Pf(x) = \frac{\mathrm{d}}{\mathrm{d}x} \int_{S^{-1}([a,x])} f\mathrm{d}m, \quad x \in [a,b] \text{ a.e.} \tag{4.2}$$

Thus, the Frobenius-Perron operator can be viewed as the composition of a kind of integral operator and a differential operator. Although integral operators are usually compact, differential operators are not bounded in general. This provides an indication that Frobenius-Perron operators are generally not compact on its natural domain.

If $S : [a,b] \to [a,b]$ is differentiable and monotonic, then it is easy to see from (4.2) that

$$Pf(x) = f(S^{-1}(x)) \left| \frac{\mathrm{d}}{\mathrm{d}x} S^{-1}(x) \right|.$$

Example 4.2.1 (*the logistic model*) Let $S : [0,1] \to [0,1]$ be the logistic model $S(x) = 4x(1-x)$. Then,

$$S^{-1}([0,x]) = \left[0, \frac{1}{2} - \frac{1}{2}\sqrt{1-x}\right] \cup \left[\frac{1}{2} + \frac{1}{2}\sqrt{1-x}, 1\right],$$

thus the Frobenius-Perron operator P has the expression

$$Pf(x) = \frac{1}{4\sqrt{1-x}} \left\{ f\left(\frac{1}{2} - \frac{1}{2}\sqrt{1-x}\right) + f\left(\frac{1}{2} + \frac{1}{2}\sqrt{1-x}\right) \right\}.$$

In 1947, Ulam and von Neumann [121] showed that the unique stationary density of P is given by

$$f^*(x) = \frac{1}{\pi\sqrt{x(1-x)}}.$$

Hence, the absolutely continuous probability measure

$$\mu_{f^*}(A) = \int_A \frac{1}{\pi\sqrt{x(1-x)}} \mathrm{d}x$$

is invariant under the quadratic mapping $S(x) = 4x(1-x)$.

Now, let X be an N-dimensional rectangle $\prod_{i=1}^{N} [a_i, b_i]$. Then, taking partial derivatives to both sides of the equality

$$\int_{a_1}^{x_1} \cdots \int_{a_N}^{x_N} Pf(s_1, s_2, \cdots, s_N)\mathrm{d}t_1 \cdots \mathrm{d}t_N$$
$$= \int_{S^{-1}([a_1,x_1]\times\cdots\times[a_N,x_N])} f(s_1, s_2, \cdots, s_N)\mathrm{d}t_1 \cdots \mathrm{d}t_N$$

with respect to $x_1, x_2 \cdots, x_N$ successively, we obtain

$$Pf(x_1, x_2, \cdots, x_N) = \frac{\partial^N}{\partial x_1 \cdots \partial x_N} \int_{S^{-1}([a_1,x_1]\times\cdots\times[a_N,x_N])} f\mathrm{d}m. \tag{4.3}$$

Example 4.2.2 (*the baker transformation*) Let $X = [0,1]^2$. Define $\boldsymbol{S} : [0,1]^2 \to [0,1]^2$ by

$$\boldsymbol{S}(x,y) = \begin{cases} \left(2x, \frac{1}{2}y\right), & 0 \leqslant x < \frac{1}{2},\ 0 \leqslant y \leqslant 1, \\ \left(2x-1, \frac{1}{2}y + \frac{1}{2}\right), & \frac{1}{2} \leqslant x \leqslant 1,\ 0 \leqslant y \leqslant 1. \end{cases}$$

Since $\boldsymbol{S}^{-1}([0,x]\times[0,y]) = [0,x/2]\times[0,2y]$ for $0 \leqslant y < 1/2$, using (4.3), we have

$$Pf(x,y) = \frac{\partial^2}{\partial x \partial y} \int_0^{\frac{x}{2}} \mathrm{d}s \int_0^{2y} f(s,t)\mathrm{d}t = f\left(\frac{1}{2}x, 2y\right),\ 0 \leqslant y < \frac{1}{2}.$$

For $1/2 \leqslant y \leqslant 1$,

$$\boldsymbol{S}^{-1}([0,x]\times[0,y]) = \left(\left[0,\frac{1}{2}x\right]\times[0,1]\right) \cup \left(\left[\frac{1}{2},\frac{1}{2}+\frac{1}{2}x\right]\times[0,2y-1]\right),$$

hence, (4.3) gives that

$$\begin{aligned} Pf(x,y) &= \frac{\partial^2}{\partial x \partial y}\left\{\int_0^{\frac{x}{2}} \mathrm{d}s \int_0^1 f(s,t)\mathrm{d}t + \int_0^{\frac{1}{2}+\frac{x}{2}} \mathrm{d}s \int_0^{2y-1} f(s,t)\mathrm{d}t\right\} \\ &= f\left(\frac{1}{2}+\frac{1}{2}x, 2y-1\right),\ \frac{1}{2} \leqslant y \leqslant 1. \end{aligned}$$

In summary, we have

$$Pf(x,y) = \begin{cases} f\left(\frac{1}{2}x, 2y\right), & 0 \leqslant y < \frac{1}{2}, \\ f\left(\frac{1}{2}+\frac{1}{2}x, 2y-1\right), & \frac{1}{2} \leqslant y \leqslant 1. \end{cases}$$

Since $P1 = 1$, the Lebesgue measure m is invariant under the baker transformation.

Example 4.2.3 (*Hénon's map*) Let $\beta > 0$. The mapping $\boldsymbol{S} : [0,1]^2 \to [0,1]^2$ defined by

$$\boldsymbol{S}(x,y) = (4x(1-x)+y, \beta x) \pmod 1$$

is called a *Hénon's map*. In Exercise 4.28, the reader is asked to find the expression of the Frobenius-Perron operator associated with a Hénon's map.

Example 4.2.4 (*the Anosov diffeomorphism*) The mapping

$$\boldsymbol{S}(x,y) = (x+y, x+2y) \pmod 1$$

is called an *Anosov diffeomorphism*. It is invertible and

$$\boldsymbol{S}^{-1}(x,y) = (2x-y, y-x) \pmod 1.$$

So,

$$Pf(x,y) = f(2x-y \pmod 1, y-x \pmod 1),$$

thus the Anosov diffeomorphism preserves the Lebesgue measure.

Example 4.2.5 This example is from the book "Chaos in Electronics" [122]. The periodic switching nature of the switched-mode DC-to-DC converter (back converter) lead Hamill *et al.* in 1988 and 1989 to study a three-segment piecewise linear mapping $S : \mathbb{R} \to \mathbb{R}$, called the *zigzag map*, given by

$$S(x) = \begin{cases} x+b-c, & \text{if } x \leqslant 1 - \dfrac{1}{a}, \\ (1-ab)x + ab - c, & \text{if } 1 - \dfrac{1}{a} < x < 1, \\ x - c, & \text{if } 1 \leqslant x, \end{cases}$$

where a, b, c are positive constants. The iteration $x_{n+1} = S(x_n)$, $\forall\, n$ of the zigzag map relates the output currents of two consecutive switching cycles. We assume that these parameters are chosen such that

$$b > 2c, \ a \leqslant \frac{1}{c}, \ \ ab = k,$$

where $k \geqslant 3$ is an integer. The corresponding Frobenius-Perron operator P has the expression

$$Pf(x) = \frac{1}{k-1} f\left(\frac{k-x-c}{k-1}\right) \chi_{[1-c, 1+(k-2)c]}(x) + f(x+c)\chi_{[1-c,1+(k-3)c]}(x).$$

It can be shown (see [122] for more details) that the stationary density f^* of P is given by

$$f^*(x) = \sum_{i=1}^{k-1} \frac{2(k-i)}{c(k-1)k} \chi_{[1+(i-2)c, 1+(i-1)c]}(x). \qquad \square$$

The remaining three examples are related to Examples 4.1.1, 4.1.3, and 4.1.4 in the previous section, respectively.

Example 4.2.6 (*solutions of evolution equations*) For each integer $i=1,2,\cdots,N$, let the function $F_i(x_1,x_2,\cdots,x_N)$ have continuous partial derivatives $\partial F_i/\partial x_j$ on its domain $\mathbb{R}^N, j=1,2,\cdots,N$. For the partial differential equation (called the *continuity equation*)

$$\frac{\partial u}{\partial t}+\sum_{i=1}^{N}\frac{\partial(uF_i)}{\partial x_i}=0,\ \ t>0,\ \boldsymbol{x}=(x_1,x_2,\cdots,x_N)^{\mathrm{T}}\in\mathbb{R}^N,$$

its solution $u(t,\boldsymbol{x})$, which satisfies the initial value condition $u(0,\boldsymbol{x})=f(\boldsymbol{x})$, is given by the *semigroup of Frobenius-Perron operators*

$$u(t,\boldsymbol{x})=P_tf(\boldsymbol{x}),$$

where P_t is defined by

$$\int_A P_tf\mathrm{d}m=\int_{\boldsymbol{S}_t^{-1}(A)}f\mathrm{d}m,\ \ \forall\ A\in\mathcal{B}$$

associated with the *semigroup of transformations* $\{\boldsymbol{S}_t\}_{t\geqslant0}$ which are defined by $S_t(\boldsymbol{x}^0)=\boldsymbol{x}(t)$, where $\boldsymbol{x}(t)$ is the solution of the corresponding system of ordinary differential equations

$$\frac{\mathrm{d}x_i}{\mathrm{d}t}=F_i(\boldsymbol{x}),\ i=1,2,\cdots,N$$

with the initial value conditions $x_i(0)=x_i^0$ for $i=1,2,\cdots,N$. See Section 7.6 of [82] for a proof of this fact.

As a special case of the above system of ordinary differential equations, consider the following *Hamiltonian system*

$$\frac{\mathrm{d}\boldsymbol{q}_i}{\mathrm{d}t}=\frac{\partial H}{\partial\boldsymbol{p}_i},\ \frac{\mathrm{d}\boldsymbol{p}_i}{\mathrm{d}t}=-\frac{\partial H}{\partial\boldsymbol{q}_i},\ \ i=1,2,\cdots,N,$$

where the *Hamiltonian* $H(\boldsymbol{p},\boldsymbol{q})$ is the *energy* of the system, and $\boldsymbol{q}$ and $\boldsymbol{p}$ are referred to as the *generalized positions* and the *generalized momenta*, respectively. In this case, the continuity equation takes the form

$$\frac{\partial u}{\partial t}+[u,H]=0,$$

where the *Poisson bracket*

$$[u,H]\equiv\sum_{i=1}^{N}\left(\frac{\partial u}{\partial\boldsymbol{q}_i}\frac{\partial H}{\partial\boldsymbol{p}_i}-\frac{\partial u}{\partial\boldsymbol{p}_i}\frac{\partial H}{\partial\boldsymbol{q}_i}\right).$$

For Hamiltonian systems, the change with time of an arbitrary function g of the variables $\boldsymbol{q}_1, \boldsymbol{q}_2, \cdots, \boldsymbol{q}_N, \boldsymbol{p}_1, \boldsymbol{p}_2, \cdots, \boldsymbol{p}_N$ is given by

$$\frac{\mathrm{d}g}{\mathrm{d}t} = \sum_{i=1}^{N} \left(\frac{\partial g}{\partial \boldsymbol{q}_i} \frac{\partial H}{\partial \boldsymbol{p}_i} - \frac{\partial g}{\partial \boldsymbol{p}_i} \frac{\partial H}{\partial \boldsymbol{q}_i} \right) = [g, H].$$

In particular, if we take g to be a function of the energy H, then

$$\frac{\mathrm{d}g}{\mathrm{d}t} = \frac{\mathrm{d}g}{\mathrm{d}H} \frac{\mathrm{d}H}{\mathrm{d}t} = \frac{\mathrm{d}g}{\mathrm{d}H}[H, H] = 0.$$

Thus, any function of the energy H is a *constant of the motion.* We shall come back to Hamiltonian systems in Section 9.3.

Example 4.2.7 (*fractal geometry*) Many fractals are generated by the so-called "iterated functions system" (IFS), i.e., $S = \{S_1, S_2, \cdots, S_r\}$ is a finite family of r *contraction mappings* with r corresponding probabilities $\{p_1, p_2, \cdots, p_r\}$ on a compact metric space $(X, d(\cdot, \cdot))$. Such an IFS gives a *random dynamical system*: in the orbit

$$x_0, x_1, x_2, \cdots, x_n, \cdots,$$

$x_{n+1} = S_{\alpha_n}(x_n)$ for some $1 \leqslant \alpha_n \leqslant r$ is determined with probability p_{α_n} for each n. The asymptotic statistical properties of such orbits can be described with an invariant probability measure μ^* of the "Markov operator on measures" $M \equiv \sum_{i=1}^{r} p_i M_i$, where for each i the Frobenius-Perron operator on measures $M_i \mu = \mu \circ S_i^{-1}$ maps a Borel measure μ to a Borel measure $\mu \circ S_i^{-1}$. Using Banach's contraction fixed point theorem to the so-called *Hutchinson metric* [6] introduced for the IFS, one can prove the existence and the uniqueness of an invariant probability measure μ^* whose support, supp μ^*, satisfies the equality

$$\operatorname{supp} \mu^* = \bigcup_{i=1}^{r} S_i(\operatorname{supp} \mu^*),$$

where the *support* of a probability measure μ is the minimal set $A \subset X$ with $\mu(A) = \mu(X)$. In other words, the support is exactly the invariant set or the *attractor* of the IFS. The invariant set of an IFS is often a fractal set. For example, the well known *Sierpiński gasket* in the theory of fractals is the invariant set of the IFS consisting of three simple affine contraction mappings. In many cases, the invariant measure of an IFS is absolutely continuous, and so its density function is a fixed point of the Markov operator $P = \sum_{i=1}^{r} p_i P_{S_i}$, where P_{S_i} is the Frobenius-Perron operator associated with S_i for $i = 1, 2, \cdots, r$.

Example 4.2.8 (*wavelets construction*) If we let

$$S_n(x) = \frac{x+n}{2}, \quad p_n = \frac{a_n}{2}, \quad \forall\, n \in \mathcal{I}$$

in Example 4.1.4, then $\{S_n\}_{n\in\mathcal{I}}$ is a sequence of nonsingular transformations on $\mathbb{R}$ with a sequence of probability distributions $\{p_n\}_{n\in\mathcal{I}}$. The Frobenius-Perron operator $P_{S_n} : L^1(\mathbb{R}) \to L^1(\mathbb{R})$ associated with S_n has the expression

$$P_{S_n} f(x) = 2f(2x - n), \ \ \forall\, x \in \mathbb{R}.$$

Therefore, the Markov operator defined in Example 4.1.4 has the representation of

$$P = \sum_{n\in\mathcal{I}} p_n P_{S_n}.$$

A sequence $\{S_n\}$ of nonsingular transformations with the corresponding sequence $\{p_n\}$ of probability distributions defines *random transformations* or *coupled maps*, in which the mapping S_n in the iteration process is chosen with the probability p_n (see Example 4.2.7 above). The coupled maps are often studied in statistical physics (see, e.g., [96]). For the current example from the wavelets theory, an absolutely continuous invariant measure of the random transformation is obtained as a stationary density of the Markov operator defined above. See [119] for more discussions.

For a general invertible transformation S, based on the following *change of variables formula*, the corresponding Frobenius-Perron operator P has an explicit expression.

Lemma 4.2.1 *Let (X, Σ, μ) be a finite measure space and let $S : X \to X$ be a nonsingular transformation. Suppose that f is a nonnegative measurable function or $f \in L^1(\mu)$. Then, for every $A \in \Sigma$,*

$$\int_{S^{-1}(A)} f \circ S \mathrm{d}\mu = \int_A f \mathrm{d}(\mu \circ S^{-1}) = \int_A f h \mathrm{d}\mu,$$

where $h = \mathrm{d}(\mu\circ S^{-1})/\mathrm{d}\mu$ is the Radon-Nikodym derivative of $\mu\circ S^{-1}$ with respect to μ, that is,

$$(\mu \circ S^{-1})(A) = \int_A h \mathrm{d}\mu, \ \ \forall\, A \in \Sigma.$$

Proof First let $f = \chi_B$ with $B \in \Sigma$. Then, $f \circ S = \chi_{S^{-1}(B)}$, and

$$\begin{aligned}\int_{S^{-1}(A)} \chi_B \circ S \mathrm{d}\mu &= \int_X \chi_{S^{-1}(A)}\chi_{S^{-1}(B)} \mathrm{d}\mu \\ &= \mu(S^{-1}(A) \cap S^{-1}(B)) = \mu(S^{-1}(A \cap B)).\end{aligned}$$

The second integral in the lemma may be written as

$$\int_A \chi_B \mathrm{d}(\mu \circ S^{-1}) = \int_X \chi_A \chi_B \mathrm{d}(\mu \circ S^{-1}) = \mu(S^{-1}(A \cap B)),$$

while the last integral of the lemma becomes

$$\int_A \chi_B h \mathrm{d}\mu = \int_{A\cap B} h \mathrm{d}\mu = \mu(S^{-1}(A \cap B)).$$

Thus, the lemma is true for all simple functions, and a limit process gives the required result. □

Proposition 4.2.2 *Let (X, Σ, μ) be a finite measure space, let $S : X \to X$ be an invertible nonsingular transformation such that S^{-1} is also nonsingular, and let P be the corresponding Frobenius-Perron operator. Then, for every $f \in L^1$,*

$$Pf(x) = f(S^{-1}(x))h(x),$$

where $h = \mathrm{d}(\mu \circ S^{-1})/\mathrm{d}\mu$.

Proof Let $A \in \Sigma$. Then, from the definition of P,

$$\int_A Pf(x)\mathrm{d}\mu(x) = \int_{S^{-1}(A)} f(x)\mathrm{d}\mu(x).$$

Changing the variable x to $y = S(x)$ to the right-hand side integral, we have

$$\int_{S^{-1}(A)} f(x)\mathrm{d}\mu(x) = \int_A f(S^{-1}(y))h(y)\mathrm{d}\mu(y).$$

Thus,

$$\int_A Pf(x)\mathrm{d}\mu(x) = \int_A f(S^{-1}(x))h(x)\mathrm{d}\mu(x).$$

Since $A \in \Sigma$ is arbitrary, it follows that

$$Pf(x) = f(S^{-1}(x))h(x).$$ □

4.3 Koopman Operators

In this section, we introduce the Koopman operator which is the dual operator of the Frobenius-Perron operator.

Definition 4.3.1 *Let (X, Σ, μ) be a measure space and let $S : X \to X$ be a nonsingular transformation. The linear operator $U : L^\infty(X) \to L^\infty(X)$ defined by*

$$Uf(x) \equiv U_S f(x) = f(S(x)), \ \forall\, x \in X, \ \forall\, f \in L^\infty(X)$$

is called the Koopman operator *with respect to S.*

Because of the nonsingularity assumption of S, the Koopman operator is well-defined since $f_1(x) = f_2(x)$, $x \in X$ μ-a.e. implies that $f_1(S(x)) = f_2(S(x))$, $x \in X$ μ-a.e. Some basic properties of U are listed below:

(i) U is a positive operator;

(ii) U is a (weak) contraction on L^∞, that is, $\|Uf\|_\infty \leqslant \|f\|_\infty$ for all $f \in L^\infty$;

(iii) $U_{S_1 \circ S_2} = U_{S_2} U_{S_1}$. In particular, $U_{S^n} = (U_S)^n$.

Proposition 4.3.1 *The Koopman operator U is the dual of the Frobenius-Perron operator P, that is,*

$$\langle Pf, g\rangle \ = \ \langle f, Ug\rangle, \ \ \forall\, f \in L^1,\ g \in L^\infty.$$

Proof Given $f \in L^1$. First, let $g = \chi_A$ with $A \in \Sigma$. Then,

$$\begin{aligned}\langle Pf, g\rangle &= \int_X (Pf)\chi_A \mathrm{d}\mu = \int_A Pf \mathrm{d}\mu = \int_{S^{-1}(A)} f \mathrm{d}\mu \\ &= \int_X f(\chi_A \circ S)\mathrm{d}\mu = \int_X f\, U\chi_A \mathrm{d}\mu = \ \langle f, Ug\rangle.\end{aligned}$$

Thus, $\langle Pf, g\rangle \ = \ \langle f, Ug\rangle$ for all simple functions $g \in L^\infty$. Now let $g \in L^\infty$. Then, $g = \lim_{n\to\infty} g_n$ for a sequence $\{g_n\}$ of simple functions in L^∞. Therefore,

$$\langle Pf, g\rangle \ = \ \lim_{n\to\infty} \langle Pf, g_n\rangle \ = \ \lim_{n\to\infty} \langle f, Ug_n\rangle \ = \ \langle f, Ug\rangle. \qquad \square$$

Using the notion of Koopman operators, we can restate Birkhoff's pointwise ergodic theorem and von Neumann's mean ergodic theorem, the two most important classic ergodic theorems, as follows, in which the domain of the Koopman operator associated with a measure preserving transformation can be the space $L^p(X, \Sigma, \mu)$ for $1 \leqslant p \leqslant \infty$. It can be shown (see Exercise 4.29) that when S is measure preserving, $U_S : L^p \to L^p$ is well-defined and is actually an *isometry*, that is, $\|U_S f\|_p = \|f\|_p$ for all $f \in L^p$.

Remark 4.3.1 The claim that $U_S : L^\infty(\mu) \to L^\infty(\mu)$ is isometric if S preserves μ is a consequence of Remark 4.4.2 in Section 4.4.

Theorem 4.3.1 (Birkhoff's pointwise ergodic theorem) *Let $S : (X, \Sigma, \mu) \to (X, \Sigma, \mu)$ be measure preserving and let U be the corresponding Koopman operator. Then, $U : L^1 \to L^1$ is well-defined and is isometric. For any $f \in L^1$, we have*

$$\lim_{n\to\infty} \frac{1}{n} \sum_{i=0}^{n-1} U^i f(x) = \tilde{f}(x), \ \ \forall\, x \in X \ \mu\text{-a.e.},$$

where $\tilde{f} \in L^1$ satisfies that $U\tilde{f} = \tilde{f}$. Moreover, if $\mu(X) < \infty$, then $\int_X \tilde{f} \mathrm{d}\mu = \int_X f \mathrm{d}\mu$. Furthermore, if S is ergodic, then $\tilde{f}$ is a constant function.

Theorem 4.3.2 (von Neumann's mean ergodic theorem) *Let $S : (X, \Sigma, \mu) \to (X, \Sigma, \mu)$ be measure preserving on a probability measure space (X, Σ, μ), let U be the corresponding Koopman operator, and let $p \in [1, \infty)$. Then, $U : L^\infty \to L^\infty$ can be extended as an isometry from L^p into itself. Moreover, if $f \in L^p$, then there is $\tilde{f} \in L^p$ such that $U\tilde{f} = \tilde{f}$ and*

$$\lim_{n\to\infty} \left\| \frac{1}{n} \sum_{i=0}^{n-1} U^i f - \tilde{f} \right\|_p = 0.$$

Similarly, one can restate the ergodic theorems for the topological dynamical systems of Section 3.4 in terms of the notation of Koopman operators. In this case, naturally, $U_S : C(X) \to C(X)$ is well-defined.

4.4 Ergodicity and Frobenius-Perron Operators

We can employ Frobenius-Perron operators to study different versions of chaos via density evolutions of the dynamical system. We first prove a theorem about the equivalence of ergodicity of a transformation and uniqueness of a stationary density of the corresponding Frobenius-Perron operator, which will also be used to characterize ergodic transformations via Frobenius-Perron operators.

Theorem 4.4.1 *Let (X, Σ, μ) be a measure space, let $S : X \to X$ be a nonsingular transformation, and let P be the Frobenius-Perron operator associated with S. If S is ergodic with respect to μ, then P has at most one stationary density. Conversely, if P has a unique stationary density f^* and f^* is strictly positive on X, then S is ergodic with respect to both μ and the invariant measure μ_{f^*} whose density with respect to μ is f^*.*

Proof If P has two different stationary densities f_1 and f_2, then for $g = f_1 - f_2$ we have $Pg = g$, $g^+ \neq 0$, and $g^- \neq 0$. Proposition 4.1.3 ensures that

$$Pg^+ = g^+ \text{ and } Pg^- = g^-. \tag{4.4}$$

Denote $A = \text{supp } g^+$ and $B = \text{supp } g^-$. Then, $\mu(A) > 0$ and $\mu(B) > 0$. From Proposition 4.2.1 and (4.4), we have

$$A \subset S^{-1}(A) \text{ and } B \subset S^{-1}(B),$$

from which

$$A \subset S^{-1}(A) \subset S^{-2}(A) \subset \cdots \subset S^{-n}(A) \subset \cdots$$

and

$$B \subset S^{-1}(B) \subset S^{-2}(B) \subset \cdots \subset S^{-n}(B) \subset \cdots$$

by induction. Since $A \cap B = \varnothing$, we see that $S^{-n}(A) \cap S^{-n}(B) = \varnothing$ for all $n = 0, 1, 2, \cdots$. Now, define two sets

$$\bar{A} = \bigcup_{n=0}^{\infty} S^{-n}(A) \text{ and } \bar{B} = \bigcup_{n=0}^{\infty} S^{-n}(B).$$

Then, $\bar{A} \cap \bar{B} = \varnothing$ and both of them have positive measures. Since

$$S^{-1}(\bar{A}) = \bigcup_{n=0}^{\infty} S^{-n-1}(A) = \bigcup_{n=0}^{\infty} S^{-n}(A) = \bar{A}$$

and the same equality is true for $\bar{B}$ in place of $\bar{A}$, we have found two nontrivial S-invariant sets, so S is not ergodic.

Conversely, suppose that $f^* > 0$ μ-a.e. is the unique stationary density of P. If S is not ergodic with respect to $\mu \cong \mu_{f^*}$, then there is a nontrivial S-invariant set $A \in \Sigma$. So $B = A^c$ is also a nontrivial S-invariant set. Since $f^* = \chi_A f^* + \chi_B f^*$,

$$\chi_A f^* + \chi_B f^* = f^* = Pf^* = P(\chi_A f^*) + P(\chi_B f^*). \tag{4.5}$$

Since f^* is positive everywhere, supp $\chi_A f^* = A$ and supp $\chi_B f^* = B$, Hence, Propositions 4.2.1 and (4.5) imply that

$$\chi_A f^* = P(\chi_A f^*) \text{ and } \chi_B f^* = P(\chi_B f^*).$$

Since $\chi_A f^*$ and $\chi_B f^*$ are nonzero, dividing them by their respective L^1-norm gives two distinct stationary densities of P, which is a contradiction to the assumption. □

Frobenius-Perron operators and Koopman operators may be used to reformulate the concepts of ergodicity, mixing, and exactness in the following theorem.

Theorem 4.4.2 *Let (X, Σ, μ) be a probability measure space, let $S : X \to X$ be a measure preserving transformation, and let P be the Frobenius-Perron operator associated with S. Then,*

(i) *S is ergodic if and only if*

$$\lim_{n\to\infty} \frac{1}{n} \sum_{i=0}^{n-1} \langle P^i f, g\rangle = \langle f, 1\rangle\langle 1, g\rangle, \quad \forall\, f \in L^1,\ g \in L^\infty. \tag{4.6}$$

(ii) *S is mixing if and only if*

$$\lim_{n\to\infty} \langle P^n f, g\rangle = \langle f, 1\rangle\langle 1, g\rangle, \quad \forall\, f \in L^1,\ g \in L^\infty. \tag{4.7}$$

(iii) *Assume further that $S\Sigma \subset \Sigma$. Then S is exact if and only if*

$$\lim_{n\to\infty} \|P^n f - \langle f, 1\rangle\| = 0, \quad \forall\, f \in L^1. \tag{4.8}$$

Proof Before we prove the theorem we note that $P1 = 1$ since S preserves μ.

(i) If S is ergodic, then, the constant function $f^* = 1$ is the unique stationary density of P from Theorem 4.4.1. We show that the strong limit

$$\lim_{n\to\infty} \frac{1}{n}\sum_{i=0}^{n-1} P^i f = 1, \ \ \forall\, f \in \mathcal{D} \tag{4.9}$$

holds in L^1. Let $A_n = n^{-1}\sum_{i=0}^{n-1} P^i$. First assume that $f \in \mathcal{D}$ is such that $f(x) \leqslant c$, $\forall\, x \in X$ for some constant $c > 0$. Then, $P^n f \leqslant P^n c = c$ for all n, so $A_n f = n^{-1}\sum_{i=0}^{n-1} P^i f \leqslant c$. Thus, the sequence $\{A_n f\}$ is precompact by Proposition 2.5.1, and it follows from Theorem 5.1.1 in the next chapter that $\{A_n f\}$ converges to a stationary density of P. Since 1 is the unique stationary density, $\lim_{n\to\infty} A_n f = 1$. This proves (4.9) for bounded densities.

If $f \in \mathcal{D}$ is not uniformly bounded on X, we define a function $f_c(x) = \min\{f(x), c\}$ on X for each $c > 0$. Then,

$$\lim_{c\to\infty} f_c(x) = f(x), \ \ \forall\, x \in X.$$

Since $f_c \leqslant f$ for all c, by Lebesgue's dominated convergence theorem, $\lim_{c\to\infty} \|f_c - f\| = 0$ and $\lim_{c\to\infty} \|f_c\| = \|f\|$. Write

$$f = \frac{1}{\|f_c\|} f_c + h_c, \tag{4.10}$$

where $h_c = f - \|f_c\|^{-1} f_c \to 0$ strongly as $c \to \infty$. In other words, given $\epsilon > 0$, there is $c > 0$ such that $\|h_c\| < \epsilon/2$. It follows that

$$\|A_n h_c\| \leqslant \|h_c\| < \frac{\epsilon}{2}. \tag{4.11}$$

On the other hand, for the fixed c, the function $f_c/\|f_c\| \in \mathcal{D}$ is bounded by the number $c/\|f_c\|$, so by what we have proved above, there is an integer $n_0 > 0$ such that for all $n \geqslant n_0$,

$$\left\| A_n\left(\frac{f_c}{\|f_c\|}\right) - 1 \right\| \leqslant \frac{\epsilon}{2}. \tag{4.12}$$

Combining inequalities (4.11) and (4.12) with the decomposition (4.10), we obtain

$$\|A_n f - 1\| \leqslant \epsilon, \ \ \forall\, n \geqslant n_0.$$

Since $\epsilon > 0$ is arbitrary, (4.9) is valid, which implies (4.6) immediately. This proves the necessity part of (i).

Conversely, if (4.6) holds, then the sequence $\{A_n f\}$ converges to 1 weakly for any $f \in \mathcal{D}$, so (4.9) is valid by Theorem 5.1.1. Applying (4.9) to any stationary density f of P gives $f = 1$. Thus, 1 is the unique stationary density of P. By Theorem 4.4.1, S is ergodic.

(ii) Assume that S is mixing. Then, the definition of mixing means that for all $A \in \Sigma$ and $B \in \Sigma$,

$$\lim_{n\to\infty} \int_X \chi_A(\chi_B \circ S^n) d\mu = \int_X \chi_A d\mu \int_X \chi_B d\mu,$$

which, via using the notion of the Koopman operator, is

$$\lim_{n\to\infty} \langle \chi_A, U^n \chi_B \rangle = \langle \chi_A, 1 \rangle \langle 1, \chi_B \rangle. \tag{4.13}$$

Since the Koopman operator is the dual of the Frobenius-Perron operator, (4.13) can be written as

$$\lim_{n\to\infty} \langle P^n \chi_A, \chi_B \rangle = \langle \chi_A, 1 \rangle \langle 1, \chi_B \rangle.$$

Thus, (4.7) is true for $f = \chi_A$ and $g = \chi_B$, and so it is true for all simple functions $f = \sum_i a_i \chi_{A_i}$ and $g = \sum_j b_j \chi_{B_j}$. Since every function $f \in L^1$ is the strong limit of a sequence of simple functions and each function $g \in L^\infty$ is the uniform limit of simple functions, and since $\|P^n\| \leqslant 1$ for all n, the convergence (4.7) is satisfied for all $f \in L^1$ and $g \in L^\infty$ by a limit argument.

Conversely, if (4.7) is satisfied, then by letting $f = \chi_A$ and $g = \chi_B$ with $A, B \in \Sigma$, we get the mixing condition (3.5).

(iii) We only prove the sufficiency part. A proof to the necessity of (4.8) for an exact transformation can be seen from [91].

Suppose that (4.8) is true. Assume $\mu(A) > 0$ for some $A \in \Sigma$. Let $f_A(x) = \chi_A(x)/\mu(A)$ for $x \in X$. Then $f_A \in \mathcal{D}$, so (4.8) ensures that $\lim_{n\to\infty} \|P^n f_A - 1\| = 0$. We have

$$\begin{aligned}
\mu(S^n(A)) &= \int_X \chi_{S^n(A)} d\mu \\
&= \int_{S^n(A)} P^n f_A d\mu - \int_{S^n(A)} (P^n f_A - 1) d\mu \\
&\geqslant \int_{S^n(A)} P^n f_A d\mu - \|P^n f_A - 1\| \\
&= \int_{S^{-n}(S^n(A))} f_A d\mu - \|P^n f_A - 1\| \\
&= 1 - \|P^n f_A - 1\| \to 1 \text{ as } n \to \infty,
\end{aligned}$$

where the last two equalities are from the definition of the Frobenius-Perron operator and the fact that $A \subset S^{-n}(S^n(A))$ and supp $f_A = A$. Thus, S is exact. This completes the proof of the theorem. □

Remark 4.4.1 As in Remark 3.3.2, the convergence equalities (4.6), (4.7), and (4.8) of Theorem 4.4.2 can be satisfied by f in a fundamental set of L^1 and g in a fundamental set of L^∞ only.

Using the dual relation of the Koopman operator and the Frobenius-Perron operator, part (i) and part (ii) of the above theorem can also be expressed by

Corollary 4.4.1 *Let (X, Σ, μ) be a probability measure space, let $S : X \to X$ be a measure preserving transformation, and let U be the Koopman operator corresponding to S. Then,*

(i) S is ergodic if and only if

$$\lim_{n\to\infty} \frac{1}{n} \sum_{i=0}^{n-1} \langle f, U^i g\rangle = \langle f, 1\rangle\langle 1, g\rangle, \quad \forall\, f \in L^1,\ g \in L^\infty;$$

(ii) S is mixing if and only if

$$\lim_{n\to\infty} \langle f, U^n g\rangle = \langle f, 1\rangle\langle 1, g\rangle, \quad \forall\, f \in L^1,\ g \in L^\infty.$$

Exactly the same idea of the proof to the necessary condition (4.9) of ergodicity in Theorem 4.4.2 (i) can establish the more general result for Markov operators as follows.

Theorem 4.4.3 *Let (X, Σ, μ) be a measure space and let $P : L^1 \to L^1$ be a Markov operator with a unique stationary density f^*. If $f^*(x) > 0$ for all $x \in X$, then*

$$\lim_{n\to\infty} \frac{1}{n} \sum_{i=0}^{n-1} P^i f = f^*, \quad \forall\, f \in \mathcal{D}.$$

Remark 4.4.2 Theorem 4.4.2 basically says that

(i) S is ergodic if and only if the sequence $\{P^n f\}$ is Cesáro convergent to 1 for all $f \in \mathcal{D}$;

(ii) S is mixing if and only if the sequence $\{P^n f\}$ is weakly convergent to 1 for all $f \in \mathcal{D}$;

(iii) S is exact if and only if the sequence $\{P^n f\}$ is strongly convergent to 1 for all $f \in \mathcal{D}$.

This motivates the extension of the notions of ergodicity, mixing, and exactness for transformations to general Markov operators in the following definition.

Definition 4.4.1 *Let (X, Σ, μ) be a probability measure space and let $P : L^1 \to L^1$ be a Markov operator with stationary density* 1. *Then, we say:*

(i) *P is* ergodic *if the sequence $\{P^n f\}$ is Cesáro convergent to* 1 *for all $f \in \mathcal{D}$;*

(ii) P *is* mixing *if the sequence* $\{P^n f\}$ *is weakly convergent to* 1 *for all* $f \in \mathcal{D}$;

(iii) P *is* exact *if the sequence* $\{P^n f\}$ *is strongly convergent to* 1 *for all* $f \in \mathcal{D}$.

4.5 Decomposition Theorem and Spectral Analysis

We are going to present some general results on the structure of Frobenius-Perron operators $P : L^1(X, \Sigma, \mu) \to L^1(X, \Sigma, \mu)$. It will be shown [35] that a Frobenius-Perron operator is *partially isometric* on its domain $L^1(\Sigma)$, that is, it is isometric on a vector subspace of $L^1(\Sigma)$. More specifically, it is isometric on the subspace $L^1(S^{-1}\Sigma)$ which is a topological complement of its null space in $L^1(\Sigma)$.

Let (X, Σ, μ) be a measure space and let $S : X \to X$ be a nonsingular transformation such that the measure space $(X, S^{-1}\Sigma, \mu)$ is also σ-finite. The nonsingularity assumption about S just says that the measure $\nu \equiv \mu \circ S^{-1}$ is absolutely continuous with respect to the measure μ. By the Radon-Nikodym theorem, there is a nonnegative Σ-measurable function h such that

$$\nu(A) = \int_A h \mathrm{d}\mu, \quad \forall\, A \in \Sigma.$$

Given any $f \in L^1(X, \Sigma, \mu)$, the set function

$$\mu_f(B) = \int_B f \mathrm{d}\mu, \quad \forall\, B \in S^{-1}\Sigma$$

defines a finite measure on the measurable space $(X, S^{-1}\Sigma)$, which is absolutely continuous with respect to μ. The Radon-Nikodym theorem ensures that there is a function $\tilde{f} \in L^1(X, S^{-1}\Sigma, \mu) \subset L^1(X, \Sigma, \mu)$ such that

$$\int_B \tilde{f} \mathrm{d}\mu = \mu_f(B) = \int_B f \mathrm{d}\mu$$

for all $B \in S^{-1}\Sigma$. This $\tilde{f}$ is unique by Proposition 2.1.1 (v), and is denoted as Ef. The operator $E : L^1(X, \Sigma, \mu) \to L^1(X, \Sigma, \mu)$, which is obviously linear, is well-defined.

Definition 4.5.1 *The operator* $E : L^1(X, \Sigma, \mu) \to L^1(X, \Sigma, \mu)$ *defined above is referred to as the* conditional expectation *associated with the sub-*σ*-algebra* $S^{-1}\Sigma$.

If $f \geqslant 0$, then $Ef \geqslant 0$ and $\|Ef\| = \|f\|$. Hence, E is a Markov operator, and so $\|Ef\| \leqslant \|f\|$ for all $f \in L^1$. Moreover, $Ef = f$ if and only if $f \in L^1$ is $S^{-1}\Sigma$-measurable. Thus, E is a positive projection operator from $L^1(X, \Sigma, \mu)$

onto $L^1(X, S^{-1}\Sigma, \mu)$ with operator norm 1. We thus have the *direct sum decomposition* $L^1(X, \Sigma, \mu) = R(E) \oplus N(E)$, where $R(E)$ is the *range* of E and $N(E)$ is the *null space* of E. More properties of E are referred to [14, 123].

Lemma 4.5.1 *$f \in L^1(X, S^{-1}\Sigma, \mu)$ if and only if $f = g \circ S$ for some $g \in L^1(X, \Sigma, \nu)$. Moreover, if $f = g \circ S$, then $\|f\|_\mu = \|g\|_\nu$. In other words, the correspondence $f \longmapsto g$ is an isometric isomorphism between $L^1(X, S^{-1}\Sigma, \mu)$ and $L^1(X, \Sigma, \nu)$.*

Proof First we show that each $f \in L^1(X, S^{-1}\Sigma, \mu)$ can be written as $g \circ S$ with $g \in L^1(X, \Sigma, \nu)$. This is true since if $f = \chi_{S^{-1}(A)}$, then $g = \chi_A$, and the general case follows from a limit process.

Suppose that $f = g \circ S$. Then it is easy to see that f is $S^{-1}\Sigma$-measurable if and only if g is Σ-measurable. $\|f\|_\mu = \|g\|_\nu$ since

$$\int_X |f| \mathrm{d}\mu = \int_X |g| \circ S \mathrm{d}\mu = \int_X |g| \mathrm{d}\nu. \qquad \square$$

Theorem 4.5.1 (Ding-Hornor's decomposition theorem) *Let (X, Σ, μ) be a measure space, let $S : X \to X$ be a nonsingular transformation such that the measure space $(X, S^{-1}\Sigma, \mu)$ is σ-finite, and let $P : L^1(\mu) \to L^1(\mu)$ be the Frobenius-Perron operator associated with S. Then,*

$$L^1(X, \Sigma, \mu) = N(P) \oplus L^1(X, S^{-1}\Sigma, \mu) \tag{4.14}$$

and P is isometric on $L^1(X, S^{-1}\Sigma, \mu)$, i.e., *$P$ is a partial isometry.*

Proof Let $f \in L^1(X, \Sigma, \mu)$. Then, by the definition of Ef,

$$\int_A P(f - Ef) \mathrm{d}\mu = \int_{S^{-1}(A)} f \mathrm{d}\mu - \int_{S^{-1}(A)} Ef \mathrm{d}\mu = 0$$

for all $A \in \Sigma$. Hence, $P(f - Ef) = 0$ by Proposition 2.1.1 (v). Thus, $N(E) \subset N(P)$.

Now we show that $\|Pf\| = \|f\|$ for all $f \in R(E) = L^1(X, S^{-1}\Sigma, \mu)$. Let $f \in L^1(X, S^{-1}\Sigma, \mu)$. From Lemma 4.5.1, there is $g \in L^1(X, \Sigma, \nu)$ such that $f = g \circ S$ and $\|f\|_\mu = \|g\|_\nu$. Hence, by Lemma 4.5.1,

$$\begin{aligned}\int_A Pf \mathrm{d}\mu &= \int_A P(g \circ S) \mathrm{d}\mu = \int_{S^{-1}(A)} g \circ S \mathrm{d}\mu \\ &= \int_A g \mathrm{d}\nu = \int_A gh \mathrm{d}\mu\end{aligned}$$

for every $A \in \Sigma$. It follows from Proposition 2.1.1 (v) that

$$Pf = gh. \tag{4.15}$$

Therefore, $\|Pf\|_\mu = \|gh\|_\mu = \|g\|_\nu = \|f\|_\mu$. Thus, $\|Pf\| = \|f\|$ for all $f \in L^1(X, S^{-1}\Sigma, \mu)$. Since $L^1(X, \Sigma, \mu) = N(E) \oplus L^1(X, S^{-1}\Sigma, \mu)$, we have $N(P) = N(E)$ and the decomposition (4.14). □

Corollary 4.5.1 *$P : L^1(\mu) \to L^1(\mu)$ is isometric if and only if $S^{-1}\Sigma = \Sigma$.*

Corollary 4.5.2 *$R(P)$ is a closed vector subspace of $L^1(X, \Sigma, \mu)$.*

Remark 4.5.1 (4.15) gives an explicit definition of the Frobenius-Perron operator in terms of the sub-σ-algebra $S^{-1}\Sigma$; see [35] for more details.

Remark 4.5.2 A general decomposition theorem for Koopman operators has been obtained in [30], which basically says that the Koopman operator U is isometric on the subspace $L^\infty(\text{supp } h)$ which is a topological complement of $N(U)$, and U is isometric on $L^\infty(X)$ if and only if the two measures μ and $\mu \circ S^{-1}$ are equivalent.

Using the above results and Banach's closed range theorem [57], we can get more relations between P and U.

Theorem 4.5.2 *Let (X, Σ, μ) be a measure space, let $S : X \to X$ be a nonsingular transformation such that the measure space $(X, S^{-1}\Sigma, \mu)$ is σ-finite, and let $P : L^1 \to L^1$ and $U : L^\infty \to L^\infty$ be the Frobenius-Perron operator and the Koopman operator associated with S, respectively. Then,*
(i) *P is one-to-one if and only if U is onto.*
(ii) *U is one-to-one if and only if P is onto.*

Proof Since U is the dual of P, the fact $R(P)$ is closed implies that $R(U)$ is closed. Thus, the theorem follows from a standard argument in functional analysis. □

Corollary 4.5.3 *Under the same conditions of the above theorem,*
(i) *P is an isometry if and only if U is onto.*
(ii) *U is an isometry if and only if P is onto.*

Finally, we give an application of the decomposition theorem to the spectral analysis of Frobenius-Perron operators [33, 35]. Let P be a Frobenius-Perron operator associated with S. We extend the definition (4.1) of P to the complex L^1 space in the natural way. Then, it is still true that $\|P\| = 1$ (see Exercise 4.31). The *spectrum* $\sigma(P)$ of P is defined to be the set of all the complex numbers λ such that the linear operator $P - \lambda I$ does not have a bounded inverse defined on $L^1(\mu)$, where I is the identity operator. The complement of $\sigma(P)$ in the complex plane $\mathbb{C}$ is called the *resolvent set* of P and is denoted by $\rho(P)$. Since $\|P\| = 1$, the spectrum $\sigma(P)$ is a compact subset of the closed unit disk $\mathbb{D} \equiv \{\lambda \in \mathbb{C} : |\lambda| \leqslant 1\}$ and is a disjoint union of the *point spectrum* $\sigma_p(P)$, the *continuous spectrum* $\sigma_c(P)$, and the *residual spectrum* $\sigma_r(P)$. The boundary of

$\sigma(P)$ is denoted by $\partial\sigma(P)$. A number $\lambda \in \mathbb{C}$ is said to be in the *approximate point spectrum* $\sigma_a(P)$ if there exists a sequence $\{f_n\}$ in L^1 such that $\|f_n\| = 1$ for all n and $\|(P - \lambda P)f_n\| \to 0$ as $n \to \infty$. Obviously, $\sigma_a(P) \subset \sigma(P)$. The following lemma is a standard result concerning $\sigma_a(P)$ (see [17]).

Lemma 4.5.2 $\sigma_a(P) \supset \sigma_p(P) \cup \sigma_c(P) \cup \partial\sigma(P)$.

Denote $\partial\mathbb{D} = \{\lambda \in \mathbb{C} : |\lambda| = 1\}$ to be the unit circle. The following general result is a basis of our spectral analysis.

Lemma 4.5.3 *Let B be a Banach space and let $T : B \to B$ be an isometry. Then, $0 \in \sigma(T)$ implies that $\sigma(T) = \mathbb{D}$ and $0 \in \rho(T)$ implies that $\sigma(T) \subset \partial\mathbb{D}$.*

Proof Let λ be such that $|\lambda| < 1$. From

$$\|Tx - \lambda x\| \geqslant \|Tx\| - |\lambda|\|x\| = (1 - |\lambda|)\|x\|,$$

we see that $T - \lambda$ is bounded from below, and thus $\lambda \notin \sigma_a(T)$. Hence, for $|\lambda| < 1, \lambda \notin \partial\sigma(T)$ by Lemma 4.5.2. In particular, $0 \notin \partial\sigma(T)$.

Consider first the case $0 \in \sigma(T)$. If there exists $|\lambda| < 1$ such that $\lambda \notin \sigma(T)$, then it is easy to see that there exists $\tilde{\lambda} \in \partial\sigma(T)$ such that $|\tilde{\lambda}| < 1$, which is a contradiction to the fact that $\lambda \notin \partial\sigma(T)$ for $|\lambda| < 1$. Therefore,

$$|\lambda| < 1 \Rightarrow \lambda \in \sigma(T).$$

Since $\sigma(T)$ is a closed subset of $\mathbb{D}$, we have that $\sigma(T) = \mathbb{D}$.

Consider now the case $0 \in \rho(T)$. If there exists $|\lambda| < 1$ such that $\lambda \in \sigma(T)$, then there exists a $\tilde{\lambda} \in \partial\sigma(T)$ with $|\tilde{\lambda}| < 1$, which also contradicts the fact that $\lambda \notin \partial\sigma(T)$ for $|\lambda| < 1$. Therefore,

$$|\lambda| < 1 \Rightarrow \lambda \notin \sigma(T).$$

In this case, $\sigma(T) \subset \partial\mathbb{D}$. □

Combining the above lemma, Corollary 4.5.1, and Remark 4.5.2, we immediately have

Theorem 4.5.3 *If $S^{-1}\Sigma = \Sigma$ or $\mu \circ S^{-1} \cong \mu$, then $0 \in \sigma(P)$ implies that $\sigma(P) = \mathbb{D}$ and $0 \in \rho(P)$ implies that $\sigma(P) \subset \partial\mathbb{D}$.*

Proof If $S^{-1}\Sigma = \Sigma$, then P is an isometry. If $\mu \circ S^{-1} \cong \mu$, then U is an isometry. Now the theorem follows from Lemma 4.5.3. □

Remark 4.5.3 $S^{-1}\Sigma = \Sigma$ if and only if S is one-to-one, and $\mu \circ S^{-1} \cong \mu$ if and only if S is onto. See [33] and [35] for more details.

Corollary 4.5.4 *If $S : X \to X$ is a measure preserving transformation, then $0 \in \sigma(P)$ implies that $\sigma(P) = \mathbb{D}$ and $0 \in \rho(P)$ implies that $\sigma(P) \subset \partial\mathbb{D}$.*

Proof $\mu(S^{-1}(A)) = \mu(A)$ for all $A \in \Sigma$ imply that $\mu \circ S^{-1} \cong \mu$. □

Corollary 4.5.5 *If P has a stationary density $\tilde{f}$ such that* supp $\tilde{f} = X$, *then $0 \in \sigma(P)$ implies that $\sigma(P) = \mathbb{D}$ and $0 \in \rho(P)$ implies that $\sigma(P) \subset \partial\mathbb{D}$.*

Proof Define $\tilde{\mu}(A) = \int_A \tilde{f} \mathrm{d}\mu$ for all $A \in \Sigma$. Then, the measure $\tilde{\mu}$ is invariant under S. Let $A \in \Sigma$ be such that $\mu(A) > 0$. Since supp $\tilde{f} = X, \tilde{\mu}(A) > 0$. So $\tilde{\mu}(S^{-1}(A)) = \tilde{\mu}(A) > 0$, which implies that $\mu(S^{-1}(A)) > 0$. Hence, $\mu \circ S^{-1} \cong \mu$. □

Remark 4.5.4 See [53] for some extensions of Theorem 4.5.3.

Exercises

4.1 The boundary value problem

$$-u'' + u = f(x),\ 0 \leqslant x \leqslant 1,\ \ u'(0) = u'(1) = 0$$

has a unique solution $u = u(x)$ defined on $[0, 1]$ for every $f \in L^1(0, 1)$. Show that the correspondence from f to u is a Markov operator on $L^1(0, 1)$.

4.2 Consider the finite set $X = \{1, 2, \cdots, k\}$ with the counting measure. Prove that any Markov operator $P : L^1(X) \to L^1(X)$ can be written as

$$(Pf)_j = \sum_{i=1}^{k} p_{ij} f_i,\ \ j = 1, 2, \cdots, k,$$

where (p_{ij}) is a $k \times k$ *stochastic matrix*, i.e.,

$$p_{ij} \geqslant 0,\ \ \sum_{j=1}^{k} p_{ij} = 1,\ \ \forall\, i = 1, 2, \cdots, k.$$

This discrete Markov operator is called a *Markov chain* and will be used in the finite approximation of Frobenius-Perron operators in the later chapters.

4.3 Let (X, Σ, μ) be a measure space and let $K(x, y)$ be a nonnegative measurable function on $X \times X$ such that $\int_X K(x, y) d\mu(y) = 1$ for each $x \in X$. Show that the linear operator $P : L^1(X) \to L^1(X)$ defined by

$$Pf(y) = \int_X K(x, y) f(x) \mathrm{d}\mu(x)$$

is a Markov operator. This is called a *Markov operator with a stochastic kernel.*

4.4 Let $P : L^1(X, \Sigma, \mu) \to L^1(X, \Sigma, \mu)$ be a Markov operator and let P^* : $L^\infty(X) \to L^\infty(X)$ be its dual operator. Define the *transition function* $P(x, A) = P^* \chi_A(x)$ for all $x \in X$ and $A \in \Sigma$. Show that this function satisfies

(i) $0 \leqslant P(x, A) \leqslant 1$;

(ii) $P(\cdot, A)$ is Σ-measurable for each fixed $A \in \Sigma$;

(iii) $P(x, \cdot)$ is a measure for each $x \in X$, and it is absolutely continuous with respect to μ.

4.5 Let (X, Σ, μ) be a measure space and suppose that $P(x, A)$, satisfying (i)–(iii) of the previous problem, is given. Using the Radon-Nikodym theorem to show that there is a unique Markov operator $P : L^1(X) \to L^1(X)$ that induces the given transition function $P(\cdot, \cdot)$.

4.6 Let (X, Σ, μ) be a measure space, let $S : X \to X$ be nonsingular, and let $P : L^1(X) \to L^1(X)$ be the corresponding Frobenius-Perron operator. Show that the associated transition function $P(x, A) = \chi_{S^{-1}(A)}(x)$ and $Ug(x) = P^* g(x) = \int_X g(y) P(x, \mathrm{d}y) = g(S(x))$.

4.7 Let (X, Σ, μ) be a measure space and let Σ_0 be a sub-σ-algebra of Σ. Show that the conditional expectation $E(\cdot\,|\Sigma_0)$ with respect to Σ_0 is a Markov operator on $L^1(X)$.

4.8 Show that any convex combination of Markov operators on $L^1(X)$ is a Markov operator.

4.9 Let $P : L^1(X) \to L^1(X)$ be a Markov operator. Let $B \in \Sigma$ with $\mu(B) > 0$ be such that $P^* \chi_{B^c} = 0$ on B. Show that supp $f \subset B$ implies that supp $Pf \subset B$. As a consequence, $P : L^1(B) \to L^1(B)$ is a Markov operator.

4.10 Find the Frobenius-Perron operator corresponding to the following mappings:

(i) $S : [0,1] \to [0,1]$, $S(x) = 4x^2(1 - x^2)$;

(ii) $S : [0,1] \to [0,1]$, $S(x) = \sin \pi x$;

(iii) $S : [0,1] \to [0,1]$, $S(x) = a \tan(bx + c)$;

(iv) $S : [0,1] \to [0,1]$, $S(x) = rx(1 - x)$, $0 < r < 4$;

(v) $S : [0,1] \to [0,1]$, $S(x) = kx (\mathrm{mod}\ 1)$, $k \geqslant 2$ is an integer;

(vi) $S : [0,1] \to [0,1]$, $S(x) = rxe^{-bx}$, $r, b > 0$.

4.11 Prove property (iv) of Frobenius-Perron operators in Section 4.2.

4.12 Let (X, Σ) be a measurable space with two equivalent σ-finite measures μ and ν, let $S : X \to X$ be a nonsingular transformation, and let $P_\mu : L^1(\mu) \to L^1(\mu)$ and $P_\nu : L^1(\nu) \to L^1(\nu)$ be the Frobenius-Perron operator with respect to μ and ν, respectively. Show that for any $f \in L^1(\mu)$,

$$P_\mu f = \frac{P_\nu(h \cdot f)}{h},$$

where $h \in L^1(\nu)$ is the Radon-Nikodym derivative of μ with respect to ν.

4.13 Let $E(\cdot|S^{-1}\Sigma)$ be the conditional expectation with respect to the sub-σ-algebra $S^{-1}\Sigma$. Show that the Frobenius-Perron operator P_S satisfies the equality

$$(P_S f) \circ S = E(f|S^{-1}\Sigma).$$

4.14 A mapping $S : [0,1] \to [0,1]$ is called a *generalized tent function* if $S(x) = S(1-x)$ for $0 \leqslant x \leqslant 1$ and if S is strictly increasing on $[0, 1/2]$. Show that there is a unique generalized tent function for which the standard Lebesgue measure is invariant.

4.15 Show that for any absolutely continuous measure μ such that $d\mu/dm > 0$ on $[0,1]$ a.e., there exists a unique generalized tent function S that preserves μ.

4.16 A Markov operator $P : L^1(X) \to L^1(X)$ is called *deterministic* if its dual operator P^* has the following property: for every $A \in \Sigma$, there is $B \in \Sigma$ such that $P^*\chi_A = \chi_B$. Show that the Frobenius-Perron operator is deterministic.

4.17 Describe a general form of the matrix (p_{ij}) in Exercise 4.2 which corresponds to a deterministic Markov operator.

4.18 If P_1 and P_2 are both deterministic Markov operators, are the operators P_1P_2 and $\alpha P_1 + (1-\alpha)P_2$, $0 < \alpha < 1$ also deterministic?

4.19 Show that $P : L^1(0,1) \to L^1(0,1)$ defined by

$$Pf(x) = \frac{1}{2}f(x) + \frac{1}{4}f\left(\frac{x}{2}\right) + \frac{1}{4}f\left(\frac{x}{2} + \frac{1}{2}\right)$$

is not a deterministic Markov operator.

4.20 Let $P : L^1 \to L^1$ be a Markov operator. Prove that for any two nonnegative functions $f,\ g \in L^1$ the condition supp $f \subset$ supp g implies that supp $Pf \subset$ supp Pg.

4.21 Show that the mapping $S : [0,1] \to [0,1]$ defined by

$$S(x) = \begin{cases} \dfrac{x}{a}, & 0 \leqslant x \leqslant a, \\ \dfrac{1-x}{1-a}, & a \leqslant x \leqslant 1 \end{cases}$$

preserves the Lebesgue measure on $[0,1]$.

4.22 Show that the function $f(x) = 4/[\pi(1+x^2)]$ is a stationary density for $S : [0,1] \to [0,1]$ given by

$$S(x) = \begin{cases} \dfrac{2x}{1-x^2}, & 0 \leqslant x \leqslant \sqrt{2}-1, \\ \dfrac{1-x^2}{2x}, & \sqrt{2}-1 \leqslant x \leqslant 1. \end{cases}$$

4.23 Show that the function $f(x) = 2x/(1+x)^2$ is a stationary density for $S : [0,1] \to [0,1]$ given by

$$S(x) = \begin{cases} \dfrac{2x}{1-x}, & 0 \leqslant x \leqslant \dfrac{1}{3}, \\ \dfrac{1-x}{2x}, & \dfrac{1}{3} \leqslant x \leqslant 1. \end{cases}$$

4.24 Show that the function $f(x) = px^{p-1}$ with $p > 1$ is a stationary density for $S : [0,1] \to [0,1]$ given by

$$S(x) = \begin{cases} 2^{\frac{1}{p}}x, & 0 \leqslant x \leqslant \left(\frac{1}{2}\right)^{\frac{1}{p}}, \\ 2^{\frac{1}{p}}(1-x^p)^{\frac{1}{p}}, & \left(\frac{1}{2}\right)^{\frac{1}{p}} \leqslant x \leqslant 1. \end{cases}$$

4.25 Show that the function $f(x) = 12(x-1/2)^2$ is a stationary density for $S : [0,1] \to [0,1]$ given by

$$S(x) = \left(\frac{1}{8} - 2\left|x - \frac{1}{2}\right|^3\right)^{\frac{1}{3}} + \frac{1}{2}.$$

4.26 Show that the function $f(x) = (1-x)/2$ is a stationary density for the *cusp map* $S : [-1,1] \to [-1,1]$ given by

$$S(x) = 1 - 2|x|^{\frac{1}{2}}.$$

4.27 Show that the Frobenius-Perron operator $P : L^1 \to L^1$ is continuous under the weak topology on L^1.

4.28 Find the expression of the Frobenius-Perron operator associated with the Hénon map in Example 4.2.3.

4.29 Let $S : (X, \Sigma, \mu) \to (X, \Sigma, \mu)$ be a measure preserving transformation and let $U : L^\infty \to L^\infty$ be the corresponding Koopman operator. Show that $U : L^p \to L^p$ is well-defined and is an isometry, where $1 \leqslant p \leqslant \infty$.

4.30 Show that the converse of Corollary 4.1.1 is not true by providing a counter example.

4.31 Let $L^1(X)$ be a complex L^1-space and let $P : L^1(X) \to L^1(X)$ be a Markov operator. Show that $\|P\| = 1$.

Chapter 5
Invariant Measures—Existence

JiuDing, Department of Mathematics, The University of Southern Mississippi in America.
Aihui Zhou, Academy of Mathematics and Systems Science, Chinese Academy of Sciences.

Abstract The existence problem of a stationary density to a Frobenius-Perron operator will be investigated. Three general existence theorems for Markov operators are presented with the help of the concepts of compactness, quasi-compactness, and constrictiveness. A powerful spectral decomposition theorem for constrictive Markov operators will be given. Then we prove the existence results for three concrete classes of transformations. They are the class of piecewise C^2 and stretching interval mappings, the class of piecewise convex mappings with a weak repellor, and the class of multi-dimensional piecewise C^2 and expanding transformations.
Keywords Kakutani-Yosida abstract ergodic theorem, quasi-compactness, constrictiveness, spectral decomposition theorem, piecewise C^2 and stretching mapping, piecewise convex mapping, piecewise C^2 and expanding transformation.

We turn to the theoretical problem on the existence of absolutely continuous invariant finite measures for some classes of nonsingular transformations. We first give an abstract approach to the general existence problem of stationary densities of Markov operators in Section 5.1 which gives three general existence results for stationary densities of Markov operators and an important spectral decomposition theorem concerning the asymptotic periodicity of the iteration sequence $\{P^n f\}$ for the class of constrictive Markov operators. Then, the existence results for several concrete classes of one or higher dimensional nonsingular transformations are presented in the subsequent sections. The presentation of the existence results in this chapter is by no means complete, but we shall show how to employ the variation technique to prove several important existence theorems, including the now classic Lasota-Yorke's theorem. For a more complete coverage and discussion of various approaches to the existence problem, see Chapters 5 and 6 of the monograph [82] by Lasota and Mackey and the textbook [14] by Boyarsky and Góra.

5.1 General Existence Results

Let (X, Σ, μ) be a measure space. As shown in Chapter 4, the problem of the existence of an absolutely continuous invariant probability measure with respect to a given nonsingular transformation $S : X \to X$ is equivalent to that of a density solution to the fixed point equation $Pf = f$ for the Frobenius-

Perron operator P associated with S. But, finding a stationary density of P is in general difficult except for some simple or special cases.

Since $P : L^1 \to L^1$ is continuous in both the strong topology and the weak topology, if the sequence $\{P^n f\}$ of the iterates converges to some function f^* strongly or weakly, then f^* must be a fixed point of P. However, because the L^1-space is not reflexive, the theory of Hilbert space is not applicable. On the other hand, since $\|P\| = 1$, we simply cannot use some standard techniques in functional analysis, such as Banach's contraction fixed point theorem, to prove the convergence of $\{P^n f\}$. As a matter of fact, the sequence $\{P^n f\}$ does not converge in most situations of interest. Instead, we must seek appropriate compactness arguments and special techniques for L^1 spaces to prove the existence of a stationary density. One successful approach is based on the ergodic theory of linear operators with uniformly bounded powers (e.g., see the last chapter of [57]). In this section, we present two general results for the existence of stationary densities of more general Markov operators along this direction.

We first state and prove a special version of the classic Kakutani-Yosida abstract ergodic theorem [57] that is adapted to the case of Markov operators. This fundamental result is an important tool for determining the convergence of a sequence of the iterates of Markov operators. Suppose that $P : L^1 \to L^1$ is a Markov operator. The following proposition about the sequence of the *Cesáro averages*

$$A_n f \equiv A_n(P) f = \frac{1}{n} \sum_{i=0}^{n-1} P^i f \tag{5.1}$$

for the iterates of $f \in L^1$ under repeated iterations of P will be used for the proof of Theorem 5.1.1.

Lemma 5.1.1 *Let $f \in L^1$. Then,*

$$\lim_{n \to \infty} \|A_n f - A_n P f\| = 0. \tag{5.2}$$

Proof By the definition (5.1) of $A_n f$, we have

$$A_n f - A_n P f = \frac{1}{n}(f - P^n f),$$

which together with the fact that $\|P^n\| \leqslant 1$ uniformly gives that

$$\|A_n f - A_n P f\| \leqslant \frac{2}{n} \|f\| \to 0$$

as $n \to \infty$. This completes the proof. □

Proposition 5.1.1 *If, for some $f \in L^1$, there is a subsequence $\{A_{n_k} f\}$ of the sequence $\{A_n f\}$ of the Cesáro averages that converges weakly to some $f^* \in L^1$, then $Pf^* = f^*$.*

Proof Since $PA_{n_k} f = A_{n_k} Pf$ and P is weakly continuous, the sequence $\{A_{n_k} Pf\}$ converges weakly to Pf^*. Since $\{A_{n_k} Pf\}$ has the same limit as $\{A_{n_k} f\}$ by Lemma 5.1.1, we have $Pf^* = f^*$. □

Theorem 5.1.1 (Kakutani-Yosida's abstract ergodic theorem) *Let (X, Σ, μ) be a measure space and let $P : L^1 \to L^1$ be a Markov operator. If for a given $f \in L^1$, the sequence $\{A_n f\}$ of the Cesáro averages given by (5.1) is weakly precompact in L^1, then it converges strongly in L^1 to some $f^* \in L^1$ which is a fixed point of P, that is,*

$$\lim_{n\to\infty} \|A_n f - f^*\| = 0 \tag{5.3}$$

and $Pf^* = f^*$. Furthermore, if $f \in \mathcal{D}$, then $f^* \in \mathcal{D}$, so that f^* is a stationary density of P.

Proof The assumption that the sequence $\{A_n f\}$ is weakly precompact means that there exists a subsequence $\{A_{n_k} f\}$ of $\{A_n f\}$ such that $\{A_{n_k} f\}$ converges weakly to some $f^* \in L^1$ as $k \to \infty$. By Proposition 5.1.1, we have $Pf^* = f^*$. If f is a density, then f^* is a density since P preserves the positivity and the L^1-norm of f.

In order to show (5.3), we first prove that $f - f^* \in \overline{R(P-I)}$, the *closure* of the range $R(P-I)$ of the operator $P - I$. If this were not the case, then, by the Hahn-Banach theorem [57], there must exist a function $g^* \in L^\infty$ such that

$$\langle f - f^*, g^* \rangle \neq 0 \tag{5.4}$$

and

$$\langle h, g^* \rangle = 0, \quad \forall\, h \in \overline{R(P-I)}.$$

It follows that

$$\langle (P-I)P^j f, g^* \rangle = 0, \;\; j = 0, 1, \cdots,$$

which implies that

$$\langle P^{j+1} f, g^* \rangle = \langle P^j f, g^* \rangle, \;\; j = 0, 1, \cdots,$$

and so we obtain

$$\langle P^j f, g^* \rangle = \langle f, g^* \rangle, \;\; j = 1, 2, \cdots.$$

As a consequence, for any n,

$$\langle A_n f, g^* \rangle = \langle f, g^* \rangle.$$

Since the sequence $\{A_{n_k} f\}$ converges weakly to f^*, we have

$$\langle f - f^*, g^* \rangle = 0,$$

which contradicts (5.4).

Thus, for any $\varepsilon > 0$, there exists $g,\ h \in L^1$ such that

$$f - f^* = Pg - g + h,$$

where $\|h\| \leqslant \varepsilon/2$. Thus, $\|A_n h\| \leqslant \varepsilon/2$ for all n. The equality $Pf^* = f^*$ implies $A_n f^* = f^*$, so

$$A_n f = A_n(Pg - g) + A_n h + f^*,$$

from which we obtain

$$\|A_n f - f^*\| \leqslant \|A_n(Pg - g)\| + \|A_n h\|.$$

However, (5.2) implies that there exists an integer $n_0 \geqslant 1$ such that

$$\|A_n(Pg - g)\| \leqslant \frac{\varepsilon}{2}, \quad \forall\, n \geqslant n_0.$$

Therefore, we have

$$\|A_n f - f^*\| \leqslant \varepsilon, \quad \forall\, n \geqslant n_0.$$

This completes the proof. □

Remark 5.1.1 The general Kakutani-Yosida abstract ergodic theorem (Theorem VIII.5.1 of [57]) says that, if T is a bounded linear operator from a Banach space B into itself such that $\lim\limits_{n\to\infty} \|T^n x\|/n = 0$ and the sequence of the Cesáro averages $A_n x = n^{-1} \sum\limits_{i=0}^{n-1} T^i x$ is weakly precompact for some $x \in B$, then $\lim\limits_{n\to\infty} \|A_n x - x^*\| = 0$ for some fixed point $x^* \in B$ of T. This remark will be used in the proof of Theorem 5.1.2 below.

Several corollaries follow immediately by the criteria for the weak precompactness in Section 2.5.

Corollary 5.1.1 *If for some density $f \in \mathcal{D}$ there is a nonnegative function $g \in L^1$ and a positive integer n_0 such that*

$$P^n f \leqslant g, \quad \forall\, n \geqslant n_0,$$

then there exists a stationary density f^* for P.

Corollary 5.1.2 *If for some density $f \in \mathcal{D}$ there are two real numbers $p > 1$ and $M > 0$ and a positive integer n_0 such that*

$$\|P^n f\|_p \leqslant M, \quad \forall\, n \geqslant n_0,$$

then there exists a stationary density f^* for P.

Another way to investigate the stationary density problem of Markov operators is based on the quasi-compactness argument.

Theorem 5.1.2 *Let (X, Σ, μ) be a measure space and let $P : L^1 \to L^1$ be a Markov operator. Suppose that V is a dense vector subspace of L^1 which is also invariant under P, and suppose that V is a Banach space under the norm $\|\cdot\|_V$. If $P : (V, \|\cdot\|_V) \to (V, \|\cdot\|_V)$ is quasi-compact and the sequence $\{\|P^n\|_V\}$ is uniformly bounded, then for any density $f \in V$, there exists a stationary density $f^* \in V$ of P such that*

$$\lim_{n\to\infty} \|A_n f - f^*\|_V = 0.$$

Proof By the definition of quasi-compactness, there is a positive integer r and a compact operator $K : (V, \|\cdot\|_V) \to (V, \|\cdot\|_V)$ such that $\|P^r - K\|_V < 1$. Denote $T = P^r - K$. Then, $\|T\|_V < 1$ implies that the bounded linear operator $I - T$ is invertible and its inverse $(I-T)^{-1} : (V, \|\cdot\|_V) \to (V, \|\cdot\|_V)$ is a bounded linear operator with the Neumann series expression

$$(I - T)^{-1} = \sum_{n=0}^{\infty} T^n.$$

It is easy to see that

$$A_n = (I - T)^{-1} K A_n + (I - T)^{-1}(A_n - P^r A_n) \tag{5.5}$$

and

$$A_n - P^r A_n = \frac{I + P + \cdots + P^{r-1}}{n}(I - P^n).$$

Since the sequence $\{\|P^n\|_V\}$ is uniformly bounded,

$$\begin{aligned} &\lim_{n\to\infty} \|A_n - P^r A_n\|_V \\ \leqslant\ &\lim_{n\to\infty} \frac{1 + \|P\|_V + \cdots + \|P^{r-1}\|_V}{n}(1 + \|P^n\|_V) = 0. \end{aligned}$$

Thus, for any density $f \in V$, we have

$$\lim_{n\to\infty} (I - T)^{-1}(A_n - P^r A_n) f = 0.$$

On the other hand, since $(I - T)^{-1}K$ is a compact operator and the sequence $\{\|A_n f\|_V\}$ is uniformly bounded, the set $\{(I-T)^{-1}KA_n f\}_{n=0}^{\infty}$ has a limit point f^* in V, which together with (5.5) means that the set $\{A_n f\}_{n=0}^{\infty}$ also has f^* as its limit point. Therefore, the Kakutani-Yosida abstract ergodic theorem (see Remark 5.1.1) applied to $(V, \|\cdot\|_V)$ gives the result. □

Remark 5.1.2 Furthermore, one can show that under the conditions of Theorem 5.1.2, $\lim_{n\to\infty} \|A_n - E(P)\|_V = 0$, where $E(P)$ is a positive projection from

V onto the vector subspace of all the fixed points of P in V; see Corollary VIII.8.4 of [57]. A more extensive analysis of this property of *uniform ergodicity* is contained in Section VIII.8 of [57].

Corollary 5.1.3 *Let (X, Σ, μ) be a measure space and let $P : L^1 \to L^1$ be a Markov operator. Suppose that V is a dense vector subspace of L^1 and is a Banach space under the norm $\|\cdot\|_V$ such that $\|f\| \leqslant c\|f\|_V$ for all $f \in V$ and some constant c. If there exist two positive numbers $\alpha < 1$ and β and a positive integer k such that*

$$\|P^k f\|_V \leqslant \alpha\|f\|_V + \beta\|f\|, \quad \forall\, f \in V,$$

then there exists a density $f^* \in V$ such that $Pf^* = f^*$, and the conclusion of Theorem 5.1.2 is true.

Proof By the Ionescu-Tulcea and Marinescu theorem, $P : V \to V$ is quasi-compact. For any $f \in V$, write $n = ik + j$ with $0 \leqslant j < k$, then

$$\begin{aligned}
\|P^n f\|_V = \|P^{ik+j} f\|_V &\leqslant \|(P^k)^i f\|_V + \|P^j f\|_V \\
&\leqslant \alpha^i \|f\|_V + \beta \sum_{l=0}^{i-1} \alpha^l \|f\| + \|P^j f\|_V \\
&\leqslant \|f\|_V + \frac{\beta}{1-\alpha}\|f\| + \|P^j f\|_V \\
&\leqslant \left(1 + \frac{\beta c}{1-\alpha} + \max_{0 \leqslant j < k} \|P^j\|_V\right) \|f\|_V.
\end{aligned}$$

Hence, the sequence $\{\|P^n\|_V\}$ is uniformly bounded, and so the conditions of Theorem 5.1.2 are satisfied. □

Now we focus on a special class of Markov operators for the investigation of the asymptotic behavior of the sequence $\{P^n f\}$, the evolution of a density $f \in \mathcal{D}$ under the repeated action of P. This class of Markov operators contains the Frobenius-Perron operators associated with the classes of the transformations to be studied in the next three sections. First, we give the definition of such operators.

Definition 5.1.1 *Let (X, Σ, μ) be a finite measure space. A Markov operator $P : L^1 \to L^1$ is said to be* constrictive *if there exist two positive numbers δ and $\epsilon < 1$ with the property that for each $f \in \mathcal{D}$, there is a positive integer $n_0(f)$ such that*

$$\int_A P^n f \mathrm{d}\mu \leqslant \epsilon, \quad \forall\, n \geqslant n_0(f)$$

for all $A \in \Sigma$ with $\mu(A) \leqslant \delta$.

Remark 5.1.3 A Markov operator $P : L^1 \to L^1$ is constrictive if and only if there exists a compact set $\mathcal{F} \subset L^1$ such that

$$\lim_{n\to\infty} \text{dist}\,(P^n f, \mathcal{F}) = 0$$

for any $f \in \mathcal{D}$, where dist $(g, \mathcal{F}) = \min\{\|g - h\| : h \in \mathcal{F}\}$ is the *distance* of $g \in L^1$ to $\mathcal{F}$ [14].

A simple sufficient condition for constrictiveness is given by the following proposition.

Proposition 5.1.2 *Let (X, Σ, μ) be a finite measure space and let $P : L^1 \to L^1$ be a Markov operator. If there are two numbers $p > 1$ and $K > 0$ such that for every $f \in \mathcal{D}$, there is a positive integer $n_1(f)$ such that $P^n f \in L^p$ for all $n \geqslant n_1(f)$ and*

$$\limsup_{n\to\infty} \|P^n f\|_p \leqslant K,$$

then P is constrictive.

Proof Given $f \in \mathcal{D}$, there is a positive integer $n_0(f)$ such that $\|P^n f\|_p \leqslant K+1$ for all $n \geqslant n_0(f)$. By Proposition 2.5.2, the set

$$\{P^n f : f \in \mathcal{D}, n \geqslant n_0(f)\}$$

is weakly precompact. Then, for any chosen $\epsilon \in (0, 1)$, Theorem 2.5.2 implies that there is $\delta > 0$ such that

$$\int_A P^n f \mathrm{d}\mu \leqslant \epsilon, \quad \forall\, n \geqslant n_0(f), \quad \forall\, A \in \Sigma \text{ with } \mu(A) < \delta.$$

So, P is constrictive by Definition 5.1.1. □

We state without proof an important spectral decomposition theorem for constrictive Markov operators due to Komornic and Lasota [77].

Theorem 5.1.3 (Komornic-Lasota's spectral decomposition theorem) *Let (X, Σ, μ) be a finite measure space and let $P : L^1 \to L^1$ be a constrictive Markov operator. Then, there are an integer $k > 0$ and functions $g_i \in \mathcal{D}$ and $h_i \in L^\infty$ for $i = 1, 2, \cdots, k$ such that the bounded linear operator R defined by*

$$Rf = Pf - \sum_{i=1}^{k} \left(\int_X f h_i \mathrm{d}\mu \right) g_i, \quad \forall\, f \in L^1 \tag{5.6}$$

satisfies the following properties:

(i) $\lim_{n\to\infty} \|P^n R f\| = 0$ *for every* $f \in L^1$;

(ii) supp $g_i \cap$ supp $g_j = \varnothing$ *for all* $i \neq j$;

(iii) $Pg_i = g_{j_i}$ *for each* $i = 1, 2, \cdots, k$, *where* $(j_1, j_2, \cdots, j_k)$ *is a permutation of* $(1, 2, \cdots, k)$.

The above theorem provides another existence result for stationary densities of Markov operators.

Corollary 5.1.4 *Let (X, Σ, μ) be a finite measure space and let $P : L^1 \to L^1$ be a constrictive Markov operator. Then, P has a stationary density.*

Proof Let a density f be defined by

$$f = \frac{1}{k}\sum_{i=1}^{k} g_i,$$

where k and g_i, $i = 1, 2, \cdots, k$ are as in Theorem 5.1.3. Because of property (iii) in that theorem, we have

$$Pf = \frac{1}{k}\sum_{i=1}^{k} g_{j_i} = \frac{1}{k}\sum_{i=1}^{k} g_i = f,$$

and thus P has a stationary density. □

Remark 5.1.4 The decomposition equality (5.6) implies the *asymptotic periodicity* of the sequence $\{P^n f\}$ for a constrictive Markov operator $P : L^1 \to L^1$. See [77, 81, 82] for more details on the asymptotic periodicity of constrictive Markov operators. Furthermore, the above spectral decomposition theorem can be used to study ergodicity, mixing, and exactness of constrictive Markov operators. In particular, applying the spectral decomposition theorem to constrictive Frobenius-Perron operators gives conditions for ergodic, mixing, or exact transformations, respectively. See Section 5.3 of [82] for more information.

Remark 5.1.5 The spectral decomposition of Theorems 5.1.3 for constrictive Markov operators is similar to that for quasi-compact Markov operators as described by Theorem 2.5.3, but the relationship between constrictiveness and quasi-compactness for Markov operators is still an open question [14].

In the remaining sections of this chapter, we shall study several classes of concrete transformations satisfying some conditions that guarantee the existence of absolutely continuous invariant finite measures. We start with Lasota-Yorke's classic result on interval mappings in the next section.

5.2 Piecewise Stretching Mappings

In 1973, Lasota and Yorke [83] proved a fundamental existence result for a class of piecewise C^2 and stretching mappings of the interval $[0, 1]$, thus they answered a question posed by Ulam in [120] on the existence of absolutely continuous invariant probability measures for "simple" interval mappings such as the line-broken ones. In this section, we state the Lasota-Yorke theorem and give its proof.

Theorem 5.2.1 (Lasota-Yorke's existence theorem) *Suppose that a mapping $S : [0,1] \to [0,1]$ satisfies the following conditions:*

(i) *there is a partition $0 = a_0 < a_1 < \cdots < a_r = 1$ of the interval $[0,1]$ such that for $i = 1, 2, \cdots, r$, the restriction $S|_{(a_{i-1},a_i)}$ of S to the open interval (a_{i-1}, a_i) can be extended to the closed interval $[a_{i-1}, a_i]$ as a C^2-function;*

(ii) *there is a constant $\lambda > 1$ such that*

$$\inf\{|S'(x)| : x \in [0,1] \setminus \{a_1, a_2, \cdots, a_{r-1}\}\} \geqslant \lambda;$$

(iii) *there is a constant s such that*

$$\sup\left\{\frac{S''(x)}{[S'(x)]^2} : x \in [0,1] \setminus \{a_1, a_2, \cdots, a_{r-1}\}\right\} \leqslant s.$$

Then, the corresponding Frobenius-Perron operator P has a stationary density.

Proof For $i = 1, 2, \cdots, r$ let $S_i = S|_{(a_{i-1},a_i)}$, $g_i = S_i^{-1}$, and $I_i = S((a_{i-1}, a_i))$, and denote

$$\Delta_i(x) = \begin{cases} (a_{i-1}, g_i(x)), & x \in I_i,\ g_i'(x) > 0, \\ (g_i(x), a_i), & x \in I_i,\ g_i'(x) < 0, \\ (a_{i-1}, a_i), & x \notin I_i. \end{cases}$$

Then, for $x \in [0,1]$,

$$S^{-1}((0,x)) = \bigcup_{i=1}^{r} \Delta_i(x),$$

and so from the explicit formula (4.2) for the Frobenius-Perron operator, we have

$$Pf(x) = \frac{\mathrm{d}}{\mathrm{d}x}\int_{S^{-1}([0,x])} f(t)\mathrm{d}t = \sum_{i=1}^{r} \frac{\mathrm{d}}{\mathrm{d}x}\int_{\Delta_i(x)} f(t)\mathrm{d}t,$$

where

$$\frac{\mathrm{d}}{\mathrm{d}x}\int_{\Delta_i(x)} f(t)\mathrm{d}t = \begin{cases} g_i'(x)f(g_i(x)), & x \in I_i,\ g_i'(x) > 0, \\ -g_i'(x)f(g_i(x)), & x \in I_i,\ g_i'(x) < 0, \\ 0, & x \notin I_i. \end{cases}$$

It follows from the above that

$$Pf(x) = \sum_{i=1}^{r} \sigma_i(x) f(g_i(x)) \chi_{I_i}(x), \tag{5.7}$$

where for each i, $\sigma_i(x) = |g_i'(x)| = 1/|S'(g_i(x))| \leqslant 1/\lambda$ for all $x \in I_i$ and σ_i' satisfies

$$|\sigma_i'(x)| = \frac{|-S'(g_i(x))g_i'(x)|}{[S'(g_i(x))]^2} \leqslant s\sigma_i(x),\ \ \forall\, x \in I_i. \tag{5.8}$$

Let $f \in \mathcal{D} \cap BV(0,1)$. Then, from the expression (5.7), Proposition 2.3.1, Yorke's inequality (2.5), and the inequality (5.8),

$$\begin{aligned}
\bigvee_0^1 Pf &\leqslant \sum_{i=1}^r \bigvee_0^1 [\sigma_i (f \circ g_i)\chi_{I_i}] \\
&\leqslant 2\sum_{i=1}^r \bigvee_{I_i} [\sigma_i (f \circ g_i)] + \sum_{i=1}^r \frac{2}{m(I_i)} \int_{I_i} \sigma_i (f \circ g_i) \mathrm{d}m \\
&\leqslant 2\sum_{i=1}^r \left(\sup_{x \in I_i} \sigma_i(x) \bigvee_{I_i} f \circ g_i + \int_{I_i} |\sigma_i'|(f \circ g_i)\mathrm{d}m \right) \\
&\quad + \sum_{i=1}^r \frac{2}{m(I_i)} \int_{I_i} \sigma_i (f \circ g_i) \mathrm{d}m \\
&\leqslant \frac{2}{\lambda} \sum_{i=1}^r \bigvee_{I_i} f \circ g_i + 2\sum_{i=1}^r \left[s + \frac{1}{m(I_i)} \right] \int_{I_i} \sigma_i (f \circ g_i)\mathrm{d}m \\
&= \frac{2}{\lambda} \sum_{i=1}^r \bigvee_{a_{i-1}}^{a_i} f + 2\sum_{i=1}^r \left[s + \frac{1}{m(I_i)} \right] \int_{a_{i-1}}^{a_i} f(y)\mathrm{d}y \\
&= \frac{2}{\lambda} \bigvee_0^1 f + \beta \int_0^1 f(y)\mathrm{d}y = \alpha \bigvee_0^1 f + \beta,
\end{aligned}$$

where the constants $\alpha = 2/\lambda$ and $\beta = \max_{i=1,\cdots,r} 2\,(s + 1/m(I_i))$.

First, we assume that $\lambda > 2$. Then, $0 < \alpha < 1$. With an induction argument, we have

$$\bigvee_0^1 P^n f \leqslant \alpha^n \bigvee_0^1 f + \beta \sum_{k=0}^{n-1} \alpha^k \leqslant \bigvee_0^1 f + \frac{\beta}{1-\alpha}, \quad \forall\, n,$$

and hence, for every $f \in \mathcal{D} \cap BV(0,1)$,

$$\bigvee_0^1 A_n f = \bigvee_0^1 \frac{1}{n} \sum_{k=0}^{n-1} P^k f \leqslant \frac{1}{n} \sum_{k=0}^{n-1} \bigvee_0^1 P^k f \leqslant \bigvee_0^1 f + \frac{\beta}{1-\alpha}, \quad \forall\, n.$$

By Helly's lemma, the sequence $\{A_n f\}$ is precompact, therefore by Theorem 5.1.1,

$$\lim_{n \to \infty} A_n f = f^*$$

in $L^1(0,1)$, where f^* is a stationary density of P.

Next, we assume only that $2 \geqslant \lambda > 1$. Then, $1 \leqslant \alpha < 2$. Choose a positive integer k such that $\gamma = \lambda^k > 2$. Then, for $\phi = S^k$ we have $\inf_{x \in [0,1]} |\phi'(x)| \geqslant \gamma > 2$.

So, from what we have proved, for any $f \in \mathcal{D} \cap BV(0,1)$, there is a constant M such that

$$\bigvee_0^1 (P_\phi)^n f \leqslant M, \ \forall\, n.$$

Write $n = jk + l$ where $0 \leqslant l < k$. Then,

$$\begin{aligned}\bigvee_0^1 P^n f = \bigvee_0^1 P^l (P^k)^j f &= \bigvee_0^1 P^l (P_\phi)^j f \\ &\leqslant \alpha^l \bigvee_0^1 (P_\phi)^j f + \beta \sum_{i=0}^k \alpha^i \\ &\leqslant \alpha^{k-1} M + \beta \sum_{i=0}^k \alpha^i, \ \forall\, n.\end{aligned}$$

So, the same argument implies that P has a stationary density. □

Remark 5.2.1 In fact, every stationary density of P in the above theorem belongs to $BV(0,1)$ (see [82]).

Remark 5.2.2 The proof of the Lasota-Yorke theorem implies that the corresponding Frobenius-Perron operator P is constrictive, so that the sequence $\{P^n f\}$ is asymptotically periodic.

Remark 5.2.3 The assumptions of the Lasota-Yorke theorem can be relaxed to that the mapping S is piecewise C^1 and stretching, and the function $1/|S'|$ is of bounded variation (see [14, 124]). It was shown that there exist constants $0 < \eta < 1$, C_1, and C_2 such that for any $f \in BV(0,1)$ and all integers $n \geqslant 1$,

$$\|P^n f\|_{BV} \leqslant C_1 \eta^n \|f\|_{BV} + C_2 \|f\|,$$

where the BV-norm $\|\cdot\|_{BV}$ is defined by (2.6).

Remark 5.2.4 If, in addition, S is piecewise *onto* in Theorem 5.2.1, then it is easy to see from the proof of the theorem that

$$\bigvee_0^1 Pf \leqslant \frac{1}{\lambda} \bigvee_0^1 f + s\|f\|, \ \forall\, f \in BV(0,1),$$

which will be used in Section 7.2.

Remark 5.2.5 The structure of the fixed point space of the Frobenius-Perron operator P under the conditions of Theorem 5.2.1 has been explored in [90], and in particular it is shown that if $r = 2$ in the theorem, then P has exactly one stationary density, which also implies the ergodicity of S by Theorem 4.4.1.

Remark 5.2.6 See [66] for more general existence results of absolutely continuous invariant measures on piecewise monotonic mappings of totally ordered sets.

Remark 5.2.7 The assumption $\inf_{x\in[0,1]\setminus\{a_1,\cdots,a_{r-1}\}} |S'(x)| > 1$ in Theorem 5.2.1 is essential for the conclusion of that theorem since it cannot be weakened to $|S'(x)| > 1$ on $(0,1)$. Even if $|S'(x)| = 1$ for only one point in the interval, the corresponding Frobenius-Perron operator may not have a nontrivial fixed point. This can be demonstrated by the following example from [88]. Let $S : [0,1] \to [0,1]$ be defined by

$$S(x) = \begin{cases} \dfrac{x}{1-x}, & x \in \left[0, \dfrac{1}{2}\right], \\ 2x-1, & x \in \left(\dfrac{1}{2}, 1\right]. \end{cases}$$

Then, $S'(x) > 1$ for all $x \in (0,1]$ and $S'(0) = 1$. A detailed analysis (see Remark 6.2.1 of [82]) shows that for any $\epsilon > 0$,

$$\lim_{n\to\infty} \int_\epsilon^1 |P_S^n f(x)|\mathrm{d}x = 0$$

for all $f \in L^1(0,1)$. Hence, the sequence $\{P_S^n f\}$ converges to 0 under the L^1-norm in $L^1(\epsilon,1)$ for each $\epsilon > 0$. Thus, the only solution to the operator equation $P_S f = f$ is the trivial solution $f = 0$, and so there is no absolutely continuous S-invariant measure at all. For each $f \in \mathcal{D}$, since $\lim_{n\to\infty} \int_0^\epsilon P_S^n f(x)\mathrm{d}x = 1$, starting with many initial points to iterate the above mapping, eventually we can see the cluster of more and more iterates near zero. This phenomenon is called the *paradox of the weak repellor* which is further discussed at the end of Section 1.3 of the book [82] by Lasota and Mackey. Note that the given S is a piecewise convex mapping with 0 as a weak repellor. In the next section, we will show that for piecewise convex mappings with 0 as a strong repellor at 0, the mentioned paradox disappears.

5.3 Piecewise Convex Mappings

Remark 5.2.7 at the end of the previous section tells us that the condition $\inf_{x\in[0,1]\setminus\{a_1,\cdots,a_{r-1}\}} |S'(x)| > 1$ in the Lasota-Yorke theorem is important for the existence of a stationary density of the Frobenius-Perron operator P corresponding to a piecewise monotonic mapping S, but it is not necessary, as the example $S(x) = 4x(1-x)$ indicates. Indeed, it has been shown (see, e.g., [5, 104]) that for the family of mappings $S_p(x) = px(1-x)$, there is an uncountable set Λ of the parameter values p near 4 such that for each $p \in \Lambda$, there is a stationary density of the corresponding Frobenius-Perron operator, even though $S_p'(1/2) = 0$ for every p.

In this section, we prove the existence of an absolutely continuous invariant probability measure for a class of mappings that are piecewise convex with a strong repellor. The result is also due to Lasota and Yorke [84].

Theorem 5.3.1 (Lasota-Yorke) *Suppose that $S : [0,1] \to [0,1]$ satisfies the conditions:*

(i) *there is a partition $0 = a_0 < a_1 < \cdots < a_r = 1$ of $[0,1]$ such that the restriction $S|_{[a_{i-1},a_i)}$ of S to $[a_{i-1}, a_i)$ is a C^2-function for each $i = 1, 2, \cdots, r$;*

(ii) *$S'(x) > 0$ and $S''(x) \geqslant 0$ for all $x \in [0,1)$, where $S'(a_i)$ and $S''(a_i)$ are understood to be the right derivatives for each i;*

(iii) *$S(a_i) = 0$ for each integer $i = 0, 1, \cdots, r-1$; and*

(iv) *$\lambda \equiv S'(0) > 1$.*

Let $P : L^1(0,1) \to L^1(0,1)$ be the corresponding Frobenius-Perron operator. Then, there exists a stationary density f^* of P. Moreover, f^* is a monotonically decreasing function.

Proof Denote $S_i = S|_{[a_{i-1},a_i)}$ and let

$$g_i(x) = \begin{cases} S_i^{-1}(x), & x \in [0, S(a_i^-)), \\ a_i, & x \in [S(a_i^-), 1] \end{cases}$$

for $i = 1, 2, \cdots, r$, where $S(a_i^-)$ is the left limit of S as x approaches a_i. Then,

$$S^{-1}([0,x]) = \bigcup_{i=1}^{r} [a_{i-1}, g_i(x)],$$

from which and (4.2)

$$Pf(x) = \sum_{i=1}^{r} g_i'(x) f(g_i(x)).$$

Since S_i is monotonically increasing, so is g_i. And, g_i' is monotonically decreasing since $g_i''(x) = -S''(x)/S'(x)^2 \leqslant 0$ on $[0,1]$. Thus, Pf is a monotonically decreasing nonnegative function if f is nonnegative and monotonically decreasing, and

$$\begin{aligned} Pf(x) &= \sum_{i=1}^{r} g_i'(x) f(g_i(x)) \leqslant \sum_{i=1}^{r} g_i'(0) f(g_i(0)) \\ &= g_1'(0) f(0) + \sum_{i=2}^{r} g_i'(0) f(a_{i-1}). \end{aligned}$$

Now, let $f \in \mathcal{D} \cap L^1(0,1)$ be monotonically decreasing. Then,

$$1 \geqslant \int_0^x f(t)dt \geqslant \int_0^x f(x)dt = xf(x), \quad \forall\, x \in [0,1],$$

which implies that

$$f(x) \leqslant \frac{1}{x}, \quad \forall\, x \in (0,1].$$

Hence, for $2 \leqslant i \leqslant r$,

$$g_i'(0) f(a_{i-1}) \leqslant \frac{g_i'(0)}{a_{i-1}}.$$

Let $\alpha = 1/\lambda$. Then, $g_1'(0) = 1/S'(0) = 1/\lambda = \alpha < 1$, so we have

$$Pf(x) \leqslant \frac{1}{\lambda} f(0) + \sum_{i=2}^{r} \frac{g_i'(0)}{a_{i-1}} = \alpha f(0) + M,$$

where the constant

$$M = \sum_{i=2}^{r} \frac{g_i'(0)}{a_{i-1}}.$$

It follows that

$$P^n f(x) \leqslant \alpha^n f(0) + \frac{M}{1-\alpha} \leqslant f(0) + K,$$

where the constant $K = M/(1-\alpha)$ is independent of the choice of f. Since $\|P\| = 1$, the set

$$\{h \in L^1(0,1) : \ 0 \leqslant h(x) \leqslant f(0) + K, \ x \in [0,1]\}$$

is weakly compact in $L^1(0,1)$ from Proposition 2.5.1. By Theorem 5.1.1, under the L^1 norm,

$$\lim_{n\to\infty} \frac{1}{n} \sum_{k=0}^{n-1} P^k f = f^*,$$

where f^* is a stationary density of P. It is obvious that f^* is monotonically decreasing on $[0,1]$ since f is monotonically decreasing on $[0,1]$, and f^* is the limit of a sequence of monotonically decreasing functions. □

Remark 5.3.1 The point $x = 0$ is called a *strong repellor* since the orbit $\{x_0, S(x_0), S^2(x_0), \cdots\}$, starting from a point $x_0 \in [0, a_1)$, will eventually leave $[0, a_1)$. Also see Remark 5.2.7.

Remark 5.3.2 It can be shown [84] that f^* is the unique stationary density of P and that the sequence $\{P^n\}$ of the iterates of P is *asymptotically stable* in the sense that

$$\lim_{n\to\infty} P^n f = f^*$$

for all $f \in \mathcal{D}$.

5.4 Piecewise Expanding Transformations

In this section, we generalize the existence result in Section 5.2 to multi-dimensional transformations. Such generalizations are based on the modern notion of variation for functions of several variables, as developed in Section 2.4.

Historically, the first existence result for multi-dimensional transformations was obtained by Krzyzewski and Szlenk [80] in 1969 by demonstrating the existence of a unique, absolutely continuous invariant probability measure for a C^1 expanding transformation S on a smooth manifold. For more general piecewise expanding transformations, the first partial result appeared in [73]. There, expanding and piecewise analytic transformations on the unit square partitioned by smooth curves were considered. A complicated definition of bounded variation is used and the method cannot be extended beyond dimension 2. Two generalizations were obtained later on. In [70], Jabłoński proved the existence of absolutely continuous invariant measures for a special class of transformations on $[0,1]^N$ with a rectangular partition, using the classic Tonnelli definition of bounded variation (Definition 2.4.4). Góra and Boyarsky [64] seem to be the first to have used the modern definition of bounded variation (Definition 2.4.1) to prove an existence result for piecewise C^2 and expanding transformations on a bounded region of $\mathbb{R}^N$. Their result has been generalized in [2] and [108].

Here, for the sake of avoiding technical difficulties of presentation, we give a unified approach [52] to the multi-dimensional existence problem, using the variation notion and the trace theorem for weakly differentiable functions. First, we prove a preliminary result.

Proposition 5.4.1 *Let $\Omega = \Omega_1 \cup \Omega_2 \cup \Gamma$ be a bounded open region of $\mathbb{R}^N$ with $\Gamma = \partial\Omega_1 \cap \partial\Omega_2$. Then, for any $f \in BV(\Omega)$, there hold*

$$\int_\Gamma \|Df\| = \int_\Gamma |\mathrm{tr}_{\Omega_1} f - \mathrm{tr}_{\Omega_2} f| \mathrm{d}H, \tag{5.9}$$

$$V(f;\Omega) = V(f;\Omega_1) + V(f;\Omega_2) + \int_\Gamma |\mathrm{tr}_{\Omega_1} f - \mathrm{tr}_{\Omega_2} f| \, \mathrm{d}H. \tag{5.10}$$

Proof Let $\boldsymbol{g} \in C_0^1(\Omega;\mathbb{R}^N)$. Then, from Theorem 2.4.3,

$$\int_{\Omega_1} f \,\mathrm{div}\, \boldsymbol{g} \,\mathrm{d}m = -\int_{\Omega_1} \langle Df, \boldsymbol{g}\rangle + \int_{\partial\Omega_1} \mathrm{tr}_{\Omega_1} f \langle \boldsymbol{g}, \boldsymbol{n}_1\rangle \mathrm{d}H,$$

$$\int_{\Omega_2} f \,\mathrm{div}\, \boldsymbol{g} \,\mathrm{d}m = -\int_{\Omega_2} \langle Df, \boldsymbol{g}\rangle + \int_{\partial\Omega_2} \mathrm{tr}_{\Omega_2} f \langle \boldsymbol{g}, \boldsymbol{n}_2\rangle \mathrm{d}H,$$

where $\boldsymbol{n}_i$ is the unit outward normal vector to $\partial\Omega_i, i = 1, 2$. From the above and using the fact that $\boldsymbol{n}_1 = -\boldsymbol{n}_2$ along Γ, we have

$$\int_\Omega f \operatorname{div} \boldsymbol{g} \, \mathrm{d}m = -\int_{\Omega_1} \langle Df, \boldsymbol{g}\rangle - \int_{\Omega_2} \langle Df, \boldsymbol{g}\rangle + \int_\Gamma (\mathrm{tr}_{\Omega_1} f - \mathrm{tr}_{\Omega_2} f)\langle \boldsymbol{g}, \boldsymbol{n}_1\rangle \mathrm{d}H.$$

On the other hand,

$$\begin{aligned}\int_\Omega f \operatorname{div} \boldsymbol{g} \, \mathrm{d}m &= -\int_\Omega \langle Df, \boldsymbol{g}\rangle \\ &= -\int_{\Omega_1} \langle Df, \boldsymbol{g}\rangle - \int_{\Omega_2} \langle Df, \boldsymbol{g}\rangle - \int_\Gamma \langle Df, \boldsymbol{g}\rangle.\end{aligned}$$

Hence,

$$-\int_\Gamma \langle Df, \boldsymbol{g}\rangle = \int_\Gamma (\mathrm{tr}_{\Omega_1} f - \mathrm{tr}_{\Omega_2} f)\langle \boldsymbol{g}, \boldsymbol{n}_1\rangle \mathrm{d}H.$$

Taking supremum over all $\boldsymbol{g} \in C_0^1(\Omega; \mathbb{R}^N)$ such that $\|\boldsymbol{g}(\boldsymbol{x})\|_2 \leqslant 1$ for all $\boldsymbol{x} \in \Omega$, we obtain

$$\int_\Gamma \|Df\| = \int_\Gamma |\mathrm{tr}_{\Omega_1} f - \mathrm{tr}_{\Omega_2} f| \, \mathrm{d}H,$$

which is (5.9). Now, from Remark 2.4.3, we see that

$$\begin{aligned}\int_\Omega \|Df\| &= \int_{\Omega_1} \|Df\| + \int_{\Omega_2} \|Df\| + \int_\Gamma \|Df\| \\ &= \int_{\Omega_1} \|Df\| + \int_{\Omega_2} \|Df\| + \int_\Gamma |\mathrm{tr}_{\Omega_1} f - \mathrm{tr}_{\Omega_2} f| \, \mathrm{d}H.\end{aligned}$$

Hence, we have (5.10). □

Corollary 5.4.1 *Let $\Omega_0 \subset \Omega$. If $f \in BV(\Omega)$, then*

$$V(f\chi_{\Omega_0}; \Omega) = V(f; \Omega_0) + \int_{\partial\Omega_0 \setminus \partial\Omega} |\mathrm{tr}_{\Omega_0} f| \, \mathrm{d}H.$$

Remark 5.4.1 Corollary 5.4.1 can be viewed as a generalization of the Yorke inequality (2.5) to multi-variable functions.

Now, we apply the above results to a class of multi-dimensional transformations. For a differentiable transformation $\boldsymbol{S}$, we denote by $J_{\boldsymbol{S}}$ the *Jacobian matrix* of $\boldsymbol{S}$ and $|J_{\boldsymbol{S}}|$ the absolute value of the determinant of $J_{\boldsymbol{S}}$. If $\boldsymbol{A} = (a_{ij})$ is an $N \times N$ matrix, then $\|\boldsymbol{A}\|_2$ is defined to be the Euclidean 2-norm of $\boldsymbol{A}$. Note that $\|\boldsymbol{A}\|_2 = \sqrt{\lambda_{\max}(\boldsymbol{A}^{\mathrm{T}}\boldsymbol{A})}$. We need the following lemma, which is basically a generalization of the familiar change of variables formula for smooth functions to weakly differentiable functions.

Lemma 5.4.1 *Let $\Omega_1, \Omega_2, \tilde{\Omega}_1, \tilde{\Omega}_2$ be bounded open sets in $\mathbb{R}^N$, $\Omega_1 \subset \tilde{\Omega}_1, \Omega_2 \subset \tilde{\Omega}_2$, let $\boldsymbol{S} = (S_1, S_2, \cdots, S_N)^{\mathrm{T}} : \tilde{\Omega}_1 \to \tilde{\Omega}_2$ be a diffeomorphism of class C^2 with $\boldsymbol{S}(\Omega_1) = \Omega_2$, and let $f \in BV(\Omega_1)$. Then, the following are valid:*

(i) *For $\boldsymbol{g} = (g_1, g_2, \cdots, g_N)^{\mathrm{T}} \in C^1(\Omega_2; \mathbb{R}^N)$,*

$$\int_{\Omega_2} (f \circ \boldsymbol{S}^{-1}) |\boldsymbol{J}_{\boldsymbol{S}^{-1}}| \operatorname{div} \boldsymbol{g} \, \mathrm{d}m$$

$$= \int_{\Omega_1} f \left\{ [\operatorname{div}(\boldsymbol{J}_{\boldsymbol{S}^{-1}} \cdot \boldsymbol{g})] \circ \boldsymbol{S} - \sum_{j,i=1}^{N} (g_j \circ \boldsymbol{S}) \frac{\partial}{\partial x_i} \left[\frac{\partial (\boldsymbol{S}^{-1})_i}{\partial y_j} \circ \boldsymbol{S} \right] \right\} \mathrm{d}m, \quad (5.11)$$

$$\int_{\Omega_2} \langle D \left[(f \circ \boldsymbol{S}^{-1}) |J_{\boldsymbol{S}^{-1}}| \right], \boldsymbol{g} \rangle = \int_{\Omega_1} \langle Df, (\boldsymbol{J}_{\boldsymbol{S}^{-1}} \cdot \boldsymbol{g}) \circ \boldsymbol{S} \rangle$$

$$+ \int_{\Omega_1} f \sum_{j,k=1}^{N} (g_j \circ \boldsymbol{S}) \frac{\partial}{\partial x_k} \left[\frac{\partial (S^{-1})_k}{\partial y_j} \circ \boldsymbol{S} \right] \mathrm{d}m, \quad (5.12)$$

where $(\boldsymbol{S}^{-1})_k$ is the kth component function of $\boldsymbol{S}^{-1}$.

(ii) *For any H-measurable set $\omega \subset \partial \Omega_2$,*

$$\int_{\omega} \left| \mathrm{tr}_{\Omega_2} \left[(f \circ \boldsymbol{S}^{-1}) |\boldsymbol{J}_{\boldsymbol{S}^{-1}}| \right] \right| \mathrm{d}H$$

$$\leqslant \sup_{z \in \Omega_2} \| \boldsymbol{J}_{\boldsymbol{S}^{-1}}(z) \|_2 \int_{\boldsymbol{S}^{-1}(\omega)} |\mathrm{tr}_{\Omega_1} f| \, \mathrm{d}H. \quad (5.13)$$

Proof (i) Since $(\boldsymbol{J}_{\boldsymbol{S}^{-1}} \cdot \boldsymbol{g}) \circ \boldsymbol{S} = (\boldsymbol{J}_{\boldsymbol{S}^{-1}} \circ \boldsymbol{S}) \cdot (\boldsymbol{g} \circ \boldsymbol{S})$,

$$[\operatorname{div} (\boldsymbol{J}_{\boldsymbol{S}^{-1}} \cdot \boldsymbol{g})] \circ \boldsymbol{S} = \sum_{i,j=1}^{N} \frac{\partial}{\partial x_i} \left[\left(\frac{\partial (S^{-1})_i}{\partial y_j} \circ \boldsymbol{S} \right) (g_j \circ \boldsymbol{S}) \right].$$

Since, for each pair of $i, j = 1, 2, \cdots, N$,

$$\frac{\partial}{\partial x_i} \left[\left(\frac{\partial (\boldsymbol{S}^{-1})_i}{\partial y_j} \circ \boldsymbol{S} \right) (g_j \circ \boldsymbol{S}) \right]$$

$$= \left[\frac{\partial (\boldsymbol{S}^{-1})_i}{\partial y_j} \circ \boldsymbol{S} \right] \frac{\partial}{\partial x_i} (g_j \circ \boldsymbol{S}) + (g_j \circ \boldsymbol{S}) \frac{\partial}{\partial x_i} \left[\frac{\partial (S^{-1})_i}{\partial y_j} \circ \boldsymbol{S} \right],$$

$$\int_{\Omega_1} f \left\{ [\operatorname{div}(\boldsymbol{J}_{\boldsymbol{S}^{-1}} \cdot \boldsymbol{g})] \circ \boldsymbol{S} - \sum_{i,j=1}^{N} (g_j \circ \boldsymbol{S}) \frac{\partial}{\partial x_i} \left[\frac{\partial (S^{-1})_i}{\partial y_j} \circ \boldsymbol{S} \right] \right\} \mathrm{d}m$$

$$= \int_{\Omega_1} \sum_{i,j=1}^{N} f \left[\frac{\partial (\boldsymbol{S}^{-1})_i}{\partial y_j} \circ \boldsymbol{S} \right] \frac{\partial}{\partial x_i} (g_j \circ \boldsymbol{S}) \mathrm{d}m$$

$$= \int_{\Omega_1} f \sum_{j=1}^{N} \left(\frac{\partial g_j}{\partial y_j} \circ \boldsymbol{S} \right) dm.$$

However, for a fixed j by change of variables, we obtain

$$\int_{\Omega_1} f \left(\frac{\partial g_j}{\partial y_j} \circ \boldsymbol{S} \right) \mathrm{d}m = \int_{\Omega_2} \left(f \circ \boldsymbol{S}^{-1}\right) |\boldsymbol{J}_{\boldsymbol{S}^{-1}}| \frac{\partial g_j}{\partial y_j} \mathrm{d}m.$$

Hence, by summing up, we have

$$\int_{\Omega_1} f \sum_{j=1}^{N} \left(\frac{\partial g_j}{\partial y_j} \circ \boldsymbol{S} \right) dm = \int_{\Omega_2} \left(f \circ \boldsymbol{S}^{-1}\right) |\boldsymbol{J}_{\boldsymbol{S}^{-1}}| \operatorname{div} \boldsymbol{g} \, dm$$

for $\boldsymbol{g} \in C^1(\Omega_2; \mathbb{R}^N)$, which gives (5.11).

If $\boldsymbol{g} \in C_0^1(\Omega_2; \mathbb{R}^N)$, then by Theorem 2.4.3 we obtain

$$\int_{\Omega_2} \left(f \circ \boldsymbol{S}^{-1}\right) |\boldsymbol{J}_{\boldsymbol{S}^{-1}}| \operatorname{div} \boldsymbol{g} \, \mathrm{d}m = - \int_{\Omega_2} \langle D \left(f \circ \boldsymbol{S}^{-1}\right) |\boldsymbol{J}_{\boldsymbol{S}^{-1}}|, \boldsymbol{g} \rangle \mathrm{d}m,$$

$$\int_{\Omega_1} f \; [\operatorname{div} (\boldsymbol{J}_{\boldsymbol{S}^{-1}} \cdot \boldsymbol{g}) \circ \boldsymbol{S}] \; \mathrm{d}m = - \int_{\Omega_1} \langle Df, (\boldsymbol{J}_{\boldsymbol{S}^{-1}} \cdot \boldsymbol{g}) \circ \boldsymbol{S} \rangle \mathrm{d}m.$$

Thus, by (5.11), we get (5.12) for $\boldsymbol{g} \in C_0^1(\Omega_2; \mathbb{R}^N)$.

Note that for any $\boldsymbol{g} \in C^1(\Omega_2; \mathbb{R}^N)$, there exists a sequence $\{\boldsymbol{g}_n\}$ in $C_0^1(\Omega_2; \mathbb{R}^N)$ such that $\|\boldsymbol{g}_n - \boldsymbol{g}\|_{1,\infty} \to 0$, which implies that (5.12) is also true for $\boldsymbol{g} \in C^1(\Omega_2; \mathbb{R}^N)$.

(ii) Given $\varepsilon > 0$, there exists $\boldsymbol{g} \in C^1(\Omega_2; \mathbb{R}^N)$ such that $\|\boldsymbol{g}(\boldsymbol{x})\|_2 \leqslant 1$ for all $\boldsymbol{x} \in \Omega_2$, $\boldsymbol{g}(\boldsymbol{x}) = \boldsymbol{0}$ for all $\boldsymbol{x} \in \partial\Omega_2 \setminus \omega$, and

$$\begin{aligned} &\int_{\omega} \left| \mathrm{tr}_{\Omega_2} \left[\left(f \circ \boldsymbol{S}^{-1}\right) |\boldsymbol{J}_{\boldsymbol{S}^{-1}}| \right] \right| \mathrm{d}H \\ &\leqslant \left| \int_{\partial\Omega_2} \mathrm{tr}_{\Omega_2} \left[\left(f \circ \boldsymbol{S}^{-1}\right) |\boldsymbol{J}_{\boldsymbol{S}^{-1}}| \right] \langle \boldsymbol{g}, \boldsymbol{n} \rangle \mathrm{d}H \right| + \varepsilon. \end{aligned}$$

Applying Theorem 2.4.3 to the right-hand side of the above inequality, we have

$$\begin{aligned} &\int_{\omega} \left| \mathrm{tr}_{\Omega_2} \left[\left(f \circ \boldsymbol{S}^{-1}\right) |\boldsymbol{J}_{\boldsymbol{S}^{-1}}| \right] \right| \mathrm{d}H \\ &\leqslant \left| \int_{\Omega_2} \left(f \circ \boldsymbol{S}^{-1}\right) |\boldsymbol{J}_{\boldsymbol{S}^{-1}}| \operatorname{div} \boldsymbol{g} \, dm + \int_{\Omega_2} \langle D \left(f \circ \boldsymbol{S}^{-1}\right) |\boldsymbol{J}_{\boldsymbol{S}^{-1}}|, \boldsymbol{g} \rangle \mathrm{d}m \right| + \varepsilon. \end{aligned}$$

Changing variables gives

$$\begin{aligned} &\int_{\omega} \left| \mathrm{tr}_{\Omega_2} \left[\left(f \circ \boldsymbol{S}^{-1}\right) |\boldsymbol{J}_{\boldsymbol{S}^{-1}}| \right] \right| \mathrm{d}H \\ &\leqslant \left| \int_{\Omega_1} f \left\{ [\operatorname{div} (|\boldsymbol{J}_{\boldsymbol{S}^{-1}}| \cdot \boldsymbol{g})] \circ \boldsymbol{S} \right\} \, dm + \int_{\Omega_1} \langle Df, [|\boldsymbol{J}_{\boldsymbol{S}^{-1}}| \cdot \boldsymbol{g}] \circ \boldsymbol{S} \rangle \mathrm{d}m \right| + \varepsilon. \end{aligned}$$

Therefore, after applying Theorem 2.4.3 again to the right-hand side above, we obtain

$$\int_{\omega} \left|\mathrm{tr}_{\Omega_2}\left[\left(f\circ \boldsymbol{S}^{-1}\right)|\boldsymbol{J}_{\boldsymbol{S}^{-1}}|\right]\right| \,\mathrm{d}H$$
$$\leqslant \int_{\partial\Omega_1} |\mathrm{tr}_{\Omega_1} f\langle[|\boldsymbol{J}_{\boldsymbol{S}^{-1}}|\cdot \boldsymbol{g}]\circ \boldsymbol{S}, \boldsymbol{n}\rangle| \,\mathrm{d}H + \varepsilon.$$

Hence,

$$\int_{\omega} \left|\mathrm{tr}_{\Omega_2}\left[\left(f\circ \boldsymbol{S}^{-1}\right)|\boldsymbol{J}_{\boldsymbol{S}^{-1}}|\right]\right| \,\mathrm{d}H$$
$$\leqslant \sup_{\|g\|_\infty\leqslant 1} \|\boldsymbol{J}_{\boldsymbol{S}^{-1}}\cdot \boldsymbol{g}\|_\infty \int_{S^{-1}(\omega)} |\mathrm{tr}_{\Omega_1} f| \,\mathrm{d}H + \varepsilon,$$

which implies (5.13). □

Definition 5.4.1 *Let $\boldsymbol{S}:\Omega\to\Omega$ and let $\{\Omega_1,\Omega_2,\cdots,\Omega_r\}$ be a partition of Ω. Denote $\boldsymbol{S}_i = \boldsymbol{S}|_{\Omega_i}$ for each i. We say that $\boldsymbol{S}$ is* piecewise C^2 and λ-expanding *if each $\boldsymbol{S}_i$ is C^2 and one-to-one on Ω_i, can be extended as a C^2 transformation on $\overline{\Omega_i}$,* i.e., *C^2 on an open neighborhood of $\overline{\Omega_i}$, and satisfies*

$$\sup_{x\in S(\Omega_i)} \left\|\boldsymbol{J}_{\boldsymbol{S}_i^{-1}}(\boldsymbol{x})\right\|_2 \leqslant \frac{1}{\lambda},\quad \forall\, i=1,2,\cdots,r.$$

Lemma 5.4.2 *If $\boldsymbol{S}:\Omega\to\Omega$ is piecewise C^2 and λ-expanding, then for any $f\in BV(\Omega)$,*

$$V(Pf;\Omega)\leqslant \frac{1}{\lambda}V(f;\Omega)+C'\|f\|+\frac{1}{\lambda}\sum_{i=1}^{r}\int_{\partial\Omega_i\setminus \boldsymbol{S}_i^{-1}(\partial\Omega)} |\mathrm{tr}_{\Omega_i} f| \,\mathrm{d}H,$$

where $C'>0$ is a constant independent of f.

Proof By the definition of P,

$$Pf=\sum_{i=1}^{r}\left(f\circ \boldsymbol{S}_i^{-1}\right)\left|\boldsymbol{J}_{\boldsymbol{S}_i^{-1}}\right|\chi_{\boldsymbol{S}(\Omega_i)},$$

from which and Corollary 5.4.1,

$$V(Pf;\Omega)\leqslant \sum_{i=1}^{r} V\left[\left(f\circ \boldsymbol{S}_i^{-1}\right)\left|\boldsymbol{J}_{\boldsymbol{S}_i^{-1}}\right|\chi_{\boldsymbol{S}(\Omega_i)};\Omega\right]$$
$$\leqslant \sum_{i=1}^{r}\left\{V\left[\left(f\circ \boldsymbol{S}_i^{-1}\right)\left|\boldsymbol{J}_{\boldsymbol{S}_i^{-1}}\right|;\boldsymbol{S}(\Omega_i)\right]\right.$$
$$\left.+\int_{\partial(\boldsymbol{S}(\Omega_i))\setminus\partial\Omega} \left|\mathrm{tr}_{\boldsymbol{S}(\Omega_i)}\left[\left(f\circ \boldsymbol{S}_i^{-1}\right)\left|\boldsymbol{J}_{\boldsymbol{S}_i^{-1}}\right|\right]\right| \,\mathrm{d}H\right\}. \qquad (5.14)$$

From the definition of variation, we have

$$V[(f \circ \boldsymbol{S}_i^{-1})|\boldsymbol{J}_{\boldsymbol{S}_i^{-1}}|; \boldsymbol{S}(\Omega_i)]$$

$$= \sup \left\{ - \int_{\boldsymbol{S}(\Omega_i)} \langle D[(f \circ \boldsymbol{S}_i^{-1})|\boldsymbol{J}_{\boldsymbol{S}_i^{-1}}|], \boldsymbol{g}\rangle : \boldsymbol{g} \in C_0^1(\boldsymbol{S}(\Omega_i); \mathbb{R}^N), \right.$$

$$\left. \|\boldsymbol{g}(\boldsymbol{x})\|_2 \leqslant 1, \boldsymbol{x} \in \boldsymbol{S}(\Omega_i) \right\}.$$

By (ii) of Lemma 5.4.1, for $\boldsymbol{g} \in C_0^1(\boldsymbol{S}(\Omega_i); \mathbb{R}^N)$ such that $\|\boldsymbol{g}(\boldsymbol{x})\|_2 \leqslant 1$, $\forall\, \boldsymbol{x} \in \boldsymbol{S}(\Omega_i)$,

$$\int_{S(\Omega_i)} \langle D[(f \circ \boldsymbol{S}_i^{-1})|J_{\boldsymbol{S}_i^{-1}}|], \boldsymbol{g}\rangle$$

$$= \int_{\Omega_i} \langle Df, (J_{\boldsymbol{S}_i^{-1}}\boldsymbol{g}) \circ \boldsymbol{S}_i\rangle + \int_{\Omega_i} f \sum_{j,k=1}^{N} (g_j \circ \boldsymbol{S}_i) \frac{\partial}{\partial x_k} \left[\frac{\partial (S_i^{-1})_k}{\partial y_j} \circ \boldsymbol{S}_i \right] dm.$$

Let $\boldsymbol{\phi} = \lambda \boldsymbol{g}$. Then, the above two equalities give

$$V[(f \circ S_i^{-1})|\boldsymbol{J}_{\boldsymbol{S}_i^{-1}}|; \boldsymbol{S}(\Omega_i)]$$

$$= \sup \left\{ - \frac{1}{\lambda} \int_{\Omega_i} \langle Df, (\boldsymbol{J}_{S_i^{-1}}\phi) \circ \boldsymbol{S}_i\rangle \right.$$

$$- \int_{\Omega_i} f \sum_{j,k=1}^{N} (g_j \circ \boldsymbol{S}_i) \frac{\partial}{\partial x_k} \left[\frac{\partial (S_i^{-1})_k}{\partial y_j} \circ \boldsymbol{S}_i \right] dm :$$

$$\left. \boldsymbol{g} \in C_0^1(\boldsymbol{S}(\Omega_i); \mathbb{R}^N), \|\boldsymbol{g}(\boldsymbol{x})\|_2 \leqslant 1, \ \boldsymbol{x} \in \boldsymbol{S}(\Omega_i) \right\}. \tag{5.15}$$

If we let C' be the L^∞-norm of

$$\sum_{j,k=1}^{N} (g_j \circ \boldsymbol{S}_i) \frac{\partial}{\partial x_k} \left[\frac{\partial (S_i^{-1})_k}{\partial y_j} \circ \boldsymbol{S}_i \right],$$

then

$$- \int_{\Omega_i} f \sum_{j,k=1}^{N} (g_j \circ \boldsymbol{S}_i) \frac{\partial}{\partial x_k} \left[\frac{\partial (S_i^{-1})_k}{\partial y_j} \circ \boldsymbol{S}_i \right] dm \leqslant C' \int_{\Omega_i} |f| dm. \tag{5.16}$$

On the other hand, since $\|(\boldsymbol{J}_{\boldsymbol{S}_i^{-1}}\boldsymbol{\phi})(\boldsymbol{x})\|_2 \leqslant \|\boldsymbol{J}_{\boldsymbol{S}_i^{-1}}\|_2 \ \|\boldsymbol{\phi}(\boldsymbol{x})\|_2 \leqslant 1$,

$$\int_{\Omega_i} \langle Df, (\boldsymbol{J}_{\boldsymbol{S}_i^{-1}}\boldsymbol{\phi}) \circ \boldsymbol{S}_i\rangle \ \leqslant \ V(f; \Omega_i). \tag{5.17}$$

Therefore, combining (5.15), (5.16), and (5.17), we obtain

$$V\left[\left(f\circ \boldsymbol{S}_i^{-1}\right)\left|\boldsymbol{J}_{\boldsymbol{S}_i^{-1}}\right|;\boldsymbol{S}(\Omega_i)\right]\leqslant \frac{1}{\lambda}V(f;\Omega_i)+C'\int_{\Omega_i}|f|\mathrm{d}m. \tag{5.18}$$

It follows from (5.14), (5.18), and (ii) of Lemma 5.4.1 that

$$\begin{aligned}V(Pf;\Omega)\leqslant&\sum_{i=1}^{r}\left(\frac{1}{\lambda}V(f;\Omega_i)+C'\int_{\Omega_i}|f|\,\mathrm{d}m\right)\\&+\sum_{i=1}^{r}\frac{1}{\lambda}\int_{\partial\Omega_i\backslash \boldsymbol{S}_i^{-1}(\partial\Omega)}|\mathrm{tr}_{\Omega_i}f|\,\mathrm{d}H\\&\leqslant\frac{1}{\lambda}V(f;\Omega)+C'\|f\|+\frac{1}{\lambda}\sum_{i=1}^{r}\int_{\partial\Omega_i\backslash \boldsymbol{S}_i^{-1}(\partial\Omega)}|\mathrm{tr}_{\Omega_i}f|\,\mathrm{d}H.\end{aligned}$$

Here we have used the fact that $\sum\limits_{i=1}^{r}V(f;\Omega_i)\leqslant V(f;\Omega)$, based on the equality (5.10). This completes the proof of the lemma. □

Lemma 5.4.3 *If $\boldsymbol{S}:\Omega\to\Omega$ is piecewise C^2 and λ-expanding, then*

$$V(Pf;\Omega)\leqslant\frac{1+\kappa_\Omega(\boldsymbol{S})}{\lambda}V(f;\Omega)+C\|f\|,\ \ \forall\, f\in BV(\Omega),$$

where $\kappa_\Omega(\boldsymbol{S})=\max\limits_{1\leqslant i\leqslant r}\kappa(\Omega_i)$, $\kappa(\Omega_i)$ is as defined in Theorem 2.4.4, and $C>0$ is a constant which is independent of the choice of f.

Proof From Lemma 5.4.2, we have

$$\begin{aligned}V(Pf;\Omega)\leqslant&\frac{1}{\lambda}V(f;\Omega)+C'\|f\|+\frac{1}{\lambda}\sum_{i=1}^{r}\int_{\partial\Omega_i\backslash \boldsymbol{S}_i^{-1}(\partial\Omega)}|\mathrm{tr}_{\Omega_i}f|\,\mathrm{d}H\\\leqslant&\frac{1}{\lambda}V(f;\Omega)+C'\|f\|+\frac{1}{\lambda}\sum_{i=1}^{r}\int_{\partial\Omega_i}|\mathrm{tr}_{\Omega_i}f|\,\mathrm{d}H.\end{aligned}$$

Note that Theorem 2.4.4 and Lemma 5.4.1 imply

$$\begin{aligned}\sum_{i=1}^{r}\int_{\partial\Omega_i}|\mathrm{tr}_{\Omega_i}f|\,\mathrm{d}H\leqslant&\sum_{i=1}^{r}\kappa(\Omega_i)\left(V(f;\Omega_i)+\int_{\Omega_i}|f|\mathrm{d}m\right)\\\leqslant&\kappa_\Omega(\boldsymbol{S})\left(V(f;\Omega)+\|f\|\right).\end{aligned}$$

Thus, we get

$$\begin{aligned}V(Pf;\Omega)\leqslant&\frac{1}{\lambda}V(f;\Omega)+C'\|f\|+\frac{\kappa_\Omega(\boldsymbol{S})}{\lambda}\left(V(f;\Omega)+\|f\|\right)\\\leqslant&\frac{1+\kappa_\Omega(\boldsymbol{S})}{\lambda}V(f;\Omega)+C\|f\|,\end{aligned}$$

where $C=C'+\kappa_\Omega(\boldsymbol{S})/\lambda$. □

Theorem 5.4.1 *Let $S : \Omega \to \Omega \subset \mathbb{R}^N$ be piecewise C^2 and λ-expanding. If $(1 + \kappa_\Omega(S))/\lambda < 1$, where $\kappa_\Omega(S)$ is as defined in Lemma 5.4.3, then S admits an absolutely continuous invariant probability measure.*

Proof From Lemma 5.4.3, the assumption implies that the sequence $\{\|P^n 1\|_{BV}\}$ is uniformly bounded. Hence, the sequence $\{P^n 1\}$ is precompact in $L^1(\Omega)$ by Theorem 2.5.1, and it follows from the Kakutani-Yosida abstract ergodic theorem (Theorem 5.1.1) that P has a stationary density which is the density of an absolutely continuous invariant probability measure. □

Corollary 5.4.2 *Let $S : \Omega \to \Omega \subset \mathbb{R}^N$ be piecewise C^2. If some iterated transformation S^k is piecewise λ-expanding and satisfies $(1 + \kappa_\Omega(S^k))/\lambda < 1$, then S admits an absolutely continuous invariant probability measure.*

Remark 5.4.2 Under the conditions of Theorem 5.4.1, P is quasi-compact on $BV(\Omega)$ and an application of the Ionescu-Tulcea and Marinescu theorem will give a spectral decomposition of P. See [64] for more details.

Remark 5.4.3 The difficult part in applying Theorem 5.4.1 is the determination or an estimation of the geometric quantity $\kappa_\Omega(S)$. The existence results obtained under different assumptions on Ω and S, e.g. [2, 64], can be viewed as consequences of the general framework after giving an upper bound for $\kappa_\Omega(S)$ in different cases.

Exercises

5.1 Let (X, Σ, μ) be a finite measure space and let $P : L^1 \to L^1$ be a Markov operator. Show that if there exists a function $h \in L^1$ and a positive number $\gamma < 1$ such that

$$\limsup_{n \to \infty} \|(P^n f - h)^+\| \leqslant \gamma, \ \ \forall\, f \in \mathcal{D},$$

then P is constrictive.

5.2 Let $S : [0,1] \to [0,1]$ be a nonsingular transformation, and let $\{S_n\}$ be a sequence of nonsingular transformations from $[0,1]$ into itself that converges to S uniformly on $[0,1]$. Suppose that f_n is a stationary density of P_{S_n} associated with S_n for each n. If $f_n \to f$ weakly in L^1, show that $P_S f = f$.

5.3 A nonsingular transformation S on a measure space (X, Σ, μ) is said to be *statistically stable* if there exists $f^* \in \mathcal{D}$ such that for any $f \in \mathcal{D}$ we have $\lim_{n \to \infty} \|P^n f - f^*\| = 0$, where P is the associated Frobenius-Perron operator. Let $k \geqslant 1$ be an integer. Show that S^k is statistically stable if and only if S is statistically stable.

5.4 Let $S : [0,1] \to [0,1]$ be a nonsingular transformation and let $h : [0,1] \to [0,1]$ be a homeomorphism. Denote $T = h \circ S \circ h^{-1}$ and $g = (f \circ h^{-1}) \cdot |(h^{-1})'|$. Prove:

(i) $P_S f = f$ if and only if $P_T g = g$.

(ii) f is a stationary density for S if and only if g is a stationary density for T.

5.5 Let $S : [0,1] \to [0,1]$ be defined by

$$S(x) = 4x\chi_{[0,\frac{1}{4})}(x) + \left(\frac{3}{2} - 2x\right)\chi_{[\frac{1}{4},\frac{1}{2})}(x) + (2x-1)\chi_{[\frac{1}{2},1]}(x).$$

Show that the corresponding Frobenius-Perron operator P has a stationary density f^* which is constant on each of the subintervals $\left[0, \frac{1}{4}\right)$, $\left[\frac{1}{4}, \frac{1}{2}\right)$, $\left[\frac{1}{2}, 1\right]$. Find the expression of f^*.

5.6 Let $T : [0,1] \to [0,1]$ be defined by $T = h \circ S \circ h^{-1}$, where $h(x) = \sqrt{x}$ and S is as in Exercise 5.5. Find a stationary density g for T.

5.7 Let $S : [0,1] \to [0,1]$ be the logistic model $S(x) = 4x(1-x)$. Define

$$h(x) = \frac{1}{\pi}\int_0^x \frac{\mathrm{d}x}{\sqrt{t(1-t)}} = \frac{1}{2} - \frac{1}{\pi}\sin^{-1}(1-2x).$$

Show that $h : [0,1] \to [0,1]$ is a homeomorphism and $T = h \circ S \circ h^{-1}$ is the tent function (1.2). Using the fact that T preserves the Lebesgue measure m, find the stationary density of S.

5.8 Let $S : [0,1] \to [0,1]$ be a transformation satisfying property (i) of Theorem 5.2.1, and let P_S be the corresponding Frobenius-Perron operator. If there exists a positive $C^1[0,1]$ function $\phi \in L^1(0,1)$ such that, for some real $\lambda > 1$ and s,

$$\frac{|S'(x)|\phi(S(x))}{\phi(x)} \geqslant \lambda, \ \ 0 < x < 1$$

and

$$\left|\frac{1}{\phi(x)}\frac{\mathrm{d}}{\mathrm{d}x}\left(\frac{1}{p(x)}\right)\right| \leqslant s, \ \ 0 < x < 1,$$

then P_S has a stationary density.

Hint: Define

$$g(x) = \frac{1}{\|\phi\|}\int_0^x \phi(t)\mathrm{d}t, \ \ \forall\, x \in [0,1],$$

apply Theorem 5.2.1 to $T \equiv g \circ S \circ g^{-1}$, and use the result of Exercise 5.4.

Chapter 6
Invariant Measures—Computation

JiuDing, Department of Mathematics, The University of Southern Mississippi in America.
Aihui Zhou, Academy of Mathematics and Systems Science, Chinese Academy of Sciences.

Abstract The computational problem of stationary densities of Frobenius-Perron operators will be studied. First we introduce the classic Ulam's piecewise constant method and its direct extension to multi-dimensional transformations. Then we study the piecewise linear Markov method proposed by the authors for one-and multi-dimensional transformations. We present Li's pioneering work solving Ulam's conjecture for the Lasota-Yorke class of interval mappings, and the convergence proofs of Ulam's method for piecewise convex mappings by Miller and for piecewise expanding transformations by Ding and Zhou.
Keywords Ulam's method, piecewise linear Markov method, Ulam's conjecture, consistency, stability, convergence.

Let (X, Σ, μ) be a measure space, let $S : X \to X$ be a nonsingular transformation, and let $P : L^1 \to L^1$ be the Frobenius-Perron operator associated with S. We know that any stationary density of P is the density of an S-invariant probability measure which is absolutely continuous with respect to the measure μ. In applications of ergodic theory to physical sciences and engineering, however, it is often the case that analytic expressions of such stationary densities are not available, complicated if available, or difficult to obtain, even though their existence may be guaranteed theoretically. Thus, it is important to be able to numerically compute a stationary density efficiently to any prescribed precision.

There are mainly two classes of numerical methods in the literature based on different approaches to the approximation problem of Frobenius-Perron operators. One kind of method is based on the idea of approximating a given transformation by the so-called *Markov transformations* for which the computation of stationary densities is equivalent to a matrix eigenvector computation problem. The other kind of method approximates directly the Frobenius-Perron operator associated with the given transformation by finite dimensional linear operators and then solves the fixed point problem of the approximating operators. The first approach was developed by Boyarsky and Góra together with their collaborators in the 1980's, and their main results have been included in the excellent textbook [14]. The second approach, which is more natural from the viewpoint of numerical analysis, originated with Ulam's famous book, "A Collection of Mathematical Problems," published in 1960 in which he proposed a piecewise constant numerical scheme to approximate a stationary density of interval mappings. In this chapter we shall follow the second approach by introducing some *structure-preserving* numerical methods for approximating Frobenius-Perron operators, such as Ulam's original piecewise constant approximation scheme and its higher order generalizations for the computation of absolutely continuous invariant probability measures. Such methods of *Markov finite approximations* for Frobenius-Perron operators and more general Markov operators are surveyed by, e.g., [42, 50]. Several other numerical methods, such as the Galerkin projection method [24, 32, 40, 47, 76], the maximum entropy method [31, 44], the Monte Carlo method [43, 46, 68], the interpolation method [45], and the minimal energy method [12], have been proposed in the literature. Except for the maximum entropy method and the minimal energy method which will be briefly introduced in the exercises of Chapter 8 on entropy, we do not study them in this book. The reader is also referred to [93] for a more delicate numerical analysis of Frobenius-Perron operators and to [37, 71, 93] for some new developments.

In Section 6.1 we introduce Ulam's original method for one dimensional mappings. We also prove its convergence for the two classes of interval mappings for which the existence problem of a stationary density has been investigated in Sections 5.2 and 5.3 respectively, based on Li's pioneering work [86] on Ulam's conjecture for the Lasota-Yorke class of piecewise C^2 and stretching mappings and Miller's paper [104] for the convergence of Ulam's method for the class of piecewise convex mappings, respectively. Its natural extension to multi-dimensional transformations will be treated in Section 6.2, in which Ulam's conjecture will be proved for the Góra-Boyarsky class of multi-dimensional transformations that were studied in Section 5.4, following the presentation of [49]. Sections 6.3 and 6.4 are devoted to the piecewise linear Markov approximation method for one-dimensional interval mappings and multi-dimensional transformations, respectively. The convergence rate analysis for all such numerical methods will be studied in the next chapter.

6.1 Ulam's Method for One-Dimensional Mappings

In his inspiring monograph [120] "A Collection of Mathematical Problems," Stanislaw Ulam proposed a piecewise constant approximation scheme for computing a stationary density of the Frobenius-Perron operator P associated with one-dimensional mappings. He also conjectured that, if P has a stationary density, then his method will produce a sequence of piecewise constant density functions that converges to a stationary density of P as the partition of the domain interval of the mapping becomes finer and finer. About fifteen years later, Tien-Yien Li[86] proved Ulam's conjecture for the class of piecewise C^2 and stretching interval mappings, for which the existence of a stationary density was proved in Section 5.2.

Let $S : [0,1] \to [0,1]$ be a nonsingular transformation and let $P : L^1(0,1) \to L^1(0,1)$ be the corresponding Frobenius-Perron operator. Our purpose is to compute an absolutely continuous invariant probability measure μ under S. From Birkhoff's pointwise ergodic theorem, if S is ergodic with respect to μ, then for $x \in [0,1]$ μ-a.e., the time average

$$\lim_{n\to\infty} \frac{1}{n} \sum_{k=0}^{n-1} \chi_A(S^k(x)) \tag{6.1}$$

equals $\mu(A)$, where A is any Borel measurable subset of $[0,1]$. Although in principle computing this average can be easily carried out on a computer by means of the direct iteration involved in the expression (6.1) and so the probability measure value $\mu(A)$ can be obtained numerically to any precision, practically the computer's round-off errors can dominate the calculation and may make the computed result far away from the theoretical expectation, due to the fact of sensitive dependence on initial conditions for chaotic mappings. Thus, the primitive computational scheme based on the repeated iteration of S is not a well-posed numerical method. An interesting and simple example, which shows the shortage of this primitive numerical approach from the direct iteration of points, can be seen from [86]. Therefore, instead of computing the time average via (6.1) directly to find $\mu(A) = \int_A f^* dm$, where $f^* = Pf^* \in \mathcal{D}$, we approximate the stationary density f^* of P by approximating P with a finite dimensional linear operator. Ulam suggested in his book the following scheme to approximate f^* via piecewise constant functions.

For a chosen positive integer n, we divide the interval $I = [0,1]$ into n subintervals $I_i = [x_{i-1}, x_i]$ for $i = 1, 2, \cdots, n$. If we denote

$$1_i = \frac{1}{m(I_i)} \chi_{I_i}, \quad \forall\, i = 1, 2, \cdots, n,$$

where, as usual, m stands for the Lebesgue measure on $[0,1]$, then each 1_i is a density function. We assume that $h \equiv \max_{1 \leqslant i \leqslant n} (x_i - x_{i-1})$ is such that $h \to 0$ as $n \to \infty$. Let Δ_n be the n-dimensional subspace of $L^1(0,1)$ which is spanned by $1_1, 1_2, \cdots, 1_n$, i.e., Δ_n is the space of all *piecewise constant functions* with respect to the partition of $[0,1]$. Now we want to see how the Frobenius-Perron operator P behaves on Δ_n from the probabilistic point of view. For this purpose, we define a linear operator $P_n \equiv P_n(S) : \Delta_n \to \Delta_n$ by

$$P_n 1_i = \sum_{j=1}^{n} p_{ij} 1_j, \quad \forall\, i = 1, 2, \cdots, n, \tag{6.2}$$

where, for each pair of $i, j = 1, 2, \cdots, n$,

$$p_{ij} = \frac{m(I_i \cap S^{-1}(I_j))}{m(I_i)}$$

is the "probability" that a point in the ith subinterval I_i is mapped into the jth subinterval I_j under the mapping S. It is noted that the matrix (p_{ij}) is a *stochastic matrix* since for each i,

$$\sum_{j=1}^{n} p_{ij} = \sum_{j=1}^{n} \frac{m(I_i \cap S^{-1}(I_j))}{m(I_i)} = 1.$$

Let $f = \sum_{i=1}^{n} a_i 1_i \in \Delta_n \cap \mathcal{D}$ be a piecewise constant density function. By the definition of the Frobenius-Perron operator, the probability distribution defined by f is, *approximately*, transferred into the probability distribution determined by the piecewise constant density function $\sum_{j=1}^{n} b_j 1_j$, where for each $j = 1, 2, \cdots, n$,

$$b_j = \sum_{i=1}^{n} a_i p_{ij} = \sum_{i=1}^{n} a_i \frac{m(I_i \cap S^{-1}(I_j))}{m(I_i)}.$$

In Ulam's method, one is to compute a fixed point $f_n \in \Delta_n \cap \mathcal{D}$ of P_n to approximate a stationary density f^* of P. It is interesting to note that, although the Frobenius-Perron operator P may not have a nonzero fixed point (for example, for $S(x) = x/2$ with $x \in [0, 1]$ the only fixed point of P_S is 0; for a nontrivial example see Remark 5.2.7 in Section 5.2 or Remark 6.2.1 of [82]), its Ulam's finite approximation P_n always has nontrivial fixed points for *any* positive integer n from the *Perron-Frobenius theorem* on nonnegative matrices [103] (see also the proof of Proposition 6.1.2 (ii) in the following). Let $f_n \in \mathcal{D}$ be a piecewise constant fixed point of P_n from Ulam's method for any $n = 1, 2, \cdots$, which is called a *stationary density* of P_n in Δ_n. Ulam stated his following famous conjecture [120] in 1960.

Ulam's conjecture: If the Frobenius-Perron operator $P : L^1(0, 1) \to L^1(0, 1)$ has a stationary density, then there is a subsequence $\{f_{n_k}\}$ of $\{f_n\}$ such that

$$\lim_{k \to \infty} f_{n_k} = f^*,$$

where f^* is a stationary density of P.

Although this conjecture is still open in general, it is true for some classes of transformations. For example, it has been proved by Li [86] in 1976 for the class of piecewise C^2 and stretching mappings and by Miller [101] in 1994 for the class of piecewise convex mappings with s strong repellor.

We are going to present and prove their convergence results. In order to study the convergence problem for Ulam's method as the number of the subintervals goes to infinity, Li [86] defined a sequence of linear operators $Q_n : L^1(0, 1) \to L^1(0, 1)$ by

$$Q_n f = \sum_{i=1}^{n} \frac{1}{m(I_i)} \int_{I_i} f \mathrm{d}m \cdot \chi_{I_i} = \sum_{i=1}^{n} \int_{I_i} f \mathrm{d}m \cdot 1_i \tag{6.3}$$

associated with a sequence of partitions of $[0, 1]$.

Proposition 6.1.1 *Q_n is both a Markov operator and a* Galerkin projection *from $L^1(0,1)$ onto Δ_n for each n. Moreover, $\|Q_n f\|_\infty \leqslant \|f\|_\infty$ for any $f \in L^\infty(0,1)$.*

Proof Q_n is clearly a positive operator. Let $f \in L^1(0,1)$. Then

$$\begin{aligned}\int_0^1 Q_n f(x)\mathrm{d}x &= \sum_{i=1}^{n} \int_{I_i} f \mathrm{d}m \cdot \int_0^1 1_i(x)\mathrm{d}x \\ &= \sum_{i=1}^{n} \int_{I_i} f \mathrm{d}m = \int_0^1 f(x)\mathrm{d}x,\end{aligned}$$

namely, Q_n preserves the integral of a function. So Q_n is a Markov operator. Q_n is also a Galerkin projection operator since for each $i = 1, 2, \cdots, n$,

$$\begin{aligned}\langle Q_n f - f, \chi_i\rangle &= \int_{I_i} Q_n f \mathrm{d}m - \int_{I_i} f \mathrm{d}m \\ &= \int_{I_i} f \mathrm{d}m - \int_{I_i} f \mathrm{d}m = 0\end{aligned}$$

for all $f \in L^1(0,1)$. The last conclusion is obvious. □

More analytic properties of Ulam's method are summarized in the following proposition.

Proposition 6.1.2 *Let n be a positive integer. Then*

(i) *$Q_n P : L^1(0,1) \to L^1(0,1)$ is a Markov operator and $Q_n P = P_n$ on Δ_n, where P_n was defined in (6.2).*

(ii) *Let $\Delta_n' = \left\{ \sum\limits_{i=1}^{n} a_i 1_i : a_i \geqslant 0, \ \forall\, i, \sum\limits_{i=1}^{n} a_i = 1 \right\}$. Then Δ_n' is a closed, bounded, and convex subset of Δ_n and $P_n(\Delta_n') \subset \Delta_n'$. Hence, there is a stationary density $f_n \in \Delta_n$ of P_n.*

(iii) *$\lim\limits_{n\to\infty} Q_n f = f$ under the L^1-norm for any $f \in L^1(0,1)$.*

(iv) *For any $f \in BV(0,1)$,*

$$\bigvee_0^1 Q_n f \leqslant \bigvee_0^1 f.$$

(v) *$\|f - Q_n f\| \leqslant h \bigvee\limits_0^1 f = O(h)$ for any $f \in BV(0,1)$.*

(vi) *The subset of monotonically increasing (or decreasing) functions of* $L^1(0,1)$ *is invariant under* Q_n.

Proof (i) Since both Q_n and P are Markov operators, their composition Q_nP is obviously a Markov operator. For each $i = 1, 2, \cdots, n$, from the definition (4.1) of P and the definition (6.2) of P_n,

$$\begin{aligned} Q_nP\,1_i &= \sum_{j=1}^{n}\left[\frac{1}{m(I_j)}\int_{I_j} P1_i dm\right]\chi_{I_j} = \sum_{j=1}^{n}\int_{I_j} P1_i dm \cdot 1_j \\ &= \sum_{j=1}^{n}\int_{S^{-1}(I_j)} 1_i dm \cdot 1_j = \sum_{j=1}^{n}\frac{m(I_i \cap S^{-1}(I_j))}{m(I_i)}1_j \\ &= \sum_{j=1}^{n} p_{ij}1_j = P_n 1_i. \end{aligned}$$

Therefore $P_n = Q_nP$ on Δ_n.

(ii) Clearly $\Delta_n' \subset \Delta_n$ is closed, bounded, and convex. Since P_n maps densities into densities and each element of Δ_n' is a density, $P_n(\Delta_n') \subset \Delta_n'$. By Brouwer's fixed point theorem [56], $P_nf_n = f_n$ for some $f_n \in \Delta_n'$.

(iii) It is enough to assume that $f \in C[0,1]$ since $\|Q_n\| \equiv 1$ and the set of all continuous functions on $[0,1]$ is dense in $L^1(0,1)$. Let $\epsilon > 0$. From the uniform continuity of f, if n is large enough,

$$|\hat{f}_i - f(x)| < \epsilon$$

for all $x \in I_i$ and all $i = 1, 2, \cdots, n$, where

$$\hat{f}_i = \frac{1}{m(I_i)}\int_{I_i} f dm$$

is the *average value* of f on I_i. Hence, from the definition (6.3) of Q_n,

$$\|Q_nf - f\| = \sum_{i=1}^{n}\int_{x_{i-1}}^{x_i} |\hat{f}_i - f(x)|dx < \sum_{i=1}^{n}\int_{x_{i-1}}^{x_i} \epsilon dx = \epsilon.$$

(iv) Since $Q_nf = \sum_{i=1}^{n} \hat{f}_i\chi_{I_i}$ by (6.3), we have

$$\bigvee_0^1 Q_nf = \sum_{i=1}^{n-1} |\hat{f}_i - \hat{f}_{i+1}|.$$

Choose v_i and y_i in I_i such that

$$f(v_i) \leqslant \hat{f}_i \leqslant f(y_i).$$

Let $\{\hat{f}_{i_k}\}_{k=1}^{s} \subset \{\hat{f}_i\}_{i=1}^{n}$ be the subsequence of all the local maxima and local minima among all these $\hat{f}_i$, and we define

$$z_k = \begin{cases} x_{i_k}, & \text{if } \hat{f}_{i_k} \text{ is a local minimum,} \\ y_{i_k}, & \text{if } \hat{f}_{i_k} \text{ is a local maximum,} \end{cases}$$

for $k = 1, 2, \cdots, s$. Then it is easy to see that

$$\bigvee_0^1 Q_n f \leqslant \sum_{k=1}^{s-1} |f(z_k) - f(z_{k+1})| \leqslant \bigvee_0^1 f.$$

(v) Since $|f(x) - \hat{f}_i| \leqslant \bigvee_{x_{i-1}}^{x_i} f$ on each I_i,

$$\|f - Q_n f\| = \sum_{i=1}^{n} \int_{x_{i-1}}^{x_i} |f(x) - \hat{f}_i| \mathrm{d}x \leqslant \sum_{i=1}^{n} \int_{I_i} \bigvee_{x_{i-1}}^{x_i} f \, \mathrm{d}m$$

$$= \sum_{i=1}^{n} (x_i - x_{i-1}) \bigvee_{x_{i-1}}^{x_i} f \leqslant h \sum_{i=1}^{n} \bigvee_{x_{i-1}}^{x_i} f = h \bigvee_0^1 f.$$

(vi) Since $Q_n f = \sum_{i=1}^{n} \hat{f}_i \chi_{I_i}$, it is clear that the ordered numbers $\hat{f}_1, \cdots, \hat{f}_n$ are monotonically increasing or decreasing depending on whether f is monotonically increasing or decreasing, respectively. □

Because of Proposition 6.1.2 (i), from now on let P_n be *defined* to be $Q_n P$ on the whole domain $L^1(0,1)$ of P. Then, Proposition 6.1.2 (iii) ensures that $\lim_{n\to\infty} P_n f = Pf$ under the L^1-norm for all $f \in L^1(0,1)$, i.e., $\lim_{n\to\infty} P_n = P$ *strongly.*

We are ready to prove the convergence of Ulam's method for the class of piecewise C^2 and stretching mappings and the class of piecewise convex mappings with a strong repellor.

Theorem 6.1.1 (Li's convergence theorem) *Let $S : [0,1] \to [0,1]$ satisfy the conditions of Theorem 5.2.1 with*

$$\lambda \equiv \inf_{x \in [0,1] \setminus \{a_1, \cdots, a_{r-1}\}} |S'(x)| > 2,$$

and let P be the corresponding Frobenius-Perron operator. Then for any sequence $\{f_n\}$ of the stationary densities of P_n in Δ_n, there is a subsequence $\{f_{n_k}\}$ such that

$$\lim_{k\to\infty} \|f_{n_k} - f^*\| = 0,$$

where f^* is a stationary density of P. If in addition f^* is the unique stationary density of P, then

$$\lim_{n\to\infty} \|f_n - f^*\| = 0.$$

Proof From the proof of Theorem 5.2.1, for any $f \in \mathcal{D} \cap BV(0,1)$,

$$\bigvee_0^1 Pf \leqslant \alpha \bigvee_0^1 f + \beta,$$

where $0 < \alpha < 1$ and $\beta > 0$ are two constants which are independent of f. Thus, from Proposition 6.1.2 (iv),

$$\bigvee_0^1 f_n = \bigvee_0^1 Q_n P f_n \leqslant \bigvee_0^1 P f_n \leqslant \alpha \bigvee_0^1 f_n + \beta,$$

which implies that

$$\bigvee_0^1 f_n \leqslant \frac{\beta}{1-\alpha}$$

uniformly for all n. Hence, from Helly's lemma, there is a subsequence $\{f_{n_k}\}$ of $\{f_n\}$ that converges to a density f^* under the L^1-norm. Since $\|P_n\| \equiv 1$ and $\lim_{n\to\infty} P_n = P$ strongly,

$$\begin{aligned}\|Pf^* - f^*\| \leqslant & \|f^* - f_{n_k}\| + \|f_{n_k} - P_{n_k} f_{n_k}\| \\ & + \|P_{n_k} f_{n_k} - P_{n_k} f^*\| + \|P_{n_k} f^* - Pf^*\| \\ \leqslant & \|f^* - f_{n_k}\| + \|f_{n_k} - f^*\| + \|P_{n_k} f^* - Pf^*\| \to 0\end{aligned}$$

as $k \to \infty$. Thus f^* is a stationary density of P. If f^* is the unique stationary density of P, then we must have $\lim_{n\to\infty} f_n = f^*$ since every convergent subsequence of $\{f_n\}$ converges to the same limit from the above analysis. □

Corollary 6.1.1 *Theorem* 6.1.1 *is still true for* $\lambda > 1$.

Proof Take an integer $k > 0$ such that $\lambda^k > 2$ and denote $\phi = S^k$. For any integer n, let $f_n^{(\phi)}$ be a stationary density of $P_n(\phi) = Q_n P_\phi$, and let

$$g_n = \frac{1}{k} \sum_{j=0}^{k-1} P^j f_n^{(\phi)}.$$

Then the sequence $\{g_n\}$ converges, by the above theorem, at

$$g = \frac{1}{k} \sum_{j=0}^{k-1} P^j f^{(\phi)},$$

where $f^{(\phi)}$ is a stationary density of P_ϕ. This g is a stationary density of P since $P^k f^{(\phi)} = P_\phi f^{(\phi)} = f^{(\phi)}$ implies that

$$Pg = \frac{1}{k} \{Pf^{(\phi)} + \cdots + P^k f^{(\phi)}\} = g.$$

□

Theorem 6.1.2 (Miller's convergence theorem) *Suppose that $S : [0,1] \to [0,1]$ is a piecewise convex mapping with a strong repellor as in Theorem 5.3.1, let P be the corresponding Frobenius-Perron operator, and let $\{P_n\}$ be a sequence of Ulam's approximations of P. Then for any n, there is a stationary density of P_n which is a monotonically decreasing function, and for any sequence of the monotonically decreasing stationary densities f_n of P_n in Δ_n,*

$$\lim_{n\to\infty} \|f_n - f^*\| = 0,$$

where f^* is the unique stationary density of P, which is also a monotonically decreasing function.

Proof From the proof of Theorem 5.3.1, Pf is monotonically decreasing for any monotonically decreasing function $f \in \mathcal{D}$ and

$$Pf(x) \leqslant \alpha f(0) + K,$$

where $\alpha = 1/S'(0) \in (0,1)$ and $K > 0$ is a constant. Let

$$\mathcal{D}_n = \{f \in \Delta_n \cap \mathcal{D} : \ f \text{ is monotonically decreasing on } [0,1]\}.$$

Then $\mathcal{D}_n$ is invariant under P_n by Proposition 6.1.2 (vi). Since $\mathcal{D}_n$ is a compact convex set, by Brouwer's fixed point theorem, P_n has a fixed point $f_n \in \mathcal{D}_n$. Thus, from the fact that $\|Q_n\|_\infty = 1$,

$$f_n(0) = Q_n P f_n(0) \leqslant \max_{x\in[0,1]} P f_n(x) \leqslant \alpha f_n(0) + K.$$

Hence, since $0 < \alpha < 1$,

$$f_n(0) \leqslant \frac{K}{1-\alpha}, \quad \forall\, n,$$

which implies that

$$\bigvee_0^1 f_n = f_n(0) - f_n(1) \leqslant \frac{K}{1-\alpha}$$

uniformly for all n, and the result follows from Helly's lemma. □

Remark 6.1.1 Ulam's method can also be used to compute absolutely continuous invariant measures for expanding circle mappings (see [72]).

6.2 Ulam's Method for N-dimensional Transformations

We may directly extend Ulam's original piecewise constant approximation scheme for interval mappings to more general multi-dimensional transformations. Let $\Omega \subset \mathbb{R}^N$ be a bounded open region, let $\boldsymbol{S} : \Omega \to \Omega$ be a nonsingular transformation, and let P be the corresponding Frobenius-Perron operator. As usual we let m denote the Lebesgue measure on $\mathbb{R}^N$.

Let $h > 0$ be a positive discretization parameter. We partition the closed region $\overline{\Omega}$ into l subregions $\mathcal{P}_h = \{\Omega_i : i = 1, 2, \cdots, l\}$ such that the *diameter*, diam $\Omega_i \equiv \sup\{\|\boldsymbol{x} - \boldsymbol{y}\|_2 : \boldsymbol{x}, \boldsymbol{y} \in \Omega_i\}$, of each Ω_i is bounded by h. We assume that the partition is *quasi-uniform* in the sense that there exists a constant $\gamma > 0$ such that

$$h \leqslant \gamma \min_{i=1}^{l} \text{ diam } B_{\Omega_i},$$

where B_{Ω_i} is the closed ball inscribed in $\overline{\Omega_i}$. In this section, we assume that Ω is an N-dimensional rectangular region, and we simply partition Ω into N-subrectangles.

Let $1_i = \chi_{\Omega_i}/m(\Omega_i)$ for $i = 1, 2, \cdots, l$. Then each $1_i \in L^1(\Omega)$ is a density function with supp $1_i = \Omega_i$. Denote by Δ_h the l-dimensional subspace of $L^1(\Omega)$ consisting of all the piecewise constant functions associated with $\mathcal{P}_h$. Let $Q_h : L^1(\Omega) \to L^1(\Omega)$ be defined by

$$Q_h f = \sum_{i=1}^{l} \left(\frac{1}{m(\Omega_i)} \int_{\Omega_i} f \mathrm{d}m \right) \chi_{\Omega_i}. \tag{6.4}$$

Define $P_h = Q_h P : L^1(\Omega) \to L^1(\Omega)$, which is clearly a Markov operator. Then, with the same argument as in the previous section, the operator $P_h|_{\Delta_h} : \Delta_h \to \Delta_h$ satisfies

$$P_h 1_i = \sum_{j=1}^{l} p_{ij} 1_j, \quad \forall\, i = 1, 2, \cdots, l,$$

where

$$p_{ij} = \frac{m(\Omega_i \cap S^{-1}(\Omega_j))}{m(\Omega_i)}, \quad \forall\, i, j = 1, 2, \cdots, l.$$

Obviously, Proposition 6.1.2 (i) and (ii) are still true, that is,

Proposition 6.2.1 *Let $\mathcal{P}_h$ be a quasi-uniform partition of Ω. Then,*

(i) *For* $\Delta'_h = \left\{ \sum_{i=1}^{l} a_i 1_i : a_i \geqslant 0, \forall\, i, \sum_{i=1}^{l} a_i = 1 \right\}$, *we have* $P_h(\Delta'_h) \subset \Delta'_h$. *Hence, there is a stationary density f_h of P_h in Δ_h.*

(ii) $\lim_{h \to 0} Q_h f = f$ *under the L^1-norm for any $f \in L^1(\Omega)$.*

Now we establish a similar *stability* result to Proposition 6.1.2 (iv) with the multi-dimensional notion of variation. For this purpose, we need the following general result, in the proof of which we shall use the notation C to represent possibly different values of constants in different occasions.

Lemma 6.2.1 *Let $\Omega \subset \mathbb{R}^N$ be an N-dimensional rectangle and let $\{\mathcal{P}_h\}$ be a family of quasi-uniform rectangular partitions of Ω. For each $\mathcal{P}_h$, let $R_h : BV(\Omega) \to \Delta_h$ be a bounded linear operator that satisfies*

$$\|R_h v - v\| \leqslant Ch \int_\Omega \|\mathbf{grad}\, v\| \, \mathrm{d}m, \quad \forall\, v \in W^{1,1}(\Omega) \tag{6.5}$$

for some constant C that is independent of h. Then there exists a constant C_{BV} such that

$$V(R_h f; \Omega) \leqslant C_{BV} V(f; \Omega), \quad \forall\, f \in BV(\Omega)$$

uniformly with respect to h.

Proof Let $f \in BV(\Omega)$. By Theorem 2.4.2, there exists a sequence $\{f_j\}$ in $C^\infty(\Omega)$ such that

$$\lim_{j\to\infty} \|f_j - f\| = 0 \tag{6.6}$$

and

$$\lim_{j\to\infty} \|\mathbf{grad}\, f_j\|_{0,1} = V(f; \Omega). \tag{6.7}$$

Since R_h is continuous, (6.6) implies $\lim\limits_{j\to\infty} \|R_h f_j - R_h f\| = 0$. Hence, from Theorem 2.4.1,

$$V(R_h f; \Omega) \leqslant \liminf_{j\to\infty} V(R_h f_j; \Omega). \tag{6.8}$$

For each $i = 1, 2, \cdots, l$ denote

$$Q_{n_1,n_2,\cdots,n_N}(\Omega_i) = \{q|_{\Omega_i} : q \in Q_{n_1,n_2,\cdots,n_N}\},$$

where

$$Q_{n_1,n_2,\cdots,n_N} = \mathrm{span}\{x_1^{i_1} \cdots x_N^{i_N} : 0 \leqslant i_j \leqslant n_j,\ 1 \leqslant j \leqslant N\}$$

with $\boldsymbol{x} = (x_1, x_2, \cdots, x_N)^{\mathrm{T}} \in \mathbb{R}^N$. Let

$$W_h = \left\{\boldsymbol{w} \in H_0^1(\Omega)^N : \boldsymbol{w}|_{\Omega_i} \in Q_{0,1,\cdots,1}(\Omega_i) \times \cdots \times Q_{1,\cdots,1,0}(\Omega_i),\ \forall\, \Omega_i \in \mathcal{P}_h\right\},$$

and define $\pi_h : H_0^1(\Omega)^N \to W_h$ by requiring that

$$\int_{\partial\Omega_i} (\boldsymbol{w} - \pi_h \boldsymbol{w}) \cdot \boldsymbol{n}_{\Omega_i} \mathrm{d}H = 0, \quad \forall\, \Omega_i \in \mathcal{P}_h,$$

where $\boldsymbol{n}_{\Omega_i}$ is the unit outward normal vector to the boundary $\partial\Omega_i$ of Ω_i and H is the $(N-1)$-dimensional Hausdorff measure on $\partial\Omega_i$. Then, a standard argument in the theory of finite element methods (see, e.g., [62]) gives that

$$\|\pi_h \boldsymbol{w}\|_\infty \leqslant C\|\boldsymbol{w}\|_\infty, \quad \forall\, \boldsymbol{w} \in C_0^1(\Omega; \mathbb{R}^N) \tag{6.9}$$

for some constant C which is independent of h. The definition of π_h implies that

$$\int_{\Omega} \operatorname{div}(\boldsymbol{w} - \pi_h \boldsymbol{w}) v \, dm = 0, \quad \forall \, v \in \Delta_h. \tag{6.10}$$

From (6.10), we have

$$\langle R_h f_j, \operatorname{div} \boldsymbol{w} \rangle = -\langle \mathbf{grad} \, f_j, \pi_h \boldsymbol{w} \rangle + \langle R_h f_j - f_j, \operatorname{div} \pi_h \boldsymbol{w} \rangle.$$

Thus, from (6.5), (6.9), and the *inverse estimate* [20]

$$\|\operatorname{div} \pi_h \boldsymbol{w}\|_\infty \leqslant C h^{-1} \|\pi_h \boldsymbol{w}\|_\infty,$$

we can find a constant C_{BV} that is independent of h and $f \in BV(\Omega)$ such that

$$|\langle R_h f_j, \operatorname{div} \boldsymbol{w} \rangle| \leqslant C_{BV} \|\mathbf{grad} \, f_j\|_{0,1} \, \|\boldsymbol{w}\|_\infty,$$

which implies that

$$\begin{aligned} V(R_h f_j; \Omega) &= \sup \left\{ \int_{\Omega} R_h f_j \operatorname{div} \boldsymbol{w} \, dm : \boldsymbol{w} \in C_0^1(\Omega; \mathbb{R}^N), \ \|\boldsymbol{w}(\boldsymbol{x})\|_2 \leqslant 1, \boldsymbol{x} \in \Omega \right\} \\ &\leqslant C_{BV} \int_{\Omega} \|\mathbf{grad} \, f_j\| dm. \end{aligned} \tag{6.11}$$

Combining (6.7), (6.8), and (6.11), we have proved the lemma. □

Proposition 6.2.2 *Suppose that $\Omega \subset \mathbb{R}^N$ is an N-dimensional rectangle. Then there is a constant C_{BV} which is independent of h such that*

$$V(Q_h f; \Omega) \leqslant C_{BV} V(f; \Omega), \quad \forall \, f \in BV(\Omega).$$

Furthermore, if $\Omega = [0,1]^N$, then $C_{BV} = \sqrt{N}$.

Proof For the first part of the proposition, by Lemma 6.2.1 it is sufficient to show that

$$\|Q_h v - v\| \leqslant C h \|\mathbf{grad} \, v\|_{0,1}, \quad \forall \, v \in W^{1,1}(\Omega)$$

for some constant C independent of h.

From formula (7.45) in [61], there is a constant C such that for each $i = 1, 2, \cdots, l$,

$$\int_{e_i} \left| v - \frac{1}{m(\Omega_i)} \int_{\Omega_i} v \, dm \right| dm \leqslant C h \int_{\Omega_i} \|\mathbf{grad} \, v\| \, dm$$

for all $v \in W^{1,1}(\Omega)$. Thus, the definition (6.4) of Q_h gives that

$$\begin{aligned} \|Q_h v - v\| &= \sum_{i=1}^{l} \int_{\Omega_i} \left| \frac{1}{m(\Omega_i)} \int_{\Omega_i} v \, dm - v \right| dm \\ &\leqslant C h \sum_{i=1}^{l} \int_{\Omega_i} \|\mathbf{grad} \, v\| \, dm = C h \|\mathbf{grad} \, v\|_{0,1}. \end{aligned}$$

The proof of the last part is referred to [102]. □

Theorem 6.2.1 (Ding-Zhou's convergence theorem) *Let $\Omega \subset \mathbb{R}^N$ be an N-dimensional rectangle and let $\boldsymbol{S} : \Omega \to \Omega$ satisfy the conditions of Theorem* 5.4.1. *If in addition*

$$\frac{C_{BV}(1+\kappa_\Omega(\boldsymbol{S}))}{\lambda} < 1,$$

where $\kappa_\Omega(\boldsymbol{S})$ and λ are given in Theorem 5.4.1 *and C_{BV} is as in Proposition* 6.2.2, *then for any sequence $\{f_{h_n}\}$ of the stationary densities of $\{P_{h_n}\}$ in Δ_{h_n} from Ulam's method such that $\lim_{n\to\infty} h_n = 0$, there is a subsequence $\{f_{h_{n_k}}\}$ such that*

$$\lim_{k\to\infty} \|f_{h_{n_k}} - f^*\| = 0,$$

where f^* is a stationary density of P. If in addition f^* is the unique stationary density of P, then

$$\lim_{n\to\infty} \|f_{h_n} - f^*\| = 0.$$

Proof The proof is basically the same as that of Theorem 6.1.1 if we use the compactness argument for $BV(\Omega)$. The reader can fill the details of the proof. □

Remark 6.2.1 The results of this section are still true if Ω is an N-dimensional polygonal region.

Remark 6.2.2 The convergence of Ulam's method for the Jabłoński class of transformations has been proved (see [10, 15, 16, 49]), and for more general hyperbolic systems (see [59, 60]).

6.3 The Markov Method for One-Dimensional Mappings

Ulam's method preserves the Markov property of the Frobenius-Perron operator, that is, the finite approximation P_n of P maps densities to densities, and so its matrix representation under any basis of Δ_n that consists of densities is a stochastic matrix which must have a nonnegative left eigenvector corresponding to the spectral radius 1 (usually called the *maximal eigenvalue*). However, since the method uses only piecewise constant functions to approximate an integrable function, fast convergence cannot be expected for this piecewise constant method. In the next chapter we shall show that the convergence rate of Ulam's method under the L^1-norm is only $O(\ln n/n)$ for the one-dimensional mappings, where n is the number of the subintervals from the partition of $[0, 1]$. Moreover, Bose and Murray [11] have constructed one-dimensional mappings (even piecewise linear ones) for which Ulam's method has the L^1-norm convergence rate exactly the same order as $\ln n/n$, that is, a constant multiple of $\ln n/n$. This means that in general we cannot expect Ulam's method to have a convergence rate faster than $\ln n/n$.

In this section and the next one, we shall introduce a higher order numerical scheme using piecewise linear function approximations. This new approach was first proposed in [28] (see also [38]) for computing one-dimensional absolutely continuous invariant measures, and was then generalized in [48, 54, 55] for approximating Frobenius-Perron operators associated with multi-dimensional transformations and more general Markov operators. This method will be called the *Markov method* since, like Ulam's method, it leads to approximations of Frobenius-Perron operators by finite dimensional Markov operators. In this section we introduce the Markov method for one-dimensional mappings, and the method for multi-dimensional transformations will be proposed in Section 6.4.

Let $S : [0,1] \to [0,1]$ be a nonsingular transformation and let $P : L^1(0,1) \to L^1(0,1)$ be the corresponding Frobenius-Perron operator. Our purpose is to approximate P through a piecewise linear approximation method. As with Ulam's method, divide $[0,1]$ into n subintervals $I_i = [x_{i-1}, x_i]$, $i = 1, 2, \cdots, n$. Here for the sake of simplicity, we assume that such subintervals have the same length $h = 1/n$. We shall use the same notation Δ_n as that for Ulam's method to denote the $(n+1)$-dimensional subspace of all *continuous piecewise linear functions* with respect to the above partition of $[0,1]$. The *canonical basis* for Δ_n consists of

$$\phi_i(x) = w\left(\frac{x - x_i}{h}\right), \quad i = 0, 1, \cdots, n,$$

where

$$w(x) = (1 - |x|)\chi_{[-1,1]}(x) = \begin{cases} 1 + x, & x \in [-1, 0], \\ 1 - x, & x \in (0, 1], \\ 0, & x \notin [-1, 1] \end{cases}$$

is the standard *tent function*. It is easy to see that the sum of the basis functions is identically equal to 1 and that if $f = \sum_{i=0}^{n} q_i\phi_i$, then $f(x_i) = q_i$ for all i. Also note that $\|\phi_0\| = \|\phi_n\| = h/2$ and $\|\phi_i\| = h$ for $i = 1, 2, \cdots, n-1$. Now we define a linear operator $Q_n : L^1(0,1) \to L^1(0,1)$ by

$$Q_n f = \hat{f}_1\phi_0 + \sum_{i=1}^{n-1} \frac{\hat{f}_i + \hat{f}_{i+1}}{2}\phi_i + \hat{f}_n\phi_n, \tag{6.12}$$

where as before,

$$\hat{f}_i = \frac{1}{m(I_i)} \int_{I_i} f \mathrm{d}m = n \int_{I_i} f \mathrm{d}m$$

is the average value of f over I_i. It is obvious to see that Q_n is a positive operator. Actually we can show that Q_n is a Markov operator for any n.

Proposition 6.3.1 $\displaystyle\int_0^1 Q_n f(x)\mathrm{d}x = \int_0^1 f(x)\mathrm{d}x$ *for any* $f \in L^1(0,1)$.

Proof Let $f \in L^1(0,1)$. Then, by using the expression (6.12) of $Q_n f$, we have

$$\begin{aligned}
&\int_0^1 Q_n f(x)\mathrm{d}x \\
=&\hat{f}_1 \int_0^1 \phi_0(x)\mathrm{d}x + \sum_{i=1}^{n-1} \frac{\hat{f}_i + \hat{f}_{i+1}}{2} \int_0^1 \phi_i(x)\mathrm{d}x + \hat{f}_n \int_0^1 \phi_n(x)\mathrm{d}x \\
=&\frac{h}{2}\left[\hat{f}_1 + \sum_{i=1}^{n-1}(\hat{f}_i + \hat{f}_{i+1}) + \hat{f}_n\right] = \sum_{i=1}^{n} \int_{x_{i-1}}^{x_i} f(x)\mathrm{d}x = \int_0^1 f(x)\mathrm{d}x.
\end{aligned}$$

□

As in Proposition 6.1.2 (iii), the operator sequence $\{Q_n\}$ in the Markov method satisfies that for any $f \in L^1(0,1)$,

$$\lim_{n\to\infty} Q_n f = f$$

under the L^1-norm. If f is second order continuously differentiable, we can even estimate its *local convergence rate.* For this purpose, we need two lemmas, the first of which is a standard result in approximation theory.

Lemma 6.3.1 *Let* $f \in C^2[0,1]$ *and let*

$$L_n f(x) = \sum_{i=0}^{n} f(x_i)\phi_i(x)$$

be the piecewise linear Lagrange interpolation *of* f. *Then*

$$\|f - L_n f\| = O\left(\frac{1}{n^2}\right).$$

Proof The integral form of Taylor's expansion gives that

$$\begin{aligned}
&\|f - L_n f\| \\
=&\sum_{i=1}^{n} \int_{x_{i-1}}^{x_i} \left| f(x) - \left[f(x_i) + \frac{f(x_i) - f(x_{i-1})}{h}(x - x_i)\right]\right| \mathrm{d}x \\
=&\sum_{i=1}^{n} \int_{x_{i-1}}^{x_i} \left| \int_x^{x_i} \int_{x_{i-1}}^{y} f''(t)\mathrm{d}t\mathrm{d}y \right| \mathrm{d}x \\
\leqslant&\sum_{i=1}^{n} \frac{h^2}{2} \int_{x_{i-1}}^{x_i} |f''(t)|\mathrm{d}t = O\left(\frac{1}{n^2}\right).
\end{aligned}$$

□

Lemma 6.3.2 *Let* $f \in C^2[0,1]$. *Then*

$$\|L_n f - Q_n f\| = O\left(\frac{1}{n^2}\right).$$

Proof Given $g \in C^2[a,b]$. Using the rectangle formula in numerical integration, we see that

$$\left|\int_a^b g(x)\mathrm{d}x - g(c)(b-a)\right| \leqslant \max_{x\in[a,b]} |f'(x)|(b-a)^2$$

for $c = a$ or $c = b$, and an application of the middle point formula in numerical integration gives that

$$\left|\int_a^b g(x)\mathrm{d}x - g(c)(b-a)\right| \leqslant \frac{1}{2}\int_a^b |g''(x)|\mathrm{d}x(b-a)^3$$

for $c = (a+b)/2$. It follows that

$$\begin{aligned}
&\|L_n f - Q_n f\| \\
\leqslant &\left|hf(0) - \int_0^h f(x)\mathrm{d}x\right| \cdot \frac{1}{h}\int_0^1 \phi_0(x)\mathrm{d}x \\
&+ \sum_{i=1}^{n-1}\left(\left|2hf(x_i) - \int_{x_{i-1}}^{x_{i+1}} f(x)\mathrm{d}x\right| \cdot \frac{1}{2h}\int_0^1 \phi_i(x)\mathrm{d}x\right) \\
&+ \left|hf(1) - \int_{1-h}^1 f(x)\mathrm{d}x\right| \cdot \frac{1}{h}\int_0^1 \phi_n(x)\mathrm{d}x \\
\leqslant &\frac{h^2}{2}\left(2\max_{x\in I_1\cup I_n} |f'(x)| + \int_0^1 |f''(x)|\mathrm{d}x\right) = O(h^2).
\end{aligned}$$

□

Lemmas 6.3.1 and 6.3.2 give immediately the following proposition, which indicates that the piecewise linear Markov method has the "local convergence" of order 2, while Ulam's method can only enjoy the local convergence of order 1 (Proposition 6.1.2 (v)).

Proposition 6.3.2 *Let* $f \in C^2[0,1]$. *Then*

$$\|f - Q_n f\| = O\left(\frac{1}{n^2}\right). \tag{6.13}$$

The sequence of the piecewise linear Markov approximations shares the same stability property as Proposition 6.1.2 (iv) for Ulam's approximation sequence.

Proposition 6.3.3 *For any* $f \in BV(0,1)$,

$$\bigvee_0^1 Q_n f \leqslant \bigvee_0^1 f.$$

Proof Let $Q_n f = \sum_{i=0}^{n} q_i \phi_i$. Then $Q_n f(x_i) = q_i$ for all $i = 0, 1, \cdots, n$. Hence,

$$\begin{aligned}
\bigvee_0^1 Q_n f &= \sum_{i=1}^{n} |q_i - q_{i-1}| \\
&= \left|\frac{1}{2}(\hat{f}_1 + \hat{f}_2) - \hat{f}_1\right| + \sum_{i=2}^{n-1}\left|\frac{1}{2}(\hat{f}_i + \hat{f}_{i+1}) - \frac{1}{2}(\hat{f}_{i-1} + \hat{f}_i)\right| \\
&\quad + \left|\hat{f}_n - \frac{1}{2}(\hat{f}_{n-1} + \hat{f}_n)\right| \\
&= \frac{1}{2}\left\{|\hat{f}_2 - \hat{f}_1| + \sum_{i=2}^{n-1}|\hat{f}_{i+1} - \hat{f}_{i-1}| + |\hat{f}_n - \hat{f}_{n-1}|\right\} \\
&\leqslant \frac{1}{2}\left\{|\hat{f}_2 - \hat{f}_1| + \sum_{i=2}^{n-1}\left[|\hat{f}_{i+1} - \hat{f}_i| + |\hat{f}_i - \hat{f}_{i-1}|\right] + |\hat{f}_n - \hat{f}_{n-1}|\right\} \\
&= \sum_{i=1}^{n-1}|\hat{f}_{i+1} - \hat{f}_i| \leqslant \bigvee_0^1 f.
\end{aligned}$$

The last inequality above is from Proposition 6.1.2 (iv). □

Proposition 6.3.4 *If $f \in L^1(0,1)$ is monotonically increasing or decreasing, then $Q_n f$ is also monotonically increasing or decreasing, respectively. Moreover $\|Q_n f\|_\infty \leqslant \|f\|_\infty$ if $f \in L^\infty(0,1)$.*

Proof Suppose that $f \in L^1(0,1)$ is monotonically increasing. Since $Q_n f(x_i) = (\hat{f}_i + \hat{f}_{i+1})/2$ for $0 < i < n$, $Q_n f(x_0) = \hat{f}_1$, and $Q_n f(x_n) = \hat{f}_n$, the finite number sequence

$$\hat{f}_1, \frac{\hat{f}_1 + \hat{f}_2}{2}, \cdots, \frac{\hat{f}_{n-1} + \hat{f}_n}{2}, \hat{f}_n$$

is monotonically increasing. Since $Q_n f$ is a continuous piecewise linear function, $Q_n f$ is monotonically increasing. Similarly we can show that $Q_n f$ is monotonically decreasing if f is monotonically decreasing. The last conclusion of the proposition is obvious. □

Now we show that $\lim_{n\to\infty} Q_n f = f$ for any $f \in BV(0,1)$ under the variation norm $\|\cdot\|_{BV}$ defined by (2.6), a property that is *not* shared by Ulam's method in general. We need two additional lemmas.

Lemma 6.3.3 *Let $f \in C^2[0,1]$ and let $\{L_n\}$ be the sequence of the Lagrange interpolation operators as defined in Lemma* 6.3.1. *Then*

$$\bigvee_0^1 (f - L_n f) \leqslant h \max_{x\in[0,1]} |f''(x)| = O\left(\frac{1}{n}\right).$$

Proof Denote $C = \max_{x\in[0,1]} |f''(x)|$. For each $i = 1, 2, \cdots, n$, consider the function $g(x) = f(x) - L_n(x)$ on $[x_{i-1}, x_i]$. Since $g(x_{i-1}) = g(x_i) = 0$, Rolle's theorem in calculus ensures that there is $z \in [x_{i-1}, x_i]$ such that $g'(z) = 0$. Fix $x \in [x_{i-1}, x_i]$. Without loss of generality, assume that $z \leqslant x$. Then from $g''(x) = f''(x)$, we have

$$|g'(x)| = \left|\int_z^x g''(t)\mathrm{d}t\right| \leqslant \int_z^x |f''(t)|\mathrm{d}t \leqslant Ch.$$

Thus,

$$\begin{aligned}\bigvee_0^1 (f - L_n f) &= \sum_{i=1}^n \bigvee_{x_{i-1}}^{x_i} g = \sum_{i=1}^n \int_{x_{i-1}}^{x_i} |g'(t)|\mathrm{d}t \\ &\leqslant \sum_{i=1}^n \int_{x_{i-1}}^{x_i} Ch\,\mathrm{d}t = Ch\int_0^1 1\mathrm{d}t = Ch.\end{aligned}$$

□

Lemma 6.3.4 *Let $f \in C^2[0,1]$. Then*

$$\bigvee_0^1 (Q_n f - L_n f) = O\left(\frac{1}{n}\right).$$

Proof Clearly $Q_n f - L_n f = \sum_{i=0}^n [q_i - f(x_i)]\phi_i$, where $q_0 = f_1$, $q_i = (\hat{f}_i + \hat{f}_{i+1})/2$, $i = 1, 2, \cdots, n-1$, and $q_n = \hat{f}_n$. From the fact that $\bigvee_0^1 \phi_i \leqslant 2$ for all i, we have, using Proposition 2.3.1 (ii),

$$\bigvee_0^1 (Q_n f - L_n f) \leqslant \sum_{i=0}^n |q_i - f(x_i)| \bigvee_0^1 \phi_i \leqslant 2\sum_{i=0}^n |q_i - f(x_i)|.$$

Now the Taylor expansion

$$f(x) = f(x_i) + f'(x_i)(x - x_i) + O(h^2)$$

gives that $|q_0 - f(0)| = O(h)$, $|q_n - f(1)| = O(h)$, and

$$|q_i - f(x_i)| = O(h^2), \quad i = 1, 2, \cdots, n-1.$$

Hence,

$$\bigvee_0^1 (Q_n f - L_n f) = O(h) + \sum_{i=1}^{n-1} O(h^2) + O(h) = O(h).$$

□

Proposition 6.3.5 *Let $f \in C^2[0,1]$. Then*

$$\|f - Q_n f\|_{BV} = O\left(\frac{1}{n}\right).$$

Proof Since $\|f - Q_n f\| = O(h^2)$ from (6.13), the proposition follows from Lemmas 6.3.3 and 6.3.4 together with

$$\bigvee_0^1 (f - Q_n f) \leqslant \bigvee_0^1 (f - L_n f) + \bigvee_0^1 (Q_n f - L_n f) = O(h). \qquad \square$$

Remark 6.3.1 Unless f is a piecewise constant function associated with the partition of $[0, 1]$, for Ulam's method,

$$\bigvee_0^1 (f - Q_n f) \geqslant \bigvee_0^1 f$$

in general, where Q_n is the discretized operator corresponding to Ulam's method. Thus, we cannot expect the convergence of Ulam's method under the BV-norm (see [11, 41]).

Remark 6.3.2 Propositions 6.3.2 and 6.3.5 are still true under a weaker assumption on f, for example, it is enough to assume that $f \in W^{2,1}(0, 1)$.

Let $P_n = Q_n P$. Then $P_n : L^1(0, 1) \to L^1(0, 1)$ is a Markov operator of finite dimensional range, and $\lim_{n\to\infty} P_n f = Pf$ under the L^1-norm for any $f \in L^1(0, 1)$. The matrix representation of P_n restricted to Δ_n under any basis consisting of density functions is given by a stochastic matrix, hence P_n has a stationary density $f_n \in \Delta_n$. This proves

Proposition 6.3.6 *$P_n : \Delta_n \to \Delta_n$ has a stationary density $f_n \in \Delta_n$ for any n.*

Proposition 6.3.6 shows that the piecewise linear Markov approximation method, which was first proposed in [28] in which a stationary density $f_n \in \Delta_n$ is computed for a given n to approximate a stationary density of the Frobenius-Perron operator P, is well-posed. Now a natural question is whether this method converges if the fixed point equation $Pf = f$ has a density solution. The proof of the following two results is basically the same as that in Section 6.1 for Ulam's method, so it is omitted; the reader can see [28, 29, 38].

Theorem 6.3.1 *Under the same conditions as in Theorem* 6.1.1, *if f^* is the unique stationary density of P, then for any sequence of the stationary densities f_n of P_n in Δ_n,*

$$\lim_{n\to\infty} \|f_n - f^*\| = 0.$$

Remark 6.3.3 In the next chapter, a stronger convergence result under the BV-norm can also be established for the Markov method.

Theorem 6.3.2 *Let S be as in Theorem* 6.1.2. *Then for any n, there is a stationary density of P_n which is monotonically decreasing, and for any sequence of the monotonically decreasing stationary densities f_n of P_n in Δ_n,*

$$\lim_{n\to\infty} \|f_n - f^*\| = 0,$$

where f^* is the unique stationary density of P, which is monotonically decreasing.

6.4 The Markov Method for N-dimensional Transformations

Now we construct continuous piecewise linear Markov finite approximations of Frobenius-Perron operators $P : L^1(\Omega) \to L^1(\Omega)$ associated with a nonsingular transformation $\boldsymbol{S} : \Omega \to \Omega$, where $\Omega \subset \mathbb{R}^N$ is an N-dimensional polygonal region.

Let $h > 0$ be a positive discretization parameter and let $\mathcal{T}_h$ be a *simplicial triangulation* of Ω with $h_e \equiv \text{diam } e \leqslant h$ for all simplices e of $\mathcal{T}_h$. We also assume that $\mathcal{T}_h$ is *shape-regular*, namely, there exists a constant $\gamma > 0$, independent of h, such that

$$\frac{\text{diam } e}{\text{diam } B_e} \leqslant \gamma, \ \ \forall\, e \in \mathcal{T}_h,$$

where B_e is the ball inscribed in e.

Associated with each $\mathcal{T}_h$ let $\Delta_h \subset W^{1,1}(\Omega)$ be the corresponding subspace of $L^1(\Omega)$ that consists of all *continuous piecewise linear functions* defined on Ω. Let $\{\boldsymbol{v}_1, \boldsymbol{v}_2, \cdots, \boldsymbol{v}_d\}$ ($d \equiv d_h$ depends on h) be the collection of all the vertices (called the *nodes*) of $\mathcal{T}_h$, and let $\{e_1, e_2, \cdots, e_l\}$ ($l \equiv l_h$ depends on h) be the set of all simplices of $\mathcal{T}_h$. For each node $\boldsymbol{v}$ in $\mathcal{T}_h$ let $\tau_{\boldsymbol{v}}$ denote the number of the simplices of $\mathcal{T}_h$ with $\boldsymbol{v}$ as a vertex. For each simplex $e \in \mathcal{T}_h$ let its vertices be $\{\boldsymbol{q}_0, \boldsymbol{q}_1, \cdots, \boldsymbol{q}_N\}$. Denote by ϕ_i the unique element in Δ_h such that

$$\phi_i(\boldsymbol{v}_j) = \delta_{ij}, \ \ i, j = 1, 2, \cdots, d,$$

where the *Kronecker symbol* $\delta_{ij} = 1$ for $i = j$ and $\delta_{ij} = 0$ otherwise. Then the set $\{\phi_i\}_{i=1}^d$ forms the *canonical basis* of Δ_h. Let $V_i = \text{supp } \phi_i$ be the support of ϕ_i for $i = 1, 2, \cdots, d$. Then V_i is the union of all the τ_i simplices of $\mathcal{T}_h$ that have $\boldsymbol{v}_i$ as a vertex. Note that

$$\|\phi_i\| = \frac{1}{N+1} m(V_i) \tag{6.14}$$

for each i. Moreover,

$$\sum_{i=1}^{d} \phi_i(\boldsymbol{x}) \equiv 1, \ \ \forall\, \boldsymbol{x} \in \Omega. \tag{6.15}$$

Furthermore, for each $g \in \Delta_h$,

$$g = \sum_{i=1}^{d} g(\boldsymbol{v}_i)\phi_i.$$

With the given triangulation $\mathcal{T}_h$ of Ω, we define a linear operator $Q_h : L^1(\Omega) \to L^1(\Omega)$ by

$$Q_h f = \sum_{i=1}^{d} \left(\frac{1}{m(V_i)} \int_{V_i} f \mathrm{d}m \right) \phi_i. \tag{6.16}$$

Note that Q_h is exactly the piecewise linear Markov approximation Q_n defined by (6.12) with $h = 1/n$ when $n = 1$.

Remark 6.4.1 In the literature of finite element methods, the operator Q_h in (6.16) is called the *Clément operator*, though its domain may not be the L^1-space. (See [21]).

Proposition 6.4.1 *$Q_h : L^1(\Omega) \to L^1(\Omega)$ is a Markov operator. Hence $\|Q_h\| = 1$.*

Proof It is clear that Q_h is a positive operator with the range $R(Q_h) = \Delta_h$. Given $f \in L^1(\Omega)$ and by (6.14), the following is obtained:

$$\begin{aligned} \int_\Omega Q_h f \mathrm{d}m &= \sum_{i=1}^{d} \frac{1}{m(V_i)} \int_{V_i} f \mathrm{d}m \int_\Omega \phi_i \mathrm{d}m \\ &= \frac{1}{N+1} \sum_{i=1}^{d} \int_{V_i} f \mathrm{d}m = \int_\Omega f \mathrm{d}m, \end{aligned}$$

where the last equality is valid since each simplex has exactly $N+1$ vertices. Thus, Q_h preserves the integral of functions and is a Markov operator. Consequently, $\|Q_h\| = 1$. □

Proposition 6.4.2 *There exists a constant C which is independent of h such that*

$$\|Q_h f - f\| \leqslant C h V(f; \Omega), \quad \forall\, f \in W^{1,1}(\Omega). \tag{6.17}$$

Proof Denote by $\hat{f}_i = \int_{V_i} f \mathrm{d}m / m(V_i)$ the average value of f over V_i for each $i = 1, 2, \cdots, d$. Then, by (7.45) of [61],

$$\int_{V_i} |f - \hat{f}_i| \mathrm{d}m \leqslant \left(\frac{\omega_N}{m(V_i)} \right)^{1 - \frac{1}{N}} (\operatorname{diam} V_i)^N \int_{V_i} \|\mathbf{grad}\, f\| \mathrm{d}m,$$

where $\omega_N = 2\pi^{N/2}/(N\Gamma(N/2))$ is the *volume* of the unit closed ball in $\mathbb{R}^N$. Since $\mathcal{T}_h$ is shape-regular, $\operatorname{diam} V_i = O(h)$ and $m(V_i) = O(h^N)$ uniformly with respect to i. Hence, there is a constant C which is independent of h such that

$$\int_{V_i} |f - \hat{f}_i| \mathrm{d}m \leqslant C h \int_{V_i} \|\mathbf{grad}\, f\| \mathrm{d}m.$$

On the other hand, by (6.15),

$$|Q_h f(\boldsymbol{x}) - f(\boldsymbol{x})| = \left|\sum_{i=1}^{d} f_i \phi_i(\boldsymbol{x}) - \sum_{i=1}^{d} f(\boldsymbol{x})\phi_i(\boldsymbol{x})\right| \leqslant \sum_{i=1}^{d} |f_i - f(\boldsymbol{x})|\phi_i(\boldsymbol{x}),$$

and it follows that

$$\begin{aligned}\|Q_h f - f\| &\leqslant \frac{1}{N+1}\sum_{i=1}^{d}\int_{V_i}|f - f_i|\phi_i \mathrm{d}m \leqslant \frac{1}{N+1}\sum_{i=1}^{d}\int_{V_i}|f - f_i|\mathrm{d}m \\ &\leqslant \frac{1}{N+1}\sum_{i=1}^{d} Ch\int_{V_i}\|\mathbf{grad}\ f\|\mathrm{d}m = Ch\int_{\Omega}\|\mathbf{grad}\ f\|\mathrm{d}m. \qquad \square\end{aligned}$$

We can strengthen Proposition 6.4.2 by proving the *consistency* result for the family $\{Q_h\}$ of Markov finite approximations under the $W^{1,1}$-norm, provided that the triangulation $\mathcal{T}_h$ of Ω is *uniform*, that is, all the simplicies of $\mathcal{T}_h$ are *congruent*. For this purpose we are in need of the following lemma.

Lemma 6.4.1 *Let e be a simplex in $\mathbb{R}^N$ with vertices $\boldsymbol{q}_0, \boldsymbol{q}_1, \cdots, \boldsymbol{q}_N$, and let $\boldsymbol{U}$ be the $N \times N$ matrix with its jth column $\boldsymbol{u}_j = \boldsymbol{q}_j - \boldsymbol{q}_{j-1}$ for $j = 1, 2, \cdots, N$. If $g(\boldsymbol{x}) = \boldsymbol{a}^{\mathrm{T}}\boldsymbol{x} + b$ on e, where $\boldsymbol{a} \in \mathbb{R}^N$ and b is a real number, then*

$$\boldsymbol{a} = (\boldsymbol{U}^{\mathrm{T}})^{-1}\boldsymbol{v}, \tag{6.18}$$

where $\boldsymbol{v} = (v_1, v_2, \cdots, v_N)^{\mathrm{T}}$ with $v_i = g(\boldsymbol{q}_i) - g(\boldsymbol{q}_{i-1})$ for each i.

Proof Since $g(\boldsymbol{q}_j) = \boldsymbol{a}^{\mathrm{T}}\boldsymbol{q}_j + b$ for each j,

$$\boldsymbol{a}^{\mathrm{T}}\boldsymbol{u}_j = \boldsymbol{a}^{\mathrm{T}}(\boldsymbol{q}_j - \boldsymbol{q}_{j-1}) = g(\boldsymbol{q}_j) - g(\boldsymbol{q}_{j-1}) = v_j,\ j = 1, 2, \cdots, N.$$

In other words, $\boldsymbol{a}^{\mathrm{T}}\boldsymbol{U} = \boldsymbol{v}^{\mathrm{T}}$. Since the simplex e is N-dimensional, $\boldsymbol{U}$ is nonsingular, and so (6.18) is valid. $\square$

Theorem 6.4.1 *If the triangulation $\mathcal{T}_h$ is uniform, then, as $h \to 0$,*

$$\|Q_h f - f\|_{BV} = O(h),\ \ \forall\, f \in C^2(\overline{\Omega}), \tag{6.19}$$

$$\|Q_h f - f\|_{BV} = o(1),\ \ \forall\, f \in W^{1,1}(\Omega). \tag{6.20}$$

Proof Let $f \in C^2(\overline{\Omega})$. As before let $\mathcal{T}_h$ consist of l simplices $\{e_1, e_2, \cdots, e_l\}$ such that each e_k has the vertices $\{\boldsymbol{q}_0^k, \boldsymbol{q}_1^k, \cdots, \boldsymbol{q}_N^k\}$. For each e_k let $\boldsymbol{U}_k$ be the corresponding matrix and $\boldsymbol{v}(k)$ the corresponding vector as in Lemma 6.4.1. Then $\mathbf{grad}\ Q_h f = (\boldsymbol{U}_k^{\mathrm{T}})^{-1}\boldsymbol{v}(k)$ on e_k. Note that $\|(\boldsymbol{U}_k^{\mathrm{T}})^{-1}\| \leqslant Ch^{-1}$ uniformly for $h > 0$ for some constant C. Since $Q_h f - f \in W^{1,1}(\Omega)$, by Lemma 6.4.1,

$$\begin{aligned}V(Q_h f - f, \Omega) &= \sum_{k=1}^{l}\int_{e_k}\|\mathbf{grad}\ (Q_h f - f)\|\mathrm{d}m \\ &= \sum_{k=1}^{l}\int_{e_k}\|(\boldsymbol{U}_k^{\mathrm{T}})^{-1}\boldsymbol{v}(k) - \mathbf{grad}\ f\|\mathrm{d}m\end{aligned}$$

$$\leqslant \sum_{k=1}^{l} \int_{e_k} \|(\boldsymbol{U}_k^{\mathrm{T}})^{-1}\| \, \|\boldsymbol{v}(k) - \boldsymbol{U}_k^{\mathrm{T}} \mathbf{grad}\, f\| \mathrm{d}m$$

$$\leqslant \sum_{k=1}^{l} Ch^{-1} \int_{e_k} \|\boldsymbol{v}(k) - \boldsymbol{U}_k^{\mathrm{T}} \mathbf{grad}\, f\| \mathrm{d}m \equiv \sum_1 + \sum_2, \tag{6.21}$$

where $\sum_1$ is the sum of the integrals over all the simplices e_k each of whose vertices is interior to Ω and $\sum_2$ is the remaining sum. Fix $k = 1, 2, \cdots, l$ and let $\boldsymbol{x} \in e_k$. Denote $Q_h f(\boldsymbol{q}_i^k) = \eta_i^k$ for $i = 0, 1, \cdots, N$. Then, after a simple calculation,

$$\boldsymbol{\alpha}_k - \boldsymbol{U}_k^{\mathrm{T}} \mathbf{grad}\, f(\boldsymbol{x}) = \begin{bmatrix} \eta_1^k - \eta_0^k \\ \vdots \\ \eta_N^k - \eta_{N-1}^k \end{bmatrix} - \begin{bmatrix} f'(x)(\boldsymbol{q}_1^k - \boldsymbol{q}_0^k) \\ \vdots \\ f'(x)(\boldsymbol{q}_N^k - \boldsymbol{q}_{N-1}^k) \end{bmatrix},$$

where $f'(\boldsymbol{x}) = (\mathbf{grad}\, f(\boldsymbol{x}))^{\mathrm{T}}$ is the Frechét derivative of f at $\boldsymbol{x}$. Let e_k be a simplex in $\sum_1$. Using Taylor's expansion, we get for each i

$$f(\boldsymbol{x}) = f(\boldsymbol{q}_i^k) + f'(\boldsymbol{q}_i^k)(\boldsymbol{x} - \boldsymbol{q}_i^k) + O(h^2),$$

from which it follows that

$$\eta_i^k = f(\boldsymbol{q}_i^k) + O(h^2)$$

because $\int_{V_i^k} f'(\boldsymbol{q}_i^k)(\boldsymbol{x} - \boldsymbol{q}_i^k)\mathrm{d}m = 0$ due to the fact that V_i^k is symmetric about $\boldsymbol{q}_i^k$ because the triangulation $\mathcal{T}_h$ is uniform. Hence the above equality combined with the Taylor expansion

$$f(\boldsymbol{q}_i^k) = f(\boldsymbol{x}) + f'(\boldsymbol{x})(\boldsymbol{q}_i^k - \boldsymbol{x}) + O(h^2)$$

gives that, for each i,

$$\eta_i^k - \eta_{i-1}^k = f'(\boldsymbol{x})(\boldsymbol{q}_i^k - \boldsymbol{q}_{i-1}^k) + O(h^2),$$

and so

$$\|\boldsymbol{\alpha}_k - \boldsymbol{U}_k^{\mathrm{T}} \mathbf{grad}\, f\| = O(h^2).$$

It follows that

$$\sum_1 = Ch^{-1} \sum_1 \int_{e_k} \|\boldsymbol{v}(k) - \boldsymbol{U}_k^{\mathrm{T}} \mathbf{grad}\, f\| \mathrm{d}m$$

$$= Ch^{-1} \sum_{k=1}^{l} \int_{e_k} O(h^2)\mathrm{d}m = O(h).$$

On the other hand, since the Lebesgue measure of the union of all the simplices in $\sum_2$ is of order $O(h)$ and since the integrand is bounded, $\sum_2 = O(h)$. Thus, from the decomposition (6.21) we see that (6.19) is true, which implies (6.20) with the help of Theorem 6.4.2 below and a dense set argument. □.

Remark 6.4.2 Theorem 6.4.1 strengthens Proposition 6.4.2 under a mild smoothness condition on f. It should be noted that although it satisfies (6.17), Ulam's method does not satisfy (6.19), which makes the piecewise linear Markov finite approximation method more appealing in the numerical analysis of Frobenius-Perron operators.

We can establish a stability result for the family $\{V(Q_h; \Omega)\}$, the proof of which is based on the following lemma whose proof is referred to [125].

Lemma 6.4.2 *Given $f, g \in \Delta_h$ and $e \in \mathcal{T}_h$ with vertices $\boldsymbol{q}_0, \boldsymbol{q}_1, \cdots, \boldsymbol{q}_N$, we have*

$$\int_e (\mathbf{grad}\ f)^{\mathrm{T}}\ \mathbf{grad}\ g\ \mathrm{d}m = -\sum_{i<j} a_{ij}^e (f(\boldsymbol{q}_i) - f(\boldsymbol{q}_j))(g(\boldsymbol{q}_i) - g(\boldsymbol{q}_j)),$$

where (a_{ij}^e) is the $(N+1) \times (N+1)$ element stiffness matrix.

Let $w_{ij}^e = -a_{ij}^e$ for $i, j = 0, 1, \cdots, N$. Then the above lemma implies the following one.

Lemma 6.4.3 *Let $g \in \Delta_h$ and $e \in \mathcal{T}_h$. Then*

$$V(g; e) = \sqrt{m(e) \sum_{i<j} w_{ij}^e (g(\boldsymbol{q}_i) - g(\boldsymbol{q}_j))^2}. \tag{6.22}$$

Proof Since g is linear on e, it is obvious that

$$V(g; e) = m(e) \|\mathbf{grad}\ g\|.$$

Putting $f = g$ in Lemma 6.4.2 gives that

$$m(e)\|\mathbf{grad}\ g\|^2 = \int_e \|\mathbf{grad}\ g\|^2 \mathrm{d}m = \sum_{i<j} w_{ij}^e (g(\boldsymbol{q}_i) - g(\boldsymbol{q}_j))^2,$$

from which we have

$$(m(e)\|\mathbf{grad}\ g\|)^2 = m(e) \sum_{i<j} w_{ij}^e (g(\boldsymbol{q}_i) - g(\boldsymbol{q}_j))^2.$$

Thus (6.22) follows by taking square root of both sides of the above equality. □

Now let $Q_hf \in \Delta_h$ be the piecewise linear Markov approximation of $f \in L^1(\Omega)$, as defined by (6.16). Then $Q_hf \in W^{1,1}(\Omega)$, so

$$V(Q_hf;\Omega) = \sum_{k=1}^{l} \sqrt{m(e_k)\sum_{i<j} w_{ij}^{e_k}(Q_hf(\boldsymbol{q}_i^k) - Q_hf(\boldsymbol{q}_j^k))^2} \tag{6.23}$$

by Lemma 6.4.3, where $e_1, e_2, \cdots, e_l$ are all the simplices of $\mathcal{T}_h$ with vertices $\boldsymbol{q}_0^k, \boldsymbol{q}_1^k, \cdots, \boldsymbol{q}_N^k$ of e_k for $k = 1, 2, \cdots, l$. Let $Q_hf(\boldsymbol{q}_i^k) = \eta_i^k$, let V_i^k be the union of all the simplices of $\mathcal{T}_h$ with vertex $\boldsymbol{q}_i^k$, and denote $w_{ij}^{e_k}$ by w_{ij}^k. Since $\mathcal{T}_h$ is shape-regular, there is a constant C, which is independent of h, such that $m(e_k) \leqslant Ch^N$ for each k. Therefore, from (6.23),

$$\begin{aligned} V(Q_hf;\Omega) &= \sum_{k=1}^{l} \sqrt{m(e_k)\sum_{i<j} w_{ij}^k(\eta_i^k - \eta_j^k)^2} \\ &\leqslant Ch^{\frac{N}{2}} \sum_{k=1}^{l} \sqrt{\sum_{i<j} w_{ij}^k(\eta_i^k - \eta_j^k)^2} \\ &\leqslant Ch^{\frac{N}{2}} \sum_{k=1}^{l}\sum_{i<j} \sqrt{w_{ij}^k}|\eta_i^k - \eta_j^k|. \end{aligned} \tag{6.24}$$

□

In the proof of the following theorem, the same symbol C may represent different constants in different places.

Theorem 6.4.2 *There is a constant C_{BV} which is independent of h such that*

$$V(Q_hf;\Omega) \leqslant C_{BV}V(f;\Omega), \quad \forall\, f \in BV(\Omega). \tag{6.25}$$

Proof It is enough to prove formula (6.25) for $f \in W^{1,1}(\Omega)$ since the formula for the general case is from Theorem 2.4.2 by a limit process. As in the proof of Proposition 6.4.2, with the help of (7.45) in the book [61], we get

$$\begin{aligned} &|\eta_i^k - \eta_j^k| \\ &= \left|\frac{1}{m(V_i^k)}\int_{V_i^k} f\mathrm{d}m - \eta_j^k\right| \leqslant \frac{1}{m(V_i^k)}\int_{V_i^k} |f(\boldsymbol{x}) - \eta_j^k|\mathrm{d}m(\boldsymbol{x}) \\ &\leqslant \frac{1}{m(V_i^k)}\int_{V_i^k\cup V_j^k} |f(\boldsymbol{x}) - \eta_j^k|\mathrm{d}m(\boldsymbol{x}) \\ &\leqslant \frac{1}{m(V_i^k)}\left(\frac{\omega_N}{m(V_j^k)}\right)^{1-\frac{1}{N}}\left(\mathrm{diam}\, V_i^k \bigcup V_j^k\right)^N \int_{V_i^k\cup V_j^k} \|\mathbf{grad}\, f\|\mathrm{d}m \\ &\leqslant Ch^{1-N}\int_{V_i^k\cup V_j^k} \|\mathbf{grad}\, f\|\mathrm{d}m. \end{aligned}$$

The above inequality and (6.24) imply that

$$V(Q_h f; \Omega) \leqslant C h^{1-\frac{N}{2}} \sum_{k=1}^{l} \sum_{i<j} \sqrt{w_{ij}^k} \int_{V_i^k \cup V_j^k} \|\mathbf{grad}\, f\| dm. \tag{6.26}$$

It has been shown in [125] that

$$w_{ij}^k = \frac{1}{N(N-1)} H(\xi_{ij}^k) \cot \theta_{ij}^k, \tag{6.27}$$

where ξ_{ij}^k is the $(N-2)$-dimensional face of the simplex e_k opposite to the edge that connects the vertices $\boldsymbol{q}_i^k$ and $\boldsymbol{q}_j^k$, $H(\xi_{ij}^k)$ is the $(N-2)$-dimensional Hausdorff measure of ξ_{ij}^k, and θ_{ij}^k is the angle between the $(N-1)$-dimensional face opposite to $\boldsymbol{q}_i^k$ and the $(N-1)$-dimensional face opposite to $\boldsymbol{q}_j^k$. It follows from (6.26) and (6.27) that

$$V(Q_h f; \Omega) \leqslant \frac{C}{\sqrt{N(N-1)}} h^{1-\frac{N}{2}} \sum_{k=1}^{l} \sum_{i<j} \sqrt{H(\xi_{ij}^k) \cot \theta_{ij}^k} \int_{V_i^k \cup V_j^k} \|\mathbf{grad}\, f\| dm.$$

Let $A^k = \bigcup_{i=0}^{N} V_i^k$. Since the number of the pairs (i,j) with $i<j$ is $N(N+1)/2$, and note that $\mathcal{T}_h$ is shape-regular implies that

$$|\kappa_{ij}^k| \leqslant C h^{N-2},$$

we have

$$V(Q_h f; \Omega) \leqslant C \sum_{k=1}^{l} \int_{A^k} \|\mathbf{grad}\, f\| dm \leqslant C_{BV} V(f; \Omega)$$

for a constant C_{BV}. This completes the proof. □

With the help of the stability result in Theorem 6.4.2, we are able to prove the convergence of the piecewise linear Markov method for the Góra-Boyarsky class of multi-dimensional transformations, as was done before in Section 6.2 for Ulam's method.

Theorem 6.4.3 *Suppose that* $\boldsymbol{S} : \Omega \to \Omega$ *satisfies the conditions of Theorem* 5.4.1. *If in addition*

$$\frac{C_{BV}(1+\kappa_\Omega(\boldsymbol{S}))}{\lambda} < 1,$$

where $\kappa_\Omega(S)$ *and* λ *are given in Theorem* 5.4.1 *and the constant* C_{BV} *is as in Theorem* 6.4.2, *then for any sequence of the stationary densities* f_{h_n} *of* P_{h_n} *in* Δ_{h_n} *with the condition that* $\lim_{n\to\infty} h_n = 0$, *there is a subsequence* $\{f_{h_{n_k}}\}$ *such that*

$$\lim_{k\to\infty} \|f_{h_{n_k}} - f^*\| = 0,$$

where f^* *is a stationary density of* P. *If in addition* f^* *is the unique stationary density of* P, *then*

$$\lim_{n\to\infty} \|f_{h_n} - f^*\| = 0.$$

Exercises

6.1 Let $\boldsymbol{A}$ be an $n \times n$ nonnegative matrix. Show that the spectral radius of $\boldsymbol{A}$ is an eigenvalue of $\boldsymbol{A}$.

6.2 Let $\boldsymbol{A}$ be a stochastic matrix. Show that its spectral radius is 1. Find a right eigenvector of $\boldsymbol{A}$ corresponding to eigenvalue 1, and show that $\boldsymbol{A}$ has a nonnegative left eigenvector associated with eigenvalue 1.

6.3 If an $n \times n$ matrix $\boldsymbol{P}$ defines a Markov operator from $\mathbb{R}^n$ into itself under the vector 1-norm of the Euclidean space $\mathbb{R}^n$. Show that $\boldsymbol{P}$ is a *column stochastic matrix*, that is, $\boldsymbol{P}$ is a nonnegative matrix with each column sums 1.

6.4 Prove Proposition 6.2.1.

6.5 Let ϕ_i, $i = 0, 1, \cdots, n$ be the canonical basis of the subspace Δ_n of all the continuous piecewise linear functions associated with the equal partition $0 = x_0 < x_1 < \cdots < x_{n-1} < x_n = 1$ with $x_i = ih$, where $h = 1/n$, $\forall\, i$. Define $Q_n : L^1(0,1) \to L^1(0,1)$ by

$$Q_n f = \frac{2}{h}\int_0^{\frac{h}{2}} f\mathrm{d}m \cdot \phi_0 + \sum_{i=1}^{n-1} \frac{1}{h}\int_{(i-\frac{1}{2})h}^{(i+\frac{1}{2})h} f\mathrm{d}m \cdot \phi_i + \frac{2}{h}\int_{1-\frac{h}{2}}^{1} f\mathrm{d}m \cdot \phi_n.$$

Prove that Q_n is a Markov operator. Compare this Q_n with that defined in Section 6.3.

6.6 Let $Q_n : L^1(0,1) \to \Delta_n$ be a bounded linear operator, where Δ_n is as in the previous exercise, so that Q_n can be represented as

$$Q_n f = \sum_{i=0}^{n} \int_0^1 f(x) w_i(x)\mathrm{d}x \cdot \phi_i,$$

where $w_i \in L^\infty(0,1)$ for all i. Show that

(i) $Q_n 1 = 1$ if and only if $\int_0^1 w_i(x)\mathrm{d}x = 1$ for $i = 0, 1, \cdots, n$.

(ii) Q_n is nonnegative if and only if $w_i(x) \geqslant 0$ a.e. for all i.

(iii) Q_n preserves integrals if and only if $w_0(x) + 2\sum_{i=1}^{n-1} w_i(x) + w_n(x) = 2n$ i.e.

(iv) Q_n is a Markov operator if and only if $w_i(x) \geqslant 0$ a.e. for all i and $w_0(x) + 2\sum_{i=1}^{n-1} w_i(x) + w_n(x) = 2n$ a.e.

6.7 Let Q be the Galerkin projection of $L^1(0,1)$ onto the subspace of linear polynomial functions on $[0,1]$. That is, for all $f \in L^1(0,1)$,

$$\int_0^1 (Qf(x) - f(x))\mathrm{d}x = 0, \quad \int_0^1 (Qf(x) - f(x))x\mathrm{d}x = 0.$$

Show that Q is not a Markov operator.

6.8 Let $Q_n : L^1(0,1) \to \Delta_n$ be defined by

$$Q_n f = \sum_{i=0}^{n} \frac{1}{m(F_i)} \int_{F_i} f(x)\mathrm{d}x \cdot \phi_i, \quad \forall\, f \in L^1(0,1),$$

where each F_i is a closed subinterval of the closure of supp ϕ_i. Show that if Q_n is a Markov operator, then Q_n is given by either (6.12) or the expression in Exercise 6.5.

Hint: You may first show that if Q_n is integral preserving and if B is a closed subinterval of $[x_k, x_{k+1}]$ and $0 \leqslant k < n$, then

$$m(B) = \frac{m(B \cap F_k)}{m(F_k)} \frac{m(\text{supp } \phi_k)}{2} + \frac{m(B \cap F_{k+1})}{m(F_{k+1})} \frac{m(\text{supp } \phi_{k+1})}{2}.$$

Then show that if $m(F_i \cap F_{i+1}) = 0$ for $i = 0, 1, \cdots, n-1$, then Q_n is as given in Exercise 6.5, and if $F_k = \text{supp}\phi_k$ for some $0 < k < n$, then Q_n is given by (6.12).

6.9 Let (X, Σ, μ) be a probability measure space and let $\psi_1, \psi_2, \cdots, \psi_n$ be linearly independent and nonnegative measurable functions on X such that $\sum_{i=1}^{n} \psi_i = 1$. Define $T_n : L^1(X) \to L^1(X)$ by

$$T_n f = \sum_{i=1}^{n} \int_X f w_i \mathrm{d}\mu \cdot \psi_i,$$

where $w_1, w_2, \cdots, w_n \in L^\infty(X)$. Show that T_n preserves integrals if and only if $\sum_{i=1}^{n} \int_X \psi_i \mathrm{d}\mu \cdot w_i = 1$.

6.10 Under the same assumption as Exercise 6.9, let $A_1, A_2, \cdots, A_n \in \Sigma$ be such that $\mu(A_i) > 0$ for each $i = 1, 2, \cdots, n$. Let $Q_n : L^1(X) \to L^1(X)$ be defined by

$$Q_n f = \sum_{i=1}^{n} \frac{1}{\mu(A_i)} \int_{A_i} f \mathrm{d}\mu \cdot \psi_i.$$

Assume that $\mu(A_i \cap A_j) = 0$ if $i \neq j$. Show that Q_n is a Markov operator if and only if $\int_X \psi_i \mathrm{d}\mu = \mu(A_i)$ for each $i = 1, 2, \cdots, n$.

Chapter 7
Convergence Rate Analysis

JiuDing, Department of Mathematics, The University of Southern Mississippi in America.
Aihui Zhou, Academy of Mathematics and Systems Science, Chinese Academy of Sciences.

Abstract After showing the convergence of the two numerical methods for Frobenius-Perron operators in the previous chapter, we further investigate the convergence rate problem for them. Keller's stochastic stability result for a class of Markov operators will be studied first, which leads to his first proof of the L^1-norm convergence rate $O(\ln n/n)$ for Ulam's method applied to the Lasota-Yorke class of mappings. Then we introduce Murrary's work on an explicit upper bound of the convergence rate for Ulam's method. The convergence rate analysis for the piecewise linear Markov method under the BV-norm will be presented in the last section.
Keywords Convergence rate, stochastic stability, cone of uniformly bounded variation, difference norm, Lasota-Yorke type inequality, Cauchy integral of linear operator.

If a numerical method is designed to produce a convergent approximation sequence for the computation of a stationary density of Frobenius-Perron operators or Markov operators, a natural question, which is of practical importance in applications, is how fast the convergence will be. In this chapter we give some convergence rate analysis to Ulam's piecewise constant method and the piecewise linear Markov method for one-dimensional mappings. In the first section we present Keller's important result [74] on the stochastic stability of Frobenius-Perron operators, from which he proved that Ulam's method has a convergence rate of order $O(\ln n/n)$ under the L^1-norm for the Lasota-Yorke class of interval mappings. In Section 7.2 we introduce Murray's delicate analysis [107] for obtaining an explicit upper bound of the error estimate for Ulam's method applied to a special subclass of Lasota-Yorke's mappings. The last section will focus on the convergence rate analysis for the Markov method under the BV-norm, the approach of which was first proposed in [19] and further investigated in [41].

7.1 Error Estimates for Ulam's Method

Following Li's pioneering work on proving the convergence of Ulam's method for the Lasota-Yorke class of interval mappings, the first serious approach to the convergence rate analysis was given by Keller [74], based on the concept of *stochastic stability* of deterministic dynamical systems that was introduced by him. He obtained the convergence rate of order $O(\ln n/n)$ under the L^1-norm for the Lasota-Yorke class of piecewise monotonic mappings, as an immediate consequence of his general theory. In this section we study Keller's work on the perturbation analysis for a class of Markov operators and its direct application to the problem of estimating the errors for the numerical solutions from Ulam's method.

To present Keller's general result, we assume that a Markov operator $P : L^1(0,1) \to L^1(0,1)$ satisfies the following conditions: $P(BV(0,1)) \subset BV(0,1)$ and $\|P\|_{BV} \equiv \max_{\|f\|_{BV}=1} \|Pf\|_{BV} < \infty$, and there are constants $0 < \alpha < 1$, $\beta > 0$, and a positive integer k such that

$$\|P^k f\|_{BV} \leqslant \alpha \|f\|_{BV} + \beta \|f\|, \quad \forall\, f \in BV(0,1). \tag{7.1}$$

The class of Markov operators P satisfying the above conditions will be denoted by $\mathcal{P}$, and the subclass of $\mathcal{P}$ with the given constants α and β will be denoted as $\mathcal{P}(\alpha,\beta)$. In this section, a sequence $\{P_n\}$ of Markov operators is called $\mathcal{P}$-*bounded* if there exist two constants $0 < \alpha < 1$ and $\beta > 0$ such that $P_n \in \mathcal{P}(\alpha,\beta)$ for all n. For the sake of the spectral analysis of Markov operators, we assume that the involved function spaces are the complex ones.

It follows from Theorem 2.5.4 (the Ionescu-Tulcea and Marinescu theorem) that any operator $P \in \mathcal{P}$ is quasi-compact as a linear operator from the Banach space $(BV(0,1), \|\cdot\|_{BV})$ into itself, so there are only finitely many *peripheral eigenvalues* $1 = \lambda_1, \lambda_2, \cdots, \lambda_r$, i.e., the eigenvalues of modules 1, with each corresponding *eigenspace* E_i finite dimensional (Theorem 2.5.4; see also Theorem VIII.8.3 of [57]). From the theory of positive operators [114], the set $\{\lambda_1, \lambda_2, \cdots, \lambda_r\}$ is *fully cyclic* and hence it is contained in a finite subgroup of the unit circle $\partial\mathbb{D}$ under the product operation. Furthermore, P has the following spectral decomposition

$$P = \sum_{i=1}^{r} \lambda_i \Phi_i + R, \tag{7.2}$$

where each Φ_i is a projection onto E_i with the properties that $\|\Phi_i\| = 1$ and $\Phi_i \Phi_j = 0$ for $i \neq j$, and $R : L^1(0,1) \to L^1(0,1)$ is a bounded linear operator such that $\sup_{n \geqslant 0} \|R^n\| \leqslant r + 1$, $R(BV(0,1)) \subset BV(0,1)$, $R\Phi_i = \Phi_i R = 0$ for each i, and $\|R^n\|_{BV} \leqslant Mq^n$, $\forall\, n$ for some constants $0 < q < 1$ and $M > 0$ (see Theorem 2.5.3).

Let λ be a complex number with $|\lambda| = 1$. Define the operator

$$\Phi(\lambda, P)f = \lim_{n\to\infty} \frac{1}{n} \sum_{j=0}^{n-1} (\bar{\lambda}P)^j f, \quad \forall\, f \in L^1(0,1),$$

where $\bar{\lambda}$ is the conjugate of λ. It can be shown (see Exercise 7.1) that

$$\Phi(\lambda, P) = \begin{cases} \Phi_i, & \text{if } \lambda = \lambda_i, \\ 0, & \text{otherwise.} \end{cases}$$

In particular, $\Phi(1, P) = \Phi_1$.

Lemma 7.1.1 *If $P,\ Q \in \mathcal{P}$ with $P = \sum_{i=1}^{r} \lambda_i \Phi_i + R$ and $\|R^n\|_{BV} \leqslant Mq^n$ for all positive integers n, then for each complex number λ satisfying $|\lambda| = 1$, the following operator*

$$A(\lambda) = \sum_{i=1, \lambda_i \neq \lambda}^{r} \frac{\bar{\lambda}\lambda_i}{\lambda - \lambda_i} \Phi_i + (1 - \bar{\lambda})\Phi(\lambda, P) + (\lambda I - R)^{-1}$$

is well-defined as a bounded linear operator on $BV(0,1)$. Moreover,

(i) $A(\lambda) = (\lambda I - (P - \Phi(\lambda, P)))^{-1}$.

(ii) $(\Phi(\lambda, P) - I)\Phi(\lambda, Q) = A(\lambda)(P - Q)\Phi(\lambda, Q)$.

Proof Since $0 < q < 1$, the fact that the inequality $\|R^n\|_{BV} \leqslant Mq^n$ is satisfied for all n implies that the spectral radius of R is strictly less than 1, so $(\lambda I - R)^{-1}$ exists and equals $\lambda^{-1} \sum_{j=0}^{\infty} (\lambda^{-1} R)^j = \bar{\lambda} \sum_{j=0}^{\infty} (\bar{\lambda} R)^j$ as a bounded linear operator from $BV(0,1)$ onto itself. This implies that $A(\lambda)$ is well-defined as a bounded linear operator from $BV(0,1)$ into itself. Now (i) and (ii) can be easily verified by a direct computation. □

We introduce another norm $|||\cdot|||$ for the given Markov operator $P : L^1(0,1) \to L^1(0,1)$ as the operator norm from $(BV(0,1), \|\cdot\|_{BV})$ into $(L^1(0,1), \|\cdot\|)$, that is, we define

$$|||P||| = \sup\left\{\|Pf\| : f \in BV(0,1), \|f\|_{BV} = 1\right\}.$$

It is obvious to see that $|||P||| \leqslant 1$ and $|||P||| \leqslant \|P\|_{BV}$.

Lemma 7.1.2 *Let $P \in \mathcal{P}$ as in Lemma 7.1.1. Suppose that $Q \in \mathcal{P}(\alpha, \beta)$ and λ is a complex number with $|\lambda| = 1$. Then there are constants C_1 that is independent of λ,*

$$C_2 = \sum_{i=1, \lambda_i \neq \lambda}^{r} \frac{1}{\lambda - \lambda_i} + |1 - \lambda| \cdot \|\Phi(\lambda, P)\|,$$

and $C_3 = \beta/(1-\alpha)$ such that if $|||P - Q||| < 1$, then

$$\|(\Phi(\lambda, P) - I)\Phi(\lambda, Q)\| \leqslant (C_1 + C_2)C_3 |||P - Q||| \left(2 + \frac{\ln |||P - Q|||}{\ln q}\right).$$

Proof The inequality (7.1) implies that for each $f \in BV(0,1)$,

$$\begin{aligned} \|\Phi(\lambda, Q)f\|_{BV} &\leqslant \limsup_{n\to\infty} \frac{1}{n} \sum_{j=0}^{n-1} \|Q^j f\|_{BV} \\ &\leqslant \max_{0 \leqslant i < k} \limsup_{n\to\infty} \|Q^{nk+i} f\|_{BV} \\ &\leqslant \max_{0 \leqslant i < k} \frac{\beta}{1-\alpha} \|Q^i f\| \leqslant C_3 \|f\|. \end{aligned}$$

Now, for any positive integer J, the Neumann series expression $\bar{\lambda}\sum_{j=0}^{\infty}(\bar{\lambda}R)^j$ for $(\lambda I - R)^{-1}$ can be written as

$$(\lambda I - R)^{-1} = \bar{\lambda}\sum_{j=0}^{J-1}(\bar{\lambda}R)^j + (\bar{\lambda}R)^J(\lambda I - R)^{-1}.$$

Hence, for $f \in L^1(0,1)$, from Lemma 7.1.1 (ii) and the spectral decomposition (7.2) of P, we obtain that

$$\begin{aligned}
&\|(\Phi(\lambda,P) - I)\Phi(\lambda,Q)f\| \leqslant \|A(\lambda)(P-Q)\Phi(\lambda,Q)f\| \\
&\leqslant [C_2 + J(r+1)]\,\|(P-Q)\Phi(\lambda,Q)f\| + \frac{Mq^J}{1-q}\|(P-\lambda I)\Phi(\lambda,Q)f\|_{BV} \\
&\leqslant \left[(C_2 + J(r+1))\,|||P-Q||| + \frac{Mq^J}{1-q}\left(\|P\|_{BV}+1\right)\right]\cdot C_3\|f\| \\
&\leqslant (C_1 + C_2)(J|||P-Q||| + q^J)\|f\|
\end{aligned}$$

for some suitable constant C_1 which depends on P only. If we choose

$$J = \frac{\ln |||P-Q|||}{\ln q} + 1,$$

the lemma follows. □

Remark 7.1.1 Denote by $\delta_\lambda(P,Q)$ the smallest right-hand side of the conclusion in Lemma 7.1.2 for the given P, Q, and λ. Then Lemma 7.1.2 implies that if $P \in \mathcal{P}$ and $\{P_n\}$ is a $\mathcal{P}$-bounded sequence, then

$$\delta_\lambda(P,P_n) = O\left(|||P-P_n|||\cdot\left|\ln|||P-P_n|||\right|\right).$$

Proposition 7.1.1 *Let P, $Q \in \mathcal{P}$. If P has a unique stationary density, then*

$$\Phi(1,P) - \Phi(1,Q) = (I - (P - \Phi(1,P)))^{-1}(P-Q)\Phi(1,Q)$$

and

$$\|\Phi(1,P) - \Phi(1,Q)\| \leqslant \delta_1(P,Q).$$

Proof Since P preserves integrals and dim $E_1 = 1$, it is clear that

$$\Phi(1,P)f = \int_0^1 f(x)\mathrm{d}x\cdot\Phi(1,P)1$$

for any $f \in L^1(0,1)$. Thus,

$$\begin{aligned}
\Phi(1,P)\Phi(1,Q)f &= \int_0^1 \Phi(1,Q)f(x)\mathrm{d}x\cdot\Phi(1,P)1 \\
&= \int_0^1 f(x)\mathrm{d}x\cdot\Phi(1,P)1 = \Phi(1,P)f,
\end{aligned}$$

and the proposition follows from Lemmas 7.1.1 and 7.1.2. □

With the above preliminary results, we can state the main result of Keller in [74].

Theorem 7.1.1 (Keller's stability theorem) *Suppose that* $P \in \mathcal{P}$ *and a sequence* $\{P_n\}$ *of Markov operators is* $\mathcal{P}$*-bounded. If* $\lim_{n\to\infty} |||P - P_n||| = 0$, *then*

$$\|(\Phi(\lambda, P) - I)\Phi(\lambda, P_n)\| = O\left(|||P - P_n||| \cdot |\ln |||P - P_n|||\,|\right) \tag{7.3}$$

for each complex number λ *with* $|\lambda| = 1$. *If in addition* P *has a unique stationary density, then*

$$\|\Phi(1, P) - \Phi(1, P_n)\| = O\left(|||P - P_n||| \cdot |\ln |||P - P_n|||\,|\right). \tag{7.4}$$

Proof Formula (7.3) follows from Lemma 7.1.2 and Remark 7.1.1, and formula (7.4) is from Proposition 7.1.1 and Remark 7.1.1. □

It is time to apply Theorem 7.1.1 to a class of *stochastic perturbations* of the Frobenius-Perron operators P associated with a nonsingular transformation in the following sense, and as an easy consequence we are able to estimate the L^1-convergence rate of Ulam's method. Let $K : [0, 1]^2 \to \mathbb{R}$ be a *doubly stochastic kernel*, i.e., the function K is measurable and nonnegative, and K satisfies the following two equalities

$$\int_0^1 K(x, y)\mathrm{d}y = 1, \ x \in [0, 1] \text{ a.e.}, \quad \int_0^1 K(x, y)\mathrm{d}x = 1, \ y \in [0, 1] \text{ a.e.}$$

Consider the following Markov operator $P_K : L^1(0, 1) \to L^1(0, 1)$ defined by

$$P_K f(x) = \int_0^1 f(y)K(S(y), x)\mathrm{d}y = \int_0^1 Pf(y)K(y, x)\mathrm{d}y.$$

Let $K_z(y) = \int_0^z K(x, y)\mathrm{d}x$ for $z \in [0, 1]$, $B(z) = \{(x, y) : x \leqslant z < y \text{ or } y \leqslant z < x\}$, $b(z) = \int_{B(z)} K\mathrm{d}m$, and $c(K) = \sup_{z\in[0,1]} b(z)$. By Remark 2.4.2, the variation of $f \in L^1(0, 1)$ as defined in Section 2.3 can be expressed as

$$\bigvee_{[0,1]} f = \sup\left\{\int_0^1 f(x)g'(x)\mathrm{d}x : g \in C_0^1(0, 1), |g(x)| \leqslant 1, x \in (0, 1).\right\}. \tag{7.5}$$

First we need another characterization of the variation for L^1 functions as the following lemma shows.

Lemma 7.1.3 *Let* $f \in L^1(0, 1)$. *Then*

$$\bigvee_{[0,1]} f = \sup\left\{\int_0^1 f\xi\mathrm{d}x, \xi : \xi \in L^\infty(0, 1), \left|\int_0^x \xi(t)\mathrm{d}t\right| \leqslant 1, \forall x, \int_0^1 \xi\mathrm{d}x = 0\right\}.$$

Proof This comes from (7.5) and a limit argument. □

Proposition 7.1.2 *With the above assumptions and notations,*
(i) $\bigvee_{[0,1]} P_K f \leqslant \sup_{z\in[0,1]} \bigvee_{[0,1]} K_z \cdot \bigvee_{[0,1]} Pf$;
(ii) $|||P - P_K||| \leqslant c(K)\|P\|_{BV}$.

Proof (i) By (7.5),

$$\bigvee_{[0,1]} P_K f = \sup\left\{\int_0^1 P_K f(x)\cdot g'(x)\mathrm{d}x : g\in C_0^1(0,1),\ |g(x)|\leqslant 1, x\in(0,1).\right\}.$$

Let $g \in C_0^1(0,1)$ be such that $\|g\|_\infty \leqslant 1$. Then

$$\int_0^1 P_K f(x)\cdot g'(x)\mathrm{d}x = \int_0^1 Pf(x)\cdot \tilde{g}(x)\mathrm{d}x,$$

where $\tilde{g}(x) = \displaystyle\int_0^1 K(x,y)g'(y)\mathrm{d}y$, and $\tilde{g}$ satisfies

$$\begin{aligned}\int_0^1 \tilde{g}(x)\mathrm{d}x &= \int_0^1\int_0^1 K(x,y)g'(y)\mathrm{d}y\mathrm{d}x = \int_0^1\int_0^1 K(x,y)\mathrm{d}xg'(y)\mathrm{d}y\\ &= \int_0^1 g'(y)\mathrm{d}y = g(1)-g(0) = 0.\end{aligned}$$

Furthermore, it follows from Fubini's theorem [109], the definition of K_z, and (7.5) that for all $z \in [0,1]$,

$$\begin{aligned}\left|\int_0^z \tilde{g}(x)\mathrm{d}x\right| &= \left|\int_0^z\int_0^1 K(x,y)g'(y)\mathrm{d}y\mathrm{d}x\right|\\ &= \left|\int_0^1\int_0^z K(x,y)\mathrm{d}xg'(y)\mathrm{d}y\right| = \left|\int_0^1 K_z(y)g'(y)\mathrm{d}y\right| \leqslant \bigvee_{[0,1]} K_z.\end{aligned}$$

Hence, by Lemma 7.1.3,

$$\bigvee_{[0,1]} P_K f \leqslant \bigvee_{[0,1]} Pf \cdot \sup_{0\leqslant z\leqslant 1} \bigvee_{[0,1]} K_z.$$

This proves (i).

(ii) Let $f \in BV(0,1)$ and denote

$$g = \frac{|Pf - P_K f|}{Pf - P_K f}.$$

Then $|g(x)| \equiv 1$ for $x \in (0,1)$ and Fubini's theorem implies that

$$\begin{aligned}\int_0^1 |Pf(x) - P_K f(x)|\mathrm{d}x &= \int_0^1 (Pf(x) - P_K f(x))g(x)\mathrm{d}x \\ &= \int_0^1 \left[Pf(x) - \int_0^1 Pf(y)K(y,x)\mathrm{d}y\right] g(x)\mathrm{d}x \\ &= \int_0^1 Pf(y)\left(g(y) - \int_0^1 K(y,x)g(x)\mathrm{d}x\right)\mathrm{d}y.\end{aligned}$$

By the assumption of K and the definitions of $B(z)$ and $b(z)$,

$$\begin{aligned}&\left|\int_0^z \left(g(y) - \int_0^1 K(y,x)g(x)\mathrm{d}x\right)\mathrm{d}y\right| \\ &\leqslant \int_0^1 \left|\chi_{[0,z]}(y) - \int_0^z K(x,y)\mathrm{d}x\right|\mathrm{d}y \\ &= \int_0^1 \left|\int_0^1 \left[\chi_{[0,z]}(y) - \chi_{[0,z]}(x)\right] K(x,y)\mathrm{d}x\right|\mathrm{d}y \\ &\leqslant \int_0^1 \int_0^1 \left|\chi_{[0,z]}(y) - \chi_{[0,z]}(x)\right| K(x,y)\mathrm{d}x\mathrm{d}y \\ &= \int_{B(z)} K\mathrm{d}m = b(z).\end{aligned}$$

On the other hand, the Fubini theorem ensures that

$$\begin{aligned}&\int_0^1 \left(g(y) - \int_0^1 K(y,x)g(x)\mathrm{d}x\right)\mathrm{d}y \\ &= \int_0^1 g(y)\mathrm{d}y - \int_0^1 \int_0^1 K(y,x)\mathrm{d}y g(x)\mathrm{d}x \\ &= \int_0^1 g(y)\mathrm{d}y - \int_0^1 g(x)\mathrm{d}x = 0.\end{aligned}$$

Therefore, by using Lemma 7.1.3, we have

$$\begin{aligned}\int_0^1 |Pf(x) - P_K f(x)|\mathrm{d}x &\leqslant c(K)\cdot \bigvee_{[0,1]} Pf \leqslant c(K)\|Pf\|_{BV} \\ &\leqslant c(K)\cdot \|P\|_{BV}\cdot \|f\|_{BV}.\end{aligned}$$ □

The following corollary gives a sufficient condition for the variation of $P_K f$ to be less than or equal to that of Pf.

Corollary 7.1.1 *Let $P \in \mathcal{P}$ be the Frobenius-Perron operator associated with a piecewise monotonic mapping of* $[0,1]$ *such that the inequality* (7.1) *is satisfied for $k=1$. Let K be a doubly stochastic kernel such that*

$$\int_0^z K(x, y_1)\mathrm{d}x \geqslant \int_0^z K(x, y_2)\mathrm{d}x \tag{7.6}$$

for all $z, y_1, y_2 \in [0,1]$ with $y_1 \leqslant y_2$. Then the conclusion of Proposition 7.1.2 *is true with*

$$\bigvee_{[0,1]} P_K f \leqslant \bigvee_{[0,1]} Pf.$$

In particular, P and P_K are both in the same class $\mathcal{P}(\alpha, \beta)$.

Proof The inequality (7.6) implies that the function $K_z(y) = \int_0^z K(x,y)\mathrm{d}x$ is monotonically decreasing. Since $0 \leqslant K_z(y) \leqslant 1$ for $0 \leqslant y \leqslant 1$, we have that $\bigvee_{[0,1]} K_z \leqslant 1$ for all $z \in [0,1]$. Therefore, the corollary follows from Proposition 7.1.2. □

We can apply Corollary 7.1.1 to the convergence rate problem for Ulam's method for which the convergence was established in Section 6.1. To fulfill this task, we need to estimate the quantity $|||P - P_n|||$, where $P_n = Q_n P$ and Q_n is the finite dimensional Markov operator for the piecewise constant approximation associated with a partition of $[0,1]$, as defined by (6.3). For the sake of simplicity of the presentation, we assume that the partition of $[0,1]$ is uniform, i.e., we assume that $x_i = i/n$ for $i = 0, 1, \cdots, n$. If we define

$$K_n(x,y) = \begin{cases} n, & \text{if } \dfrac{i-1}{n} \leqslant x,\ y < \dfrac{i}{n} \text{ for some } i, \\ 0, & \text{otherwise,} \end{cases}$$

then it is easy to see that $P_n = P_{K_n}$. In other words,

$$Q_n P f(x) = P_{K_n} f(x) = \int_0^1 Pf(y) K_n(y,x)\mathrm{d}y, \quad \forall\, f \in L^1(0,1).$$

Under the same conditions of Theorem 6.1.1 that guarantee the convergence of Ulam's method as n goes to infinity, Corollary 7.1.1 together with Theorem 7.1.1 can be directly used to establish the following error estimate result for Ulam's method.

Theorem 7.1.2 (Keller's estimate theorem) *Under the conditions of Theorem* 6.1.1, *if f^* is a unique stationary density of the Frobenius-Perron operator P and $\{f_n\}$ is a sequence of the stationary densities of P_n from Ulam's method, then*

$$\|f_n - f^*\| = O\left(\frac{\ln n}{n}\right). \tag{7.7}$$

Proof We can easily check that $c(K_n) \leqslant 1/(2n)$. Hence the sequence $\{P_n\}$ is $\mathcal{P}$-bounded and $|||P - P_n||| = O(1/n)$ by Corollary 7.1.1. Finally Theorem 7.1.1 gives rise to (7.7). □

Remark 7.1.2 A Markov operator $P \in \mathcal{P}$ is said to be *stochastically pre-stable* if the condition $\lim_{n\to\infty} |||P_n - P||| = 0$ implies that $\lim_{n\to\infty} \|(\Phi(1, P_n) - I)\Phi(1, P)\| = 0$ for each $\mathcal{P}$-bounded sequence $\{P_n\}$ of Markov operators. Similarly, $P \in \mathcal{P}$ is called *stochastically stable* if $\lim_{n\to\infty} |||P_n - P||| = 0$ implies that $\lim_{n\to\infty} \|\Phi(1, P_n) - \Phi(1, P)\| = 0$ for any $\mathcal{P}$-bounded sequence $\{P_n\}$. The results of this section show that each $P \in \mathcal{P}$ is stochastically pre-stable, and if P has only one stationary density, then $P \in \mathcal{P}$ is stochastically stable.

Remark 7.1.3 Keller's general stability result, Theorem 7.1.1, can also be used for deterministic perturbations of piecewise monotonic mappings of an interval; see Chapter 11 of the book [14] by Boyarsky and Góra.

Remark 7.1.4 The idea of Keller's analysis seems not to be able to provide a better upper bound of the convergence rate under the L^1-norm for the piecewise linear Markov method than Ulam's method (see Exercise 7.6). A convergence rate of order $O(1/n)$ under the even stronger BV-norm for the piecewise linear Markov method will be presented in Section 7.3.

7.2 More Explicit Error Estimates

In this section we continue the error estimate of Ulam's method for one-dimensional mappings, but our purpose is to obtain an *explicit* expression in the error upper bound $O(\ln n/n)$. Murray began to investigate this problem in his dissertation [106] and obtained an explicit expression for the symbol $O(\ln n/n)$ after a more delicate analysis. This section is devoted to presenting his idea and method for more explicit error bounds of Ulam's method.

We restrict our analysis to a special subclass of Lasota-Yorke's interval mappings $S : [0, 1] \to [0, 1]$. That is, in our discussion that follows, we assume that the mapping is piecewise C^2, stretching, and *onto*. For the simplicity of presentation we also assume that in Ulam's method the interval $[0, 1]$ is divided into n equal subintervals, so that $x_i - x_{i-1} = 1/n$ for $i = 1, 2, \cdots, n$.

To obtain an explicit expression for the constant in the upper bound of the convergence rate, Murray [106, 107] used the concept of *cones of uniformly bounded variation*, which is similar to that introduced by Liverani in [92] and is related to the notion of *Hilbert metric*. The advantage of this approach is that the explicit constant is available in the expression of the upper bound $O(\ln n/n)$, which is more useful in the practical computation of absolutely continuous invariant probability measures.

A set $\mathcal{C}$ in a vector space is called a *convex cone* or just a *cone* if

$$f \in \mathcal{C} \Rightarrow af \in \mathcal{C}, \ \forall\, a \geqslant 0$$

and

$$f+g \in \mathcal{C}, \ \forall f, g \in \mathcal{C}.$$

Definition 7.2.1 *Let a real number $a \geqslant 0$. The convex cone*

$$\mathcal{C}_a = \left\{ 0 \leqslant f \in L^1(0,1) : \bigvee_{[0,1]} f \leqslant a\|f\| \right\}$$

is called a cone of uniformly bounded variation.

Definition 7.2.2 *Let $a \geqslant 0$. The cone*

$$\Gamma_a = \{ f \in L^1(0,1) : f = f_1 - f_2, \ f_1, f_2 \in \mathcal{C}_a, \ \|f_1\| = \|f_2\| \}$$

is called the difference cone *for $\mathcal{C}_a$, and for each $f \in \Gamma_a$ let*

$$\|f\|_a = \inf \{ \|f_1\| : f = f_1 - f_2, \ f_1, f_2 \in \mathcal{C}_a, \ \|f_1\| = \|f_2\| \}. \tag{7.8}$$

Then $\|\cdot\|_a$ is called the difference norm *for Γ_a.*

It is easy to see that if $f \in \Gamma_a$, then automatically $\int_0^1 f(x)\mathrm{d}x = 0$, so $f \in BV_0(0,1)$ and $\|f\|/2 = \|f^+\| = \|f^-\|$. Proposition 7.2.1 below shows that $f \in \Gamma_a$ if $f \in BV_0(0,1)$. As an example let $f(x) = \sin 2\pi x$. Then both $f^+, f^- \in \mathcal{C}_a$ whenever $a \geqslant 2\pi$. One can check that (Exercise 7.9)

$$\|f\|_a = \begin{cases} \dfrac{2}{a}, & \text{if } a < 2\pi, \\ \dfrac{1}{\pi}, & \text{if } a \geqslant 2\pi. \end{cases}$$

Lemma 7.2.1 *$(\Gamma_a, \|\cdot\|_a)$ is a normed vector space.*

Proof It is enough to verify the triangle inequality. Let $f, g \in \Gamma_a$ and $\epsilon > 0$. Choose $f_1, f_2, g_1, g_2 \in \mathcal{C}_a$ such that

$$f = f_1 - f_2, \ \|f_1\| = \|f_2\| < \|f\|_a + \epsilon,$$

$$g = g_1 - g_2, \ \|g_1\| = \|g_2\| < \|g\|_a + \epsilon.$$

Then

$$f + g = (f_1 + g_1) - (f_2 + g_2)$$

and, since L^1-norm $\|\cdot\|$ satisfies the property (2.2),

$$\|f_1 + g_1\| = \|f_2 + g_2\| = \|f_1\| + \|g_1\| < \|f\|_a + \|g\|_a + 2\epsilon.$$

Hence the triangle inequality follows from the definition of $\|\cdot\|_a$. □

We list some useful properties of Γ_a in the following.

Proposition 7.2.1 *For each real number $a \geqslant 0$:*

(i) *The infimum in the definition of $\|\cdot\|_a$ is attained, that is, the infimum can be replaced by the minimum in (7.8).*

(ii) *If $a < b$, then $\Gamma_a \subset \Gamma_b$ and $\|f\|_b \leqslant \|f\|_a$ for any $f \in \Gamma_a$.*

(iii) *If P is a Markov operator and n is a positive integer, then*

$$P^n C_b \subset C_a \Rightarrow \|P^n f\|_a \leqslant \|f\|_b, \ \forall f \in \Gamma_b.$$

(iv) *If $f \in \Gamma_a$, then*

$$\|f\| \leqslant 2\|f\|_a \quad \text{and} \quad \bigvee_{[0,1]} f \leqslant 2a\|f\|_a.$$

(v) *If $f \in \Gamma_a$ and $a > 0$, then*

$$\|f\|_a \leqslant \max\left\{\frac{1}{2}\|f\|, \frac{1}{a}\bigvee_{[0,1]} f\right\}.$$

(vi) *If $f \in \Gamma_a$, then*

$$\min\{2, a\}\|f\|_a \leqslant \|f\|_{BV} \leqslant 2(1+a)\|f\|_a,$$

so that $f \in BV_0(0,1)$ if and only if $f \in \Gamma_a$ and $\|f\|_a < \infty$ for every $a > 0$.

Proof (i) The infimum is attained because the collection of the functions with uniformly bounded integrals and variation is compact in $L^1(0,1)$.

(ii) If $a < b$ and $f \in \Gamma_a$, then for $f = f_1 - f_2$ with $\|f_1\| = \|f\|_a$ we have $\bigvee_{[0,1]} f_1 \leqslant a\|f_1\| < b\|f\|_b$. So $f_1 \in C_b$ and $\|f\|_b \leqslant \|f_1\| = \|f\|_a$.

(iii) Let $f = f_1 - f_2 \in \Gamma_b$ be such that $\|f_1\| = \|f\|_b$. Then $P^n f_1, P^n f_2 \in P^n C_b \subset C_a$ and $P^n f = P^n f_1 - P^n f_2$. Hence,

$$\|P^n f\|_a \leqslant \|P^n f_1\| \leqslant \|f_1\| = \|f\|_b.$$

(iv) Write $f = f_1 - f_2$ with $\|f_1\| = \|f\|_a$. Then, since $f_1 \geqslant f^+$ and $f_2 \geqslant f^-$,

$$\|f\| = 2\|f^+\| \leqslant 2\|f_1\| = 2\|f\|_a.$$

Since $f_1, f_2 \in C_a$,

$$\bigvee_{[0,1]} f \leqslant \bigvee_{[0,1]} f_1 + \bigvee_{[0,1]} f_2 \leqslant a\|f_1\| + a\|f_2\| = 2a\|f\|_a.$$

(v) Suppose first that

$$\bigvee_{[0,1]} f \leqslant a\frac{\|f\|}{2} = a\|f^+\| = a\|f^-\|.$$

Then $f^+, f^- \in \mathcal{C}_a$, and so $\|f\|_a \leqslant \|f^+\| = \|f\|/2$. Now suppose that $\bigvee_{[0,1]} f > a\|f\|/2$. Let

$$g_1 = f^+ + \left(\frac{1}{a}\bigvee_{[0,1]} f - \frac{1}{2}\|f\|\right), \quad g_2 = f^- + \left(\frac{1}{a}\bigvee_{[0,1]} f - \frac{1}{2}\|f\|\right).$$

Then $g_1 \geqslant f^+$ and $g_2 \geqslant f^-$. It follows that

$$\bigvee_{[0,1]} g_1 = \bigvee_{[0,1]} f^+ \leqslant \bigvee_{[0,1]} f = a\|g_1\|, \quad \bigvee_{[0,1]} g_2 = \bigvee_{[0,1]} f^- \leqslant \bigvee_{[0,1]} f = a\|g_2\|.$$

Thus $g_1, g_2 \in \mathcal{C}_a$. Since $f = f^+ - f^- = g_1 - g_2$, the definition of $\|\cdot\|_a$ implies that

$$\|f\| \leqslant \|g_1\| = \frac{1}{a}\bigvee_{[0,1]} f.$$

(vi) The inequalities are consequences of (iv) and (v). The other part is obvious. □

Theorem 7.2.1 *Let $P : L^1(0,1) \to L^1(0,1)$ be a Markov operator such that $P\mathcal{C}_a \subset \mathcal{C}_a$ for some a. Suppose that $0 < \alpha < 1$ and that for every $f, g \in \mathcal{C}_a$ with $\|f\| = \|g\|$, there exists a nonnegative function $\psi_{f,g} \in L^1(0,1)$ such that*

$$\psi_{f,g} \leqslant Pf, \quad \psi_{f,g} \leqslant Pg, \quad \|\psi_{f,g}\| \geqslant \alpha\|f\|$$

and

$$Pf - \psi_{f,g}, \quad Pg - \psi_{f,g} \in \mathcal{C}_a.$$

Then for every $\phi \in \Gamma_a$,

$$P\phi \in \Gamma_a \text{ and } \|P\phi\|_a \leqslant (1-\alpha)\|\phi\|_a.$$

Remark 7.2.1 $\psi_{f,g}$ in the theorem is called a *lower bound function* corresponding to f, g.

Proof Given $\epsilon > 0$, let $f_1, f_2 \in \mathcal{C}_a$ such that

$$\phi = f_1 - f_2, \quad \|f_1\| = \|f_2\| < \epsilon.$$

Denote by $\psi = \psi_{f_1,f_2}$ a lower bound function corresponding to f_1, f_2 and put

$$\phi_1 = Pf_1 - \psi_1, \quad \phi_2 = Pf_2 - \psi_2.$$

Then $\phi_1, \phi_2 \in \mathcal{C}_a$ and $P\phi = \phi_1 - \phi_2$. Since $0 \leqslant \phi_1 \leqslant Pf_1$,

$$\|f_1\| = \|Pf_1\| = \|\phi_1\| + \|\psi\|.$$

The same equality holds with f_2, ϕ_2 replacing f_1, ϕ_1, respectively. Thus,

$$\begin{aligned}\|\phi_1\| = \|\phi_2\| &= \|f_1\| - \|\psi\| \\ &\leqslant \|f_1\| - \alpha\|f_1\| < (1-\alpha)\|\phi\|_a + \epsilon.\end{aligned}$$

Since $\epsilon > 0$ is arbitrary, this proves the theorem. □

It is time to consider the Lasota-Yorke class of interval mappings such that each monotonic piece is onto. For this subclass of mappings, by Remark 5.2.4, we have the following inequality for the corresponding Frobenius-Perron operator as the conclusion of Lemma 7.2.2.

Lemma 7.2.2 *Let $S : [0,1] \to [0,1]$ be a piecewise C^2 and onto mapping associated with the partition $0 = a_0 < a_1 < \cdots < a_{r-1} < a_r = 1$ such that*

$$|S'(x)| \geqslant \lambda, \quad \frac{|S''(x)|}{(S'(x))^2} \leqslant s, \quad \forall\, x \in [0,1] \setminus \{a_1, a_2, \cdots, a_{r-1}\},$$

where $\lambda > 1$ and $s > 0$ are two constants, and let P be the Frobenius-Perron operator corresponding to S. If $f \in L^1(0,1)$ satisfies the inequality $\bigvee_{[0,1]} f \leqslant a\|f\|$ for some $a > 0$, then

$$\bigvee_{[0,1]} Pf \leqslant \frac{1}{\lambda} \bigvee_{[0,1]} f + s\|f\| \leqslant \left(\frac{a}{\lambda} + s\right)\|f\|.$$

It follows immediately from the above inequality that

$$P\mathcal{C}_a \subset \mathcal{C}_{\frac{a}{\lambda}+s}.$$

In particular, if $a \geqslant \lambda/(\lambda - 1)$, then

$$P(\mathcal{C}_a \cap \mathcal{D}) \subset \mathcal{C}_a \cap \mathcal{D}.$$

Since $\mathcal{C}_a \cap \mathcal{D}$ is compact and convex, Brouwer's fixed point theorem implies that P has a fixed point

$$f^* \in \mathcal{C}_{\frac{s\lambda}{\lambda-1}}.$$

Moreover, since $\bigvee_{[0,1]} P_n f \leqslant \bigvee_{[0,1]} Pf$, where P_n is the Ulam approximation of P, and since $P_n\mathcal{D} \subset \mathcal{D}$, we see that

$$P\mathcal{C}_a \subset \mathcal{C}_a \ \Rightarrow\ P_n\mathcal{C}_a \subset \mathcal{C}_a.$$

Thus each P_n has a fixed point

$$f_n \in \mathcal{C}_{\frac{s\lambda}{\lambda-1}}. \tag{7.9}$$

The above claim (7.9) together with Helly's lemma gives an alternative proof of Li's theorem on the convergence of Ulam's method under the additional assumption that S is piecewise onto.

In the following we give the asymptotic estimate of $\|f_n - f^*\|$. For this purpose we need a series of lemmas.

Lemma 7.2.3 *Suppose that $f \in \mathcal{C}_a$ and $\epsilon > 0$. Let $0 = x_0 < x_1 < \cdots < x_{n-1} < x_n = 1$ be a partition of $[0,1]$ such that the length of each subinterval is less than or equal to ϵ. If $\tilde{f} = \sum_{i=1}^{r} c_i \chi_{I_i}$ is a piecewise constant function corresponding to the partition such that*

$$\operatorname*{ess\,inf}_{x\in I_i} f(x) \leqslant c_i \leqslant \operatorname*{ess\,sup}_{x\in I_i} f(x)$$

for each subinterval $I_i = [x_{i-1}, x_i]$ with $i = 1, 2, \cdots, n$, then

$$\|f - \tilde{f}\| \leqslant \epsilon a \|f\|.$$

Proof Since $\bigvee_{I_i} f \geqslant \operatorname*{ess\,sup}_{x\in I_i} f(x) - \operatorname*{ess\,inf}_{x\in I_i} f(x) \geqslant |f(x) - \tilde{f}(x)|$ for $x \in I_i$, we have

$$\int_{I_i} |f - \tilde{f}| \mathrm{d}m \leqslant \bigvee_{I_i} f \cdot m(I_i) \leqslant \epsilon \bigvee_{I_i} f$$

for each i. Therefore,

$$\|f - \tilde{f}\| = \int_0^1 |f - \tilde{f}| \mathrm{d}m = \sum_{i=1}^{n} \int_{I_i} |f - \tilde{f}| \mathrm{d}m$$

$$\leqslant \epsilon \sum_{i=1}^{n} \bigvee_{I_i} f = \epsilon \bigvee_{[0,1]} f \leqslant \epsilon a \|f\|. \qquad \square$$

For the class of piecewise onto mappings, denote $\eta = \{I_i\}_{i=1}^{r}$ the collection of the intervals $\{[a_{i-1}, a_i]\}_{i=1}^{r}$ of monotonicity of S. For the iterates of S, denote $\eta^{(n)} = \{I_{n,i}\}_{i=1}^{r_n}$ the collection of the intervals of monotonicity of S^n, and the inverse mappings of S^n restricted to such intervals of monotonicity are denoted by $S_{n,i}^{-n} : [0,1] \to I_{n,i}$ since $S^n(I_{n,i}) = [0,1]$.

It will be useful to express the bounded distortion property of the Lasota-Yorke mapping in terms of the *Rényi condition*:

$$\frac{|(S_{n,i}^{-n})'(x)|}{|(S_{n,i}^{-n})'(y)|} \leqslant K, \quad \forall\, x, y \in [0,1] \tag{7.10}$$

for some constant K and all $n \geqslant 1$. It is easy to see that the constant K can be taken to be less than or equal to $e^{s\lambda}/(\lambda - 1)$.

Lemma 7.2.4 *Let $I_{n,i}$ be any member of $\eta^{(n)}$. Then*

$$\left|(S^n)'(S_{n,i}^{-n}(x))\right| \leqslant \frac{K}{m(I_{n,i})}, \quad \forall\, x \in I_{n,i}.$$

Proof By the mean value theorem, there exists $y \in [0,1]$ such that

$$\left|(S^n)'(S_{n,i}^{-n}(y))\right| m(I_{n,i}) = m([0,1]) = 1.$$

(7.10) implies that for any $x \in [0,1]$,

$$\left|(S^n)'(S_{n,i}^{-n}(x))\right| \leqslant K \left|(S^n)'(S_{n,i}^{-n}(y))\right| = \frac{K}{m(I_{n,i})}. \qquad \square$$

Proposition 7.2.2 *Let $f \in C_a$ and $n \geqslant 1$. Then*

$$P^n f \geqslant \frac{1 - a\lambda^{-n}}{K} \|f\|.$$

Proof Since $|(S^n)'(S_{I_{n,i}}^{-n}(x))| \geqslant \lambda^n$, $\forall\, x \in [0,1]$ for all i, each subinterval in $\eta^{(n)}$ has length at most λ^{-n}. Let

$$\tilde{f} = \sum_{i=1}^{r_n} \left(\operatorname*{ess\ inf}_{x \in I_{n,i}} f(x) \right) \chi_{I_{n,i}}.$$

Then Lemma 7.2.3 implies that $\|f - \tilde{f}\| \leqslant a\lambda^{-n}\|f\|$. Since $\|f\| = \|f - \tilde{f}\| + \|\tilde{f}\|$ by (2.2), it follows that $\|\tilde{f}\| \geqslant (1 - a\lambda^{-n})\|f\|$.

The explicit expression (4.2) of the Frobenius-Perron operator and Lemma 7.2.4 imply that

$$\begin{aligned} P^n f(x) &= \sum_{i=1}^{r_n} \frac{f(S_{n,i}^{-n}(x))}{\left|(S^n)'(S_{n,i}^{-n}(x))\right|} \geqslant \sum_{i=1}^{r_n} \frac{\operatorname{ess\,inf}_{x \in I_{n,i}} f(x)}{\left|(S^n)'(S_{n,i}^{-n}(x))\right|} \\ &\geqslant \sum_{i=1}^{r_n} \operatorname*{ess\ inf}_{x \in I_{n,i}} f(x) \frac{m(I_{n,i})}{K} = \frac{\|\tilde{f}\|}{K} \geqslant \frac{1 - a\lambda^{-n}}{K} \|f\|. \end{aligned} \qquad \square$$

Definition 7.2.3 *We define the following two numbers*

$$a^* = \frac{\dfrac{1}{2} + \dfrac{s\lambda}{\lambda - 1}}{1 - \dfrac{1}{2K}}, \quad n^* = \left\lceil \frac{\log 2a^*}{\log \lambda} \right\rceil,$$

where $\lceil t \rceil$ denotes the integral part of a real number t.

Then we have the following result on the contraction property of P^{n^*} under the norm $\|\cdot\|_{a^*}$.

Theorem 7.2.2 *Let $f \in \Gamma_{a^*}$. Then*

$$\|P^{n^*} f\|_{a^*} \leqslant \left(1 - \frac{1}{2K}\right) \|f\|_{a^*}.$$

Proof Let $g \in \mathcal{C}_{a^*}$. Then, by Proposition 7.2.2,

$$P^{n^*} g \geqslant \frac{1 - a^* \lambda^{-n^*}}{K} \|g\| \geqslant \frac{\|g\|}{2K}. \tag{7.11}$$

It follows that

$$\left\| P^{n^*} g - \frac{\|g\|}{2K} \right\| = \|P^{n^*} g\| - \frac{\|g\|}{2K} = \left(1 - \frac{1}{2K}\right) \|g\|. \tag{7.12}$$

Hence,

$$\begin{aligned} \bigvee_{[0,1]} \left(P^{n^*} g - \frac{\|g\|}{2K}\right) = \bigvee_{[0,1]} P^{n^*} g \leqslant \frac{1}{\lambda^{n^*}} \bigvee_{[0,1]} g + \frac{s\lambda}{\lambda - 1} \|g\| \\ \leqslant \left(\frac{1}{2} + \frac{s\lambda}{\lambda - 1}\right) \|g\| = \left(1 - \frac{1}{2K}\right) a^* \|P^{n^*} g\|. \end{aligned} \tag{7.13}$$

Combining (7.12) and (7.13) gives that

$$P^{n^*} g - \frac{\|g\|}{2K} \in \mathcal{C}_{a^*},$$

from which, (7.11), and Theorem 7.2.1, the theorem follows. □

A direct consequence of Proposition 7.2.1 and Theorem 7.2.2 is the following corollary.

Corollary 7.2.1 *Let $f \in BV_0(0,1)$. Then for any positive integer n,*

$$\|P^n f\| \leqslant 2\|P^n f\|_{a^*} \leqslant 2 \left(1 - \frac{1}{2K}\right)^{\lceil \frac{n}{n^*} \rceil} \|f\|_{a^*}.$$

Lemma 7.2.5 *The operator $I - P : (\Gamma_{a^*}, \|\cdot\|_{a^*}) \to (\Gamma_{a^*}, \|\cdot\|_{a^*})$ is invertible, and for every $f \in \Gamma_{a^*}$,*

$$\|(I - P)|_{\Gamma_{a^*}}^{-1} f\|_{a^*} \leqslant 2Kn^* \|f\|_{a^*}.$$

Proof By Corollary 7.2.1, using the Banach lemma [57], we see that $(I-P)|_{\Gamma_{a^*}}$ is an invertible linear operator from Γ_{a^*} onto itself, and the norm estimate is from the Neumann series representation of $(I - P)|_{\Gamma_{a^*}}^{-1}$. □

Proposition 7.2.3 *Let $f \in BV_0(0,1)$. Then for any $a \geqslant a^*$,*

$$\|(I - P)|_{\Gamma_{a^*}}^{-1} f\| \leqslant \left\lceil \frac{\ln(2a)}{\ln \lambda} \right\rceil \|f\| + 2(2K - 1)n^* \|f\|_a.$$

Proof Let $n_a = \ln(2a)/\ln\lambda$. Since $f \in BV_0(0,1)$, by Proposition 7.2.1, $f \in \Gamma_{a^*}$. Note that

$$(I-P)|_{\Gamma_{a^*}}^{-1} f = \sum_{k=0}^{\infty} P^k f = \sum_{k=0}^{n_a-1} P^k f + (I-P)|_{\Gamma_{a^*}}^{-1} P^{n_a} f.$$

Then, by Proposition 7.2.1 (iv) and Lemma 7.2.5,

$$\begin{aligned}\|(I-P)|_{\Gamma_{a^*}}^{-1} f\| &\leqslant n_a\|f\| + 2\|(I-P)|_{\Gamma_{a^*}}^{-1} P^{n_a} f\|_{a^*}\\ &\leqslant n_a\|f\| + 2\times 2Kn^*\|P^{n_a} f\|_{a^*}.\end{aligned}$$

Now let $f = f_1 - f_2$ such that $f_1, f_2 \in \mathcal{C}_a$ and $\|f_1\| = \|f_2\| = \|f\|_a$. Then, as in the proof to Theorem 7.2.2,

$$g_1 \equiv P^{n_a} f_1 - \frac{\|f_1\|}{2K} \in \mathcal{C}_{a^*}, \quad g_2 \equiv P^{n_a} f_2 - \frac{\|f_2\|}{2K} \in \mathcal{C}_{a^*}.$$

Since $P^{n_a} f = P^{n_a} f_1 - P^{n_a} f_2 = g_1 - g_2$,

$$\|P^{n_a} f\|_{a^*} \leqslant \|g_1\| = \|g_2\| = \left(1 - \frac{1}{2K}\right)\|f\|_a,$$

and the proposition follows from (7.13). □

Corollary 7.2.2 *Let f^* be a stationary density of P and let $g \in BV(0,1)\cap\mathcal{D}$. If $a \geqslant a^*$, then*

$$\|g - f^*\| \leqslant \left\lceil \frac{\ln(2a)}{\ln\lambda} \right\rceil \|g - Pg\| + 2(2K-1)n^*\|g - Pg\|_a.$$

Proof It is immediate from Proposition 7.2.3 since

$$(I-P)(g - f^*) = g - Pg.$$ □

We are ready to apply the above results to get an explicit error bound for Ulam's method by letting $g = f_n = P_n f_n$, where f_n is a piecewise constant density function. What we need now is to find an a such that $f_n - Pf_n \in \Gamma_a$ and $\|f_n\|_a$ is small enough.

Lemma 7.2.6 *For every $n \geqslant 0$,*

$$\|f_n - Pf_n\| \leqslant \frac{1}{n}\frac{s\lambda}{\lambda-1}, \quad \|f_n - Pf_n\|_{4n} \leqslant \frac{1}{2n}\frac{s\lambda}{\lambda-1}.$$

Proof Since $f_n,\ Pf_n \in \mathcal{C}_{s\lambda/(\lambda-1)}$, by Proposition 6.1.1 (v),

$$\|f_n - Pf_n\| = \|P_nf_n - Pf_n\| = \|(Q_n - I)Pf_n\| \leqslant \frac{1}{n}\frac{s\lambda}{\lambda - 1}.$$

On the other hand,

$$\bigvee_{[0,1]} (f_n - Pf_n) \leqslant \bigvee_{[0,1]} f_n + \bigvee_{[0,1]} Pf_n \leqslant 2\frac{s\lambda}{\lambda - 1}.$$

By Proposition 7.2.1 (v),

$$\|f_n - Pf_n\|_{4n} \leqslant \max\left\{\frac{\|f_n - Pf_n\|}{2}, \frac{\bigvee_{[0,1]}(f_n - Pf_n)}{4n}\right\} \leqslant \frac{1}{2n}\frac{s\lambda}{\lambda - 1}. \qquad \square$$

Therefore, by letting $a = 4n$ in Corollary 7.2.2, we have

Theorem 7.2.3 (Murray's estimate theorem) *Let $S : [0,1] \to [0,1]$ satisfy the assumptions of Lemma 7.2.2, let P be the Frobenius-Perron operator associated with S, let $f^* \in \mathcal{C}_{s\lambda/(\lambda-1)}$ be such that $Pf^* = f^*$ and $\|f^*\| = 1$, and let a^* and n^* be defined by Definition 7.2.3. For each $n \geqslant a^*/4$ let $f_n \in \mathcal{C}_{s\lambda/(\lambda-1)}$, $\|f_n\| = 1$ be such that $f_n = P_nf_n$, where P_n is the Ulam approximation of P. Then*

$$\|f_n - f^*\| \leqslant \left(\left\lceil\frac{\ln(8n)}{\ln\lambda}\right\rceil + (2K-1)n^*\right)\frac{1}{n}\frac{s\lambda}{\lambda - 1} = O\left(\frac{\ln n}{n}\right).$$

Remark 7.2.2 A similar result to Theorem 7.2.3 can also be obtained for non-onto Lasota-Yorke mappings and for some multi-dimensional transformations (see [106, 108] for more details).

Remark 7.2.3 One example has been found in [11] to show that $O(\ln n/n)$ is the *optimal* error bound for Ulam's method even for piecewise linear mappings, which means that the convergence rate of $\ln n/n$ *cannot* be improved in general. In the next section we shall show that a higher order of $O(1/n)$ can be achieved for the piecewise linear Markov method even under the stronger BV-norm, which indicates that the Markov method is much better as far as the convergence rate is concerned.

Remark 7.2.4 Using the concepts of *spectral stability* and the Hilbert metric, new results on the upper bounds of the convergence rate and the speed of the decay of correlations have been given in [77, 93].

7.3 Error Estimates for the Markov Method

This section will be devoted to the convergence rate analysis of the piecewise linear Markov approximation method for solving the Frobenius-Perron operator equation, in which the operator satisfies the Lasota-Yorke type variation inequality (7.14) below. Again, we only treat one-dimensional mappings.

Although such Frobenius-Perron operators are quasi-compact when they are restricted to the vector subspace of functions of bounded variation under the BV-norm, we do not explore the concept of quasi-compactness further to accomplish our purpose. The approach to the convergence rate analysis using the notion of quasi-compactness [57] and the spectral approximation technique [17] can be seen in [35]. Here we present a more direct approach with the help of the powerful analytic tool of the Cauchy integral of linear operators, which was first introduced in [19] and further investigated in [41]. The idea of Cauchy integrals was also employed independently in [69] for a similar purpose. The main result of the works [19, 41] is that the convergence rate of the Markov method is of the order $O(1/n)$ under the BV-norm, where n is the number of the equal subintervals of the partition of $[0, 1]$. The main content of this section comes from the two related papers.

First we give the basis of *Cauchy integrals of bounded linear operators* for our needs. Let T be a bounded linear operator on a complex Banach space B. Recall that the *resolvent set* $\rho(T)$ of T consists of all complex numbers z such that the inverse operator

$$R(z, T) \equiv (zI - T)^{-1}$$

is well-defined and is bounded with domain B. The resolvent set is an open subset of the complex plane $\mathbb{C}$. The bounded linear operator $R(z, T)$ is called the *resolvent* of T, which is analytic in its domain. The complement of the resolvent set in the complex plane is the *spectrum* $\sigma(T)$ of T.

Now let λ be an *isolated* point spectrum point of T, that is, λ is an *eigenvalue* of T and is the only spectral point of T in a neighborhood of λ. Then the resolvent $R(z, T)$ has the *Laurent expansion* in a deleted neighborhood of λ:

$$R(z, T) = \sum_{n=-\infty}^{\infty} (z - \lambda)^n A_n, \quad 0 < |z - \lambda| < \delta,$$

where $\delta \leqslant \infty$ is a positive number or infinity and for each n, the bounded linear operator A_n is defined by the Cauchy-type integral

$$A_n = \frac{1}{2\pi \mathrm{i}} \int_C \frac{R(z, T)}{(z - \lambda)^{n+1}} \mathrm{d}z,$$

where $\mathrm{i} = \sqrt{-1}$ is the imaginary unit of $\mathbb{C}$. Here the integral contour C is any closed Jordan curve which lies in the resolvent set of T and its interior contains no other spectral points but λ. The coefficient operator

$$A_{-1} = \frac{1}{2\pi \mathrm{i}} \int_C R(z, T) \mathrm{d}z$$

is most important for us because of the following result [57].

Proposition 7.3.1 *A_{-1} is a* projection *onto the* generalized eigenspace *corresponding to λ and the restriction $T|_{R(A_{-1})}$ of T to the range $R(A_{-1})$ of A_{-1} maps $R(A_{-1})$ into itself and its spectrum is exactly the single-point set $\{\lambda\}$.*

Let a nonsingular transformation $S : [0,1] \to [0,1]$ be such that the corresponding Frobenius-Perron operator $P : L^1(0,1) \to L^1(0,1)$ satisfies the condition that there exist two real numbers $0 < \alpha < 1$ and $\beta > 0$ such that

$$\|Pf\|_{BV} \leqslant \alpha \|f\|_{BV} + \beta \|f\|, \quad \forall\, f \in BV(0,1). \tag{7.14}$$

For the spectral analysis of P, in this section we assume that the space $L^1(0,1)$ is a complex one, and thus the Frobenius-Perron operator $P : L^1(0,1) \to L^1(0,1)$ is a complex bounded linear operator which still has operator norm 1 as in the real L^1-space case.

Remark 7.3.1 Any mapping $S : [0,1] \to [0,1]$ satisfying the conditions of Theorem 5.2.1 with $\inf_{x \in [0,1] \setminus \{a_1, a_2, \cdots, a_{r-1}\}} |S'(x)| > 2$ satisfies the Lasota-Yorke type inequality (7.14).

For the simplicity of the convergence rate analysis, we assume that f^* is a unique stationary density of P in $L^1(0,1)$. Because of (7.14), from the proof of Theorem 5.2.1, $f^* \in BV(0,1)$. Let $\{Q_n\}$ be the sequence of the finite approximation operators associated with the Markov piecewise linear method, as defined by (6.12), and let $P_n = Q_n P$. Then for all $f \in BV(0,1)$,

$$\|P_n f\|_{BV} \leqslant \alpha \|f\|_{BV} + \beta \|f\|. \tag{7.15}$$

First, Proposition 6.3.5 immediately gives the "local convergence" of the Markov approximations under the BV-norm.

Proposition 7.3.2 *If $f \in C^2[0,1]$, then*

$$\lim_{n \to \infty} \|Q_n f - f\|_{BV} = 0.$$

Then, we have the following "global convergence" under the BV-norm for the Markov approximation method.

Theorem 7.3.1 *Suppose that the Frobenius-Perron operator P satisfies* (7.14) *and $f^* \in C^2[0,1]$ is a unique stationary density of P. Let $\{f_n\}$ be a sequence of the stationary densities of P_n in Δ_n from the Markov method. Then*

$$\lim_{n \to \infty} \|f_n - f^*\|_{BV} = 0.$$

Proof It has been shown before (see Theorem 6.3.1) that

$$\lim_{n \to \infty} \|f_n - f^*\| = 0. \tag{7.16}$$

Writing

$$f^* - f_n = f^* - P_n f^* + P_n f^* - P_n f_n = P_n(f^* - f_n) + f^* - Q_n f^*$$

and using (7.14), we obtain

$$\begin{aligned}\|f^* - f_n\|_{BV} &= \|P_n(f^* - f_n) + f^* - Q_n f^*\|_{BV} \\ &\leqslant \alpha\|f^* - f_n\|_{BV} + \beta\|f^* - f_n\| + \|f^* - Q_n f^*\|_{BV}.\end{aligned}$$

It follows that, since $0 < \alpha < 1$,

$$\|f^* - f_n\|_{BV} \leqslant \frac{1}{1-\alpha}(\beta\|f^* - f_n\| + \|f^* - Q_n f^*\|_{BV}). \tag{7.17}$$

The theorem follows from (7.16), Proposition 7.3.2, and (7.17). □

Now we begin to investigate the speed of the above convergence under the BV-norm. Several lemmas will be needed first.

Lemma 7.3.1 *Suppose that* $\lambda \neq 1$. *If* $Pf = \lambda f$, *then* $\int_0^1 f(x)\mathrm{d}x = 0$. *The same conclusion is also true for* P_n.

Proof Integrating both sides of the equality $Pf = \lambda f$ over $[0, 1]$ yields

$$\lambda \int_0^1 f(x)\mathrm{d}x = \int_0^1 Pf(x)\mathrm{d}x = \int_0^1 f(x)\mathrm{d}x.$$

Thus $\int_0^1 f(x)\mathrm{d}x = 0$ since $\lambda \neq 1$. The proof for P_n is similar. □

Lemma 7.3.2 *If* n *is large enough, then the eigenspace* E_n *of* P_n *corresponding to the eigenvalue* $\lambda = 1$ *is one-dimensional.*

Proof $E_n \subset \Delta_n$ always contains a density function f_n for each n. If the lemma is not true, then there is a subsequence $\{f_{n_i}\}$ of the sequence $\{f_n\}$ such that for each f_{n_i}, there is a fixed point $g_{n_i} \in \Delta_{n_i}$ of P_{n_i} with L^1-norm 1 which is linearly independent of f_{n_i}. Let $c_{n_i} = \int_0^1 g_{n_i}(x)\mathrm{d}x$ and

$$h_{n_i} = \frac{c_{n_i} f_{n_i} - g_{n_i}}{\|c_{n_i} f_{n_i} - g_{n_i}\|}.$$

Then $\int_0^1 h_{n_i}(x)\mathrm{d}x = 0$, $P_{n_i} h_{n_i} = h_{n_i}$, and $\|h_{n_i}\| = 1$. By (7.15),

$$\|h_{n_i}\|_{BV} = \|P_{n_i} h_{n_i}\|_{BV} \leqslant \alpha\|h_{n_i}\|_{BV} + \beta\|h_{n_i}\| = \alpha\|h_{n_i}\|_{BV} + \beta,$$

from which we have

$$\|h_{n_i}\|_{BV} \leqslant \frac{\beta}{1-\alpha}, \quad \forall\, i.$$

Thus, Helly's lemma ensures that the sequence $\{h_{n_i}\}$ has a convergent subsequence, which is still denoted as $\{h_{n_i}\}$, such that it converges in the L^1-norm to some $h \in L^1(0,1)$. Now $\|h\| = \lim_{i\to\infty} \|h_{n_i}\| = 1$. Moreover h is a fixed point of P since

$$\begin{aligned}\|h - Ph\| \leqslant & \|h - h_{n_i}\| + \|h_{n_i} - P_{n_i}h_{n_i}\| \\ & + \|P_{n_i}h_{n_i} - P_{n_i}h\| + \|P_{n_i}h - Ph\| \to 0.\end{aligned}$$

Since f^* is the unique stationary density of P by assumption, the null space $N(P-I)$ is one-dimensional from Proposition 4.1.3. Thus $h = cf^*$, where c is a complex number such that $|c| = 1$. Hence $\int_0^1 h(x)\mathrm{d}x = c \neq 0$, which contradicts the fact that $\int_0^1 h_{n_i}(x)\mathrm{d}x = 0$ for all i imply $\int_0^1 h(x)\mathrm{d}x = 0$. □

Lemma 7.3.3 *$N(P_n - I) = N((P_n - I)^k)$ for all positive integers k and n. Thus, for n large enough, the generalized eigenspace of P_n is one-dimensional.*

Proof Given n, since

$$\lim_{j\to\infty} \frac{\|P_n^j\|}{j} \leqslant \lim_{j\to\infty} \frac{1}{j} = 0,$$

and since P_n is compact, the lemma follows from Theorem VII.4.5 and Lemma VIII.8.1 of [57]. □

We define two subsets

$$\Gamma = \{\lambda \neq 1 : P_n g = \lambda g,\ g \neq 0 \text{ for some } n\}$$

and

$$\gamma = \{\lambda \neq 1 : Pg = \lambda g,\ g \in BV(0,1),\ g \neq 0\}$$

of the complex plane.

Lemma 7.3.4 *$1 \notin \overline{\Gamma} \cup \overline{\gamma}$, where $\overline{\Gamma}$ and $\overline{\gamma}$ are the closures of Γ and γ.*

Proof Suppose that $1 \in \overline{\Gamma}$. Then there is a sequence $\{\lambda_{n_i}\}$ of numbers in Γ such that $\lambda_{n_i} \to 1$. Let g_{n_i} be an eigenvector with L^1-norm 1 corresponding to the eigenvalue λ_{n_i} of P_{n_i}. Then, by Lemma 7.3.1,

$$\int_0^1 g_{n_i}(x)\mathrm{d}x = 0.$$

From (7.15),

$$\|g_{n_i}\|_{BV} = \left\|\frac{1}{\lambda_{n_i}} P_{n_i} g_{n_i}\right\|_{BV} \leqslant \frac{1}{|\lambda_{n_i}|}\left(\alpha\|g_{n_i}\|_{BV} + \beta\right).$$

Since for i large enough, $|\lambda_{n_i}| > \alpha$, so

$$\|g_{n_i}\|_{BV} \leqslant \frac{\beta}{|\lambda_{n_i}| - \alpha}.$$

Thus, the sequence $\{g_{n_i}\}$ has a convergent subsequence, which is still denoted as $\{g_{n_i}\}$, such that it converges in the L^1-norm to some g of L^1-norm 1. We must have $Pg = g$ since

$$\begin{aligned}\|g - Pg\| \leqslant & \|g - g_{n_i}\| + |1 - \lambda_{n_i}|\|g_{n_i}\| + \|\lambda_{n_i} g_{n_i} - P_{n_i} g_{n_i}\| \\ & + \|P_{n_i} g_{n_i} - P_{n_i} g\| + \|P_{n_i} g - Pg\| \to 0.\end{aligned}$$

Hence $g = cf^*$ with $|c| = 1$. This leads to a contradiction that $1 = |c| \int_0^1 f^*(x)\mathrm{d}x$ $= \left|\lim_{i\to\infty} \int_0^1 g_{n_i}(x)\mathrm{d}x\right| = 0.$

Now suppose that $1 \in \overline{\gamma}$. Then $Pg_n = \lambda_n g_n$ for a sequence $\{\lambda_n\}$ in γ such that $g_n \in BV$, $\|g_n\| = 1$, and $\lim_{n\to\infty} \lambda_n = 1$. The same contradiction will appear from Lemma 7.3.1 and (7.14). □

Lemma 7.3.5 *Let $\sigma(P_n)$ be the spectrum of $P_n : (BV, \|\cdot\|_{BV}) \to (BV, \|\cdot\|_{BV})$. Then $\sigma(P_n) \setminus \{0\} \subset \Gamma$.*

Proof Let $\lambda \neq 0$. Since $R(P_n)$ is finite dimensional, P_n is a compact operator. Thus the spectral theory for compact linear operators [57] shows that $\lambda \in \sigma(P_n)$ if and only if λ is an eigenvalue of P_n. □

Denote

$$d = \frac{1}{3}\min\{\mathrm{dist}(1, \Gamma \cup \gamma), 1 - \alpha\}.$$

Then $d > 0$. Let a region G of the complex plane be defined as

$$G = \left\{z : \frac{d}{2} \leqslant |z - 1| \leqslant d\right\}.$$

By Lemma 7.3.5, for all n large enough,

$$(P_n - zI)^{-1} : (BV(0,1), \|\cdot\|_{BV}) \to (BV(0,1), \|\cdot\|_{BV})$$

are well-defined and bounded for all $z \in G$. From now on we fix a circle C centered at 1 with radius ϵ such that $C \subset G$.

Lemma 7.3.6 *There exists a constant M such that for any sequence $\{f_n\}$ of functions in $BV(0,1)$,*

$$\|(P_n - zI)^{-1} f_n\| \leqslant M\|f_n\|_{BV}, \ \forall\, z \in C.$$

Proof If the lemma is not true, then there is a subsequence $\{n_i\}$ of the natural numbers such that

$$\lim_{i\to\infty} \|g_{n_i}\| = \infty,$$

where

$$g_{n_i} = \frac{(P_{n_i} - z_{n_i} I)^{-1} f_{n_i}}{\|f_{n_i}\|_{BV}}$$

and $z_{n_i} \in C$. Let

$$\tilde{g}_{n_i} = \frac{g_{n_i}}{\|g_{n_i}\|} \text{ and } \tilde{f}_{n_i} = \frac{f_{n_i}}{\|g_{n_i}\| \, \|f_{n_i}\|_{BV}}.$$

Then $\tilde{f}_{n_i} = (P_{n_i} - z_{n_i} I)\tilde{g}_{n_i}$ and

$$\lim_{i\to\infty} \|\tilde{f}_{n_i}\|_{BV} = \lim_{i\to\infty} \frac{1}{\|g_{n_i}\|} = 0.$$

Since the set $\{z_{n_i}\}_{i=1}^{\infty} \subset C$ is precompact, without loss of generality, we assume that $z_{n_i} \to z \in C$. Then,

$$\begin{aligned}\|\tilde{g}_{n_i}\|_{BV} &= \left\| \frac{1}{z_{n_i}} (P_{n_i}\tilde{g}_{n_i} - \tilde{f}_{n_i}) \right\|_{BV} \leqslant \frac{1}{|z_{n_i}|} (\|P_{n_i}\tilde{g}_{n_i}\|_{BV} + \|\tilde{f}_{n_i}\|_{BV}) \\ &\leqslant \frac{\alpha}{1-2d}\|\tilde{g}_{n_i}\|_{BV} + \frac{\beta}{1-2d} + 1\end{aligned}$$

for i large enough. Hence

$$\|\tilde{g}_{n_i}\|_{BV} \leqslant \frac{\beta + 1 - 2d}{1 - 2d - \alpha}, \ \forall\, i \geqslant i_0,$$

where i_0 is some positive integer. It follows from Helly's lemma that there is a subsequence of $\{\tilde{g}_{n_i}\}$, which is again denoted by $\{\tilde{g}_{n_i}\}$, such that

$$\lim_{i\to\infty} \|\tilde{g}_{n_i} - g\| = 0$$

for some $g \in BV(0,1)$ and $\|g\| = \lim_{i\to\infty} \|\tilde{g}_{n_i}\| = 1$. From

$$\begin{aligned}\|zg - Pg\| \leqslant\, & |z - z_{n_i}|\|g\| + |z_{n_i}|\|g - \tilde{g}_{n_i}\| + \|(z_{n_i} - P_{n_i})\tilde{g}_{n_i}\| \\ & + \|P_{n_i}(\tilde{g}_{n_i} - g)\| + \|P_{n_i}g - Pg\| \to 0,\end{aligned}$$

we see that $z \in \gamma$, a contradiction to the fact that $G \cap \gamma = \varnothing$. □

We are in the position to use the tool of Cauchy integrals of bounded linear operators. For all natural numbers n large enough, the resolvents $R(z, P_n) = (zI - P_n)^{-1}$ of $P_n : (BV(0,1), \|\cdot\|_{BV}) \to (BV(0,1), \|\cdot\|_{BV})$ are well-defined and analytic for z in G with the Laurent expansion

$$R(z, P_n) = \sum_{k=-\infty}^{\infty} A_k(P_n)(z-1)^k$$

at $z = 1$, where for each integer k, the bounded linear operator $A_k(P_n) : (BV(0,1), \|\cdot\|_{BV}) \to (BV(0,1), \|\cdot\|_{BV})$ is given by the Cauchy integral

$$A_k(P_n) = \frac{1}{2\pi \mathrm{i}} \int_C \frac{R(z, P_n)}{(z-1)^{k+1}} \mathrm{d}z.$$

By Lemma 7.3.3, $z = 1$ is a simple pole of $R(z, P_n)$, thus $A_k(P_n) = 0$ for $k \leqslant -2$. From Proposition 7.3.1 we have

Lemma 7.3.7 *$A_{-1}(P_n) = (2\pi\mathrm{i})^{-1} \int_C R(z, P_n)\mathrm{d}z$ is a projection operator from $BV(0,1)$ onto $N(P_n - I)$ for n large enough.*

Lemma 7.3.8 *For n large enough,*

$$A_{-1}(P_n)(f^* - f_n) = 0.$$

Proof Since $\dim N(P_n - I) = 1$ for large n, $A_{-1}(P_n)(f^* - f_n) = c_n f_n$, where

$$\begin{aligned} c_n &= \int_0^1 A_{-1}(P_n)(f^* - f_n)(x)\mathrm{d}x \\ &= \int_0^1 \frac{1}{2\pi\mathrm{i}} \left(\int_C (zI - P_n)^{-1}(f^* - f_n)\mathrm{d}z \right)(x)\mathrm{d}x \\ &= \frac{1}{2\pi\mathrm{i}} \int_C \int_0^1 (zI - P_n)^{-1}(f^* - f_n)(x)\mathrm{d}x\mathrm{d}z \end{aligned}$$

since $\int_0^1 f_n(x)\mathrm{d}x = 1$. Let $g_n(z) = (zI - P_n)^{-1}(f^* - f_n)$. Then $(zI - P_n)g_n(z) = f^* - f_n$, thus, since P_n preserves integrals,

$$\begin{aligned} (z-1)\int_0^1 [g_n(z)](x)\mathrm{d}x &= \int_0^1 [(zI - P_n)g_n(z)](x)\mathrm{d}x \\ &= \int_0^1 (f^* - f_n)(x)\mathrm{d}x = 0, \end{aligned}$$

from which we have $\int_0^1 [g_n(z)](x)\mathrm{d}x = 0$ for all $z \in C$. Hence,

$$c_n = \frac{1}{2\pi \mathrm{i}} \int_C \int_0^1 [g_n(z)](x)\mathrm{d}x\mathrm{d}z = 0. \qquad \square$$

With all the above preliminary results, we can state the main theorem of this section.

Theorem 7.3.2 (Chiu-Du-Ding-Li's estimate theorem) *Let $\{f_n\}$ be a sequence of the stationary densities of P_n in Δ_n from the Markov method for approximating the unique stationary density f^* of P. Then for n large enough,*

$$\|f^* - f_n\| \leqslant M\|f^* - Q_n f^*\|_{BV}, \tag{7.18}$$

$$\|f^* - f_n\|_{BV} \leqslant \frac{\beta M + 1}{1 - \alpha}\|f^* - Q_n f^*\|_{BV}. \tag{7.19}$$

Moreover, if $f^* \in C^2[0,1]$, then

$$\|f_n - f^*\|_{BV} = O\left(\frac{1}{n}\right). \tag{7.20}$$

Proof From (7.17), (7.18) implies (7.19) which implies (7.20) by Proposition 6.3.5. So we just need to show (7.18). Since

$$(I - P_n)(f^* - f_n) = f^* - Q_n f^*,$$

for any $z \in C$,

$$\frac{1}{z-1}(f^* - f_n) = (zI - P_n)^{-1}(f^* - f_n) + \frac{1}{z-1}(zI - P_n)^{-1}(f^* - Q_n f^*).$$

Integrating both sides of the above equality over C, noting that $(2\pi\mathrm{i})^{-1}\int_C (z-1)^{-1}\mathrm{d}z = 1$, and using Lemma 7.3.8, we have

$$f^* - f_n = \frac{1}{2\pi\mathrm{i}}\int_C \frac{1}{z-1}(zI - P_n)^{-1}(f^* - Q_n f^*)\mathrm{d}z.$$

Thus, it follows from Lemma 7.3.6 that

$$\begin{aligned}\|f^* - f_n\| &\leqslant \frac{1}{2\pi}\frac{2\pi\epsilon}{\epsilon}\max_{z\in C}\|(zI - P_n)^{-1}(f^* - Q_n f^*)\| \\ &\leqslant M\|f^* - Q_n f^*\|_{BV}.\end{aligned}$$

This proves the theorem. $\square$

Remark 7.3.2 While Ulam's piecewise constant approximations method has the L^1-norm error bound no better than $O(\ln n/n)$, the piecewise linear Markov approximations method does converge under the BV-norm with the convergence rate $O(1/n)$ for the class of Frobenius-Perron operators satisfying (7.14). This theoretical result has been verified by the numerical results in [38].

Remark 7.3.3 The same idea can be used for the convergence rate analysis in the more general setting of the Markov method for multi-dimensional transformations and for quasi-compact Markov operators (see [36, 51]).

Remark 7.3.4 Although Theorem 7.3.2 implies immediately that the convergence rate of the piecewise linear Markov method under the L^1-norm is at least $O(1/n)$, it seems that its higher order L^1-norm error estimate has not been seen in the literature, although an order $O(\ln n/n^2)$ of L^1 convergence has been observed from some numerical experiments [28, 38].

Exercises

7.1 Let $\Phi(\lambda, P)$ be defined as in Section 7.1. Show that $\Phi(\lambda, P)f$ exists for all $f \in L^1(0,1)$ and

$$\Phi(\lambda, P) = \begin{cases} \Phi_i, & \text{if } \lambda = \lambda_i, \\ 0, & \text{otherwise.} \end{cases}$$

7.2 Prove (i) and (ii) of Lemma 7.1.1.
7.3 Show that if $P \in \mathcal{P}$ with a unique stationary density, then for any $f \in L^1(0,1)$,

$$\Phi(1, P)f = \int_0^1 f(x)\mathrm{d}x \cdot \Phi(1, P)1.$$

7.4 Prove Lemma 7.1.3 in more detail.
7.5 Construct a doubly stochastic symmetric kernel $K(\cdot,\cdot)$ for which $K_z(\cdot)$ as defined in Section 7.1 is not monotonically decreasing.
7.6 Construct a sequence of doubly stochastic kernels K_n corresponding to the sequence $\{P_n\}$ of the piecewise linear Markov approximation method introduced in Section 6.3, and show that Keller's approach only gives the same L^1-norm convergence rate for the Markov method as the Ulam method.
7.7[82] Consider the following discrete dynamical system with constantly applied stochastic perturbations

$$x_{n+1} = S(x_n) + \xi_n, \quad n = 0, 1, \cdots,$$

where $S : \mathbb{R} \to \mathbb{R}$ is a nonsingular transformation and $\xi_0, \xi_1, \cdots$ are independent random variables each having the same density g. Let the density of the random variable x_n be denoted by f_n. Find the Markov operator P defined by a stochastic kernel that maps f_n to f_{n+1} for all n.

7.8[82] Apply the solution of Exercise 7.7 to the following small stochastic perturbations of the discrete dynamical system

$$x_{n+1} = S(x_n) + \epsilon \xi_n,$$

where $S : \mathbb{R} \to \mathbb{R}$ is a nonsingular transformation and $\xi_0, \xi_1, \cdots$ are independent random variables each having the same density g. Find the Markov operator P_ϵ that describes the density evolution of the dynamical system.

7.9 Let $\|\cdot\|_a$ be the difference norm for $\mathcal{C}_a$ as in Section 7.2. Verify that for $f(x) = \sin(2\pi x)$,

$$\|f\|_a = \begin{cases} \dfrac{2}{a}, & \text{if } a < 2\pi, \\ \dfrac{1}{\pi}, & \text{if } a \geqslant 2\pi. \end{cases}$$

Chapter 8
Entropy

JiuDing, Department of Mathematics, The University of Southern Mississippi in America.
Aihui Zhou, Academy of Mathematics and Systems Science, Chinese Academy of Sciences.

Abstract We introduce the concepts of Shannon's entropy for discrete sample spaces, the Kolmogorov entropy for measurable transformations, the topological entropy for continuous transformations on compact metric spaces, and the Boltzmann entropy of density functions. We also study some relationship between the Boltzmann entropy and the iteration of a Frobenius-Perron operator.
Keywords Shannon entropy, Kolmogorov entropy, topological entropy, Boltzmann entropy, Gibbs inequality, Jensen's inequality.

Various entropy concepts, such as the Shannon entropy that first appeared in the 1940's for the need of information theory, the Kolmogorov entropy that emerged for solving a fundamental problem of ergodic theory in the 1950's, and the topological entropy for the study of topological dynamical systems in the 1960's, are all the mathematical measures of uncertainty. They have the origin in the classic Boltzmann entropy in thermodynamics. In this chapter we introduce such important concepts and study their mutual relationship. More detailed theoretical analysis of entropy and its important applications to ergodic theory and dynamical systems can be found in, e.g., [97].

In the first section we introduce Shannon's entropy, the idea of which will be used to define the Kolmogorov entropy of Section 8.2 and the topological entropy in Section 8.3. The classic Boltzmann entropy will be studied in Section 8.4, and in the final section we explore its relation with the concept of Frobenius-Perron operators. Part of the material in this chapter is presented by following an elementary introduction as given in [39].

8.1 Shannon Entropy

The Shannon entropy is a mathematical quantification about the degree of uncertainty out of experiments with several possible outcomes. Uncertainty is everywhere in our real world. For example, it is apparent that the claim "there is a 90% chance to rain this evening" has less degree of uncertainty than the statement "the chance to rain this evening is 60%." In the following we first use elementary calculus to develop a mathematical approach to the problem of quantization of uncertainty.

Suppose that there is given a sample space X with n basic events $w_1, w_2, \cdots, w_n$, whose probabilities are $p_1, p_2, \cdots, p_n$, respectively. Since we are only interested in the numerical values of the probabilities, this sample space $(X; p_1, p_2, \cdots, p_n)$ will simply be denoted as an n-tuple $(p_1, p_2, \cdots, p_n)$. The basic relation

$$\sum_{i=1}^{n} p_i = 1, \quad p_i \geqslant 0, \ i = 1, 2, \cdots, n$$

holds. We want to define a nonnegative function H whose domain consists of all such finite sample spaces $(p_1, p_2, \cdots, p_n)$ and positive integers n. The value of H at a sample space $(p_1, p_2, \cdots, p_n)$ is denoted by $H(p_1, p_2, \cdots, p_n)$. This number will be used to characterize mathematically the *degree of uncertainty* of the outcome of the events $w_1, w_2, \cdots, w_n$ with the corresponding probabilities $p_1, p_2, \cdots, p_n$. For example, $H(1/2, 1/2)$ should be the measure of the degree of uncertainty for tossing a regular smooth coin to see whether the Head or the Tail is up since each event has probability $1/2$.

In order to reflect the degree of uncertainty of the experimental result, $H(p_1, p_2, \cdots, p_n)$ should satisfy the following three basic conditions.

(i) For every fixed integer $n > 0$, H is a continuous function of its variables $p_1, p_2, \cdots, p_n$.

To motivate the second condition, we do the experiment of throwing a dice to see whether one specified side is up or not. The probability for each individual side of the dice to be up is obviously $1/6$. Intuitively, the degree of uncertainty for a specified side of a dice to be up is bigger than that for the Head or the Tail of a coin to be up. Using the function H to express this fact, we should have that $H(1/6, 1/6, 1/6, 1/6, 1/6, 1/6) > H(1/2, 1/2)$. In general, H should satisfy that

(ii) If $p_i = 1/n$ for all $i = 1, 2, \cdots, n$, then the corresponding numeric value $H(1/n, \cdots, 1/n)$ is monotonically increasing with respect to n.

To motivate the third condition, we consider the following simple problem. Suppose that a research grant is to be distributed to Professor A of the physics department or one of Professor B and Professor C of the mathematics department. Assume that the probability for Professor A to receive this grant is $1/2$, for Professor B the probability is $1/3$, and for Professor C it is $1/6$. Suppose that either department has the same probability to receive the grant. If the physics department obtains the grant, Professor A is the only recipient of the grant. If the grant is given to the mathematics department, Professor B has the probability $2/3$ to receive it, while Professor C's probability to get it is only $1/3$. The above "absolute uncertainty" analysis and "conditional uncertainty" analysis should give the same result concerning the final degree of uncertainty. That is, the degree of uncertainty for Professors A, B, C to receive the grant, $H(1/2, 1/3, 1/6)$, should equal the degree of uncertainty for the grant to be first assigned to the physics department or the mathematics department, $H(1/2, 1/2)$, plus the degree of uncertainty that under the condition of mathematics department's obtaining the grant, Professor B or C receives it, $H(2/3, 1/3)$, multiplied by $1/2$ because the probability for the mathematics department to receive the grant is $1/2$. Thus,

$$H\left(\frac{1}{2}, \frac{1}{3}, \frac{1}{6}\right) = H\left(\frac{1}{2}, \frac{1}{2}\right) + \frac{1}{2} H\left(\frac{2}{3}, \frac{1}{3}\right).$$

To generalize the above argument, we have the third condition:

(iii) If an experiment is decomposed into several successive ones, then the original value of H is the weighted sum of the corresponding values of H.

We show that basically the above three conditions are enough to uniquely determine the required entropy function H.

Theorem 8.1.1 *Any real-valued nonnegative function H satisfying the above conditions* (i), (ii), *and* (iii) *must have the expression*

$$H(p_1, p_2, \cdots, p_n) = -K \sum_{i=1}^{n} p_i \ln p_i, \tag{8.1}$$

where K is a positive constant.

Proof Denote $A(n) = H(1/n, \cdots, 1/n)$. Let $n = s^k$, where s and k are two positive integers. Then condition (iii) implies that

$$\begin{aligned} H\left(\frac{1}{s^k}, \cdots, \frac{1}{s^k}\right) &= \sum_{i=0}^{k-1} \left[\sum_{j=1}^{s^i} \frac{1}{s^i} H\left(\frac{1}{s}, \cdots, \frac{1}{s}\right) \right] \\ &= kH\left(\frac{1}{s}, \cdots, \frac{1}{s}\right). \end{aligned} \tag{8.2}$$

Next, suppose that four positive integers s, t, n, and k satisfy

$$s^k \leqslant t^n < s^{k+1}.$$

Then $k \ln s \leqslant n \ln t < (k+1) \ln s$, which implies that

$$\frac{k}{n} \leqslant \frac{\ln t}{\ln s} < \frac{k}{n} + \frac{1}{n}.$$

Thus,

$$\left| \frac{k}{n} - \frac{\ln t}{\ln s} \right| < \frac{1}{n}. \tag{8.3}$$

On the other hand, condition (ii) and (8.2) imply that $kA(s) \leqslant nA(t) < (k+1)A(s)$, so we have

$$\left|\frac{k}{n} - \frac{A(t)}{A(s)}\right| < \frac{1}{n}. \tag{8.4}$$

Combining (8.3) and (8.4) yields

$$\left|\frac{A(t)}{A(s)} - \frac{\ln t}{\ln s}\right| < \frac{2}{n}.$$

Since n is arbitrary and the left-hand side of the above inequality is independent of n, and since t and s are independent of each other, we must have

$$\frac{A(t)}{A(s)} = \frac{\ln t}{\ln s} \equiv K,$$

where K is a fixed positive constant. Therefore,

$$A(t) = K \ln t,$$

which means that

$$H\left(\frac{1}{n}, \cdots, \frac{1}{n}\right) = K \ln n = -K \sum_{i=1}^{n} \frac{1}{n} \ln \frac{1}{n}.$$

That is, (8.1) is true for the special case when $p_i = 1/n$, $i = 1, 2, \cdots, n$.

Next, we show that (8.1) is still true when $p_1, \cdots, p_n$ are positive rational numbers satisfying $\sum\limits_{i=1}^{n} p_i = 1$. There exist natural numbers $k_1, k_2, \cdots, k_n$ such that

$$p_i = \frac{k_i}{\sum\limits_{j=1}^{n} k_j}, \quad i = 1, 2, \cdots, n.$$

Using condition (iii), we see that

$$A\left(\sum_{i=1}^{n} k_i\right) = H(p_1, p_2, \cdots, p_n) + \sum_{i=1}^{n} p_i A(k_i).$$

Substituting $A(k_i) = K \ln k_i$, $\forall\, i$ into the above equality leads to

$$\begin{aligned} H(p_1, p_2, \cdots, p_n) &= K \ln \sum_{i=1}^{r} k_i - \sum_{i=1}^{n} p_i (K \ln k_i) \\ &= K \left(\sum_{i=1}^{n} p_i \ln \sum_{j=1}^{n} k_j \right) \end{aligned}$$

$$= -K\sum_{i=1}^{n} p_i \ln \frac{k_i}{\sum\limits_{j=1}^{n} k_j} = -K\sum_{i=1}^{n} p_i \ln p_i.$$

That is, (8.1) is satisfied by all positive rational numbers $p_1, p_2, \cdots, p_n$ satisfying $\sum\limits_{i=1}^{n} p_i = 1$. By a limit process thanks to condition (i), we see that (8.1) is still valid for all nonnegative real numbers $p_1, p_2, \cdots, p_n$ satisfying $\sum\limits_{i=1}^{n} p_i = 1$. □

From the expression of H in the theorem, if we let $p_i = 1$ for some i, then $H(p_1, p_2, \cdots, p_n) = 0$. This coincides with our expectation: $p_i = 1$ means that the corresponding event always happens, and so there is no uncertainty of the outcome.

We are ready to give the following definition of entropy.

Definition 8.1.1 *The number*

$$H(p_1, p_2, \cdots, p_n) = -\sum_{i=1}^{n} p_i \ln p_i \tag{8.5}$$

is called the entropy *or the* Shannon entropy *corresponding to the sample space* $(p_1, p_2, \cdots, p_n)$.

Since entropy is a mathematical measure of uncertainty, it is natural to expect that when $p_1 = \cdots = p_n = 1/n$, the corresponding value of H is maximal. Although this is not required in the development of the expression for H, it is actually true.

Proposition 8.1.1 *The Shannon entropy has the following extreme value property:*

$$\ln n = H\left(\frac{1}{n}, \cdots, \frac{1}{n}\right) = \max\left\{H(p_1, p_2, \cdots, p_n) : p_i \geqslant 0, \sum_{i=1}^{n} p_i = 1\right\}.$$

Proof Clearly $H(1/n, \cdots, 1/n) = \ln n$. The function $\ln t$ is strictly concave on $(0, \infty)$ since $(\ln t)'' = -1/t^2 < 0$ for $t > 0$. Thus, for any positive numbers $p_1, p_2, \cdots, p_n$ with $\sum\limits_{i=1}^{n} p_i = 1$,

$$\begin{aligned} H(p_1, p_2, \cdots, p_n) &= -\sum_{i=1}^{n} p_i \ln p_i = \sum_{i=1}^{n} p_i \ln \frac{1}{p_i} \\ &\leqslant \ln \sum_{i=1}^{n} \frac{p_i}{p_i} = \ln n = H\left(\frac{1}{n}, \cdots, \frac{1}{n}\right). \end{aligned}$$

When $p_i = 0$ for some i, say $i = 1$, and other p_i's are positive, then from the above proof,

$$H(0, p_2, \cdots, p_n) \leqslant H\left(\frac{1}{n-1}, \cdots, \frac{1}{n-1}\right) < H\left(\frac{1}{n}, \cdots, \frac{1}{n}\right),$$

where the last inequality is from condition (ii). □

Remark 8.1.1 Definition 8.1.1 of entropy was introduced by C. Shannon [118] in 1948 for the study of information theory.

8.2 Kolmogorov Entropy

In order to motivate the concept of the Kolmogorov entropy, we do the mathematical experiment of tossing a regular smooth coin n times. We use 0 and 1 to represent the Head and the Tail of the coin, respectively. Then, each outcome of the experiment can be represented by a finite sequence consisting of n digits of 0 and 1. This sequence can be represented as a binary number

$$0.a_1 a_2 \cdots a_n, \quad a_i = 0 \text{ or } 1, \quad i = 1, 2, \cdots, n$$

which is the binary expansion of a rational number $x \in [0, 1]$. That is,

$$x = 0.a_1 a_2 \cdots a_n = \frac{a_1}{2} + \frac{a_2}{2^2} + \cdots + \frac{a_n}{2^n}.$$

There are exactly 2^n possible outcomes for n tosses, so we have a sample space of 2^n basic events each of which has probability 2^{-n}. From the previous section, the Shannon entropy of this sample space equals $n \ln 2$.

Now we ask the following question: suppose that we know the results of the first $n-1$ tosses, what is the degree of uncertainty for the outcome of the nth toss? To answer this question, we let

$$\bar{A} = \left\{ I_1 = \left[0, \frac{1}{2}\right), \; I_2 = \left[\frac{1}{2}, 1\right] \right\} \tag{8.6}$$

be a partition of $I = [0, 1]$. Let $0.a_1 a_2 \cdots a_n$ be a binary representation of some $x \in [0, 1]$. Clearly, if $a_1 = 0$, then $x \in [0, 1/2)$ and $a_1 = 1$ implies that $x \in [1/2, 1]$. Consider the dyadic mapping

$$S(x) = 2x \pmod 1, \quad \forall\, x \in [0, 1]. \tag{8.7}$$

Then it is easy to see that the action of S to a binary expansion of x simply moves the digits to the left by one place. Namely, if

$$x = 0.a_1 a_2 \cdots a_n = \frac{a_1}{2} + \frac{a_2}{2^2} + \cdots + \frac{a_n}{2^n},$$

then,

$$S(x) = 0.a_2 \cdots a_n = \frac{a_2}{2} + \cdots + \frac{a_n}{2^{n-1}}.$$

Thus, $a_2 = 0$ implies $S(x) \in [0, 1/2)$ and $a_2 = 1$ implies $S(x) \in [1/2, 1]$. By the same token, in general $S^{i-1}(x) \in [0, 1/2)$ if $a_i = 0$ and $S^{i-1}(x) \in [1/2, 1]$ if $a_i = 1$. Therefore, the outcome of a consecutive toss can be represented by the addresses of $x, S(x), S^2(x), \cdots$ in the partition $\bar{A}$. Knowing the results of the first $n-1$ tosses is equivalent to knowing the locations of $x, S(x), \cdots, S^{n-2}(x)$ in the partition $\bar{A}$, and whether the nth toss is Head up or Tail up is the same as whether the value $S^{n-1}(x)$ belongs to I_1 or I_2 respectively. Therefore, our problem can be reformulated as follows: for the mapping S in (8.7), suppose that we know the addresses of $x, S(x), \cdots, S^{n-2}(x)$ in the partition (8.6), what is the degree of uncertainty for the location of $S^{n-1}(x)$ in $\bar{A}$?

With the above intuitive motivation, we can define the Kolmogorove entropy for general transformations. Let (X, Σ, μ) be a probability measure space and let $S : X \to X$ be a measurable transformation. Denote by $\bar{A} = \{A_1, A_2, \cdots, A_k\}$ an arbitrary finite measurable partition of X, where each $A_i \in \Sigma$. Then $\bar{A}$ can be viewed as a sample space consisting of k basic events $A_1, A_2, \cdots, A_k$ with probabilities $\mu(A_1), \mu(A_2), \cdots, \mu(A_k)$. The Shannon entropy of $\bar{A}$ is then given by

$$H(\bar{A}) = -\sum_{i=1}^{k} \mu(A_i) \ln \mu(A_i).$$

For the given S, the class of the measurable subsets

$$S^{-1}(\bar{A}) = \{S^{-1}(A_1), \cdots, S^{-1}(A_k)\}$$

is also a finite measurable partition of X. We ask the question: on the premise that the experimental result $\bar{A}$ is known, what is the degree of uncertainty of the experimental result $S^{-1}(\bar{A})$? That is, we want to know the degree of uncertainty where $S(x)$ is when the location of x in $\bar{A}$ is known. This problem can be dealt with from the viewpoint of *conditional probability.* Notice that $S(x) \in A_i$ if and only if $x \in S^{-1}(A_i)$. Thus, for each pair of $i, j = 1, 2, \cdots, k$, the set of all $x \in A_i$ such that $S(x) \in A_j$ is $A_i \cap S^{-1}(A_j)$ and its conditional probability is

$$\frac{\mu(A_i \cap S^{-1}(A_j))}{\mu(A_i)}.$$

By the definition of Shannon's entropy, we know that under the condition $x \in A_i$, the degree of uncertainty where $S(x)$ lies in $\bar{A}$ should be

$$H_i = -\sum_{j=1}^{k} \frac{\mu(A_i \cap S^{-1}(A_j))}{\mu(A_i)} \ln \frac{\mu(A_i \cap S^{-1}(A_j))}{\mu(A_i)}, \quad i = 1, 2, \cdots, k.$$

From condition (iii) for the Shannon entropy, under the condition that the experimental result $\{A_1, A_2, \cdots, A_k\}$ is known, the degree of uncertainty of the experimental result $S^{-1}(\bar{A}) = \{S^{-1}(A_1), \cdots, S^{-1}(A_k)\}$, denoted as $H(S^{-1}(\bar{A})|\bar{A})$, is the weighted sum of $H_1, \cdots, H_k$, i.e.,

$$\begin{aligned} H(S^{-1}(\bar{A})|\bar{A}) &= \sum_{i=1}^{k} \mu(A_i) H_i \\ &= -\sum_{i=1}^{k}\sum_{j=1}^{k} \mu(A_i \cap S^{-1}(A_j)) \ln \frac{\mu(A_i \cap S^{-1}(A_j))}{\mu(A_i)}. \end{aligned} \tag{8.8}$$

The next proposition gives an equivalent expression for $H(S^{-1}(\bar{A})|\bar{A})$. Given two finite measurable partitions $\bar{A}$, $\bar{B}$ of X, the notation $\bar{A} \bigvee \bar{B}$ stands for the finite measurable partition of X defined by

$$\bar{A} \bigvee \bar{B} = \{A \cap B : A \in \bar{A},\ B \in \bar{B}\}.$$

Proposition 8.2.1 $H(S^{-1}(\bar{A})|\bar{A}) = H\left(\bar{A} \bigvee S^{-1}(\bar{A})\right) - H(\bar{A})$.

Proof After a direct computation, we have

$$\begin{aligned} H(S^{-1}(\bar{A})|\bar{A}) &= -\sum_{i=1}^{k}\sum_{j=1}^{k} \mu(A_i \cap S^{-1}(A_j)) \ln \frac{\mu(A_i \cap S^{-1}(A_j))}{\mu(A_i)} \\ &= -\sum_{i=1}^{k}\sum_{j=1}^{k} \mu(A_i \cap S^{-1}(A_j)) \left[\ln \mu(A_i \cap S^{-1}(A_j)) - \ln \mu(A_i)\right] \\ &= -\sum_{i=1}^{k}\sum_{j=1}^{k} \mu(A_i \cap S^{-1}(A_j)) \ln \mu(A_i \cap S^{-1}(A_j)) \\ &\quad + \sum_{i=1}^{k}\sum_{j=1}^{k} \mu(A_i \cap S^{-1}(A_j)) \ln \mu(A_i) \\ &= H(\bar{A} \bigvee S^{-1}(\bar{A})) + \sum_{i=1}^{k} \mu(A_i) \ln \mu(A_i) \\ &= H(\bar{A} \bigvee S^{-1}(\bar{A})) - H(\bar{A}). \end{aligned}$$

□

Motivated by (8.8), we give the formal definition of the conditional entropy of one finite measurable partition of X with respect to another finite measurable partition of X.

Definition 8.2.1 *Let (X, Σ, μ) be a probability measure space. Suppose that $\bar{A} = \{A_1, A_2, \cdots, A_k\}$ and $\bar{B} = \{B_1, B_2, \cdots, B_l\}$ are two finite measurable partitions of X. The* conditional Shannon entropy *of $\bar{A}$ with respect to $\bar{B}$, $H(\bar{A}|\bar{B})$,* *is defined as*

$$H(\bar{A}|\bar{B}) = -\sum_{j=1}^{l} \mu(B_j) \left[\sum_{i=1}^{k} \frac{\mu(A_i \cap B_j)}{\mu(B_j)} \ln \frac{\mu(A_i \cap B_j)}{\mu(B_j)} \right].$$

Then, using the same idea of the proof to Proposition 8.2.1, we can prove the following useful result (see Exercise 8.1 (iii)).

Proposition 8.2.2 $H(\bar{A}|\bar{B}) = H\left(\bar{A} \bigvee \bar{B}\right) - H(\bar{B})$.

Remark 8.2.1 Proposition 8.2.2 is intuitively obvious: the degree of uncertainty $H\left(\bar{A} \bigvee \bar{B}\right)$ of the experimental result $\bar{A} \bigvee \bar{B}$ should be the degree of uncertainty $H(\bar{B})$ of the experimental result $\bar{B}$ plus the degree of uncertainty $H(\bar{A}|\bar{B})$ of the experimental result $\bar{A}$ under the condition that the experimental result $\bar{B}$ is known.

From the above discussion we know that, for a measurable transformation $S : X \to X$, the degree of uncertainty of the experimental result $S^{-1}(\bar{A})$ when the experimental result $\bar{A}$ is known characterizes the degree of uncertainty of the location of $S(x)$ in $\bar{A}$ when the location of x in $\bar{A}$ is known. In general, the degree of uncertainty of the location of $S^n(x)$ in $\bar{A}$ when the locations of $x, S(x), \cdots, S^{n-1}(x)$ in $\bar{A}$ are known is the degree of uncertainty of the experimental result $S^{-n}(\bar{A})$ when the experimental result $\bigvee_{i=0}^{n-1} S^{-i}(\bar{A}) = \bar{A} \bigvee S^{-1}(\bar{A}) \bigvee \cdots \bigvee S^{-(n-1)}(\bar{A})$ is known. The Kolmogorov entropy basically describes the asymptotic property of this degree of uncertainty when n approaches infinity. Using Proposition 8.2.2, we easily see that when the experimental result $\bigvee_{i=0}^{n-1} S^{-i}(\bar{A})$ is known, the degree of uncertainty of the experimental result $S^{-n}(\bar{A})$ is

$$H\left(S^{-n}(\bar{A})| \bigvee_{i=0}^{n-1} S^{-i}(\bar{A})\right) = H\left(\bigvee_{i=0}^{n} S^{-i}(\bar{A})\right) - H\left(\bigvee_{i=0}^{n-1} S^{-i}(\bar{A})\right).$$

Definition 8.2.2 *Let* $\bar{A} = \{A_1, A_2, \cdots, A_k\}$ *be a finite measurable partition of* X. *Then the* entropy *of the measurable transformation* $S : X \to X$ *with respect to* $\bar{A}$ *is defined as*

$$h_\mu(S, \bar{A}) = \limsup_{n\to\infty} H\left(S^{-n}(\bar{A})| \bigvee_{i=0}^{n-1} S^{-i}(\bar{A})\right).$$

Definition 8.2.3 *Let (X, Σ, μ) be a probability measure space and let $S : X \to X$ be a measurable transformation. Then the* Kolmogorov entropy *of S is defined as*

$$h_\mu(S) = \sup\{h_\mu(S, \bar{A}) : \bar{A} \text{ is a finite measurable partition of } X\}.$$

For a general measurable transformation $S : X \to X$, the upper limit in Definition 8.2.2 cannot be changed to a regular limit. However, for the class of measure preserving transformations, it can be shown that $\lim_{n\to\infty} H\left(S^{-n}(\bar{A})| \bigvee_{i=0}^{n-1} S^{-i}(\bar{A})\right)$ does exist, and in this particular case $h_\mu(S, \bar{A})$ has an equivalent definition. To prove this, we need the following lemma. Suppose that $\bar{C}$ and $\bar{D}$ are two finite measurable partitions of X. If each member of $\bar{C}$ is a union of some members of $\bar{D}$, then we write $\bar{C} \leqslant \bar{D}$ and say that $\bar{D}$ is a *refinement* of $\bar{C}$.

Lemma 8.2.1 *Let $\bar{A}, \bar{C}, \bar{D}$ be three finite measurable partitions of X. If $\bar{C} \leqslant \bar{D}$, then $H(\bar{A}|\bar{C}) \geqslant H(\bar{A}|\bar{D})$.*

Proof Let $\bar{A} = \{A_i\}$, $\bar{C} = \{C_j\}$, $\bar{D} = \{D_l\}$. We must show that

$$\begin{aligned}
&-\sum_j \sum_i \mu(C_j) \frac{\mu(A_i \cap C_j)}{\mu(C_j)} \ln \frac{\mu(A_i \cap C_j)}{\mu(C_j)} \\
&\leqslant \sum_l \sum_i \mu(D_l) \frac{\mu(A_i \cap D_l)}{\mu(D_l)} \ln \frac{\mu(A_i \cap D_l)}{\mu(D_l)} \\
&= -\sum_l \sum_i \sum_j \mu(C_j \cap D_l) \frac{\mu(A_i \cap D_l)}{\mu(D_l)} \ln \frac{\mu(A_i \cap D_l)}{\mu(D_l)}.
\end{aligned}$$

It is enough to show that for each i and j,

$$\begin{aligned}
&\mu(C_j) \frac{\mu(A_i \cap C_j)}{\mu(C_j)} \ln \frac{\mu(A_i \cap C_j)}{\mu(C_j)} \\
&\leqslant \sum_l \mu(C_j \cap D_l) \frac{\mu(A_i \cap D_l)}{\mu(D_l)} \ln \frac{\mu(A_i \cap D_l)}{\mu(D_l)}.
\end{aligned}$$

Define $\phi(x) = x \ln x$ for $x > 0$ and $\phi(0) = 0$. Then the above inequality can be written as

$$\phi\left(\frac{\mu(A_i \cap C_j)}{\mu(C_j)}\right) \leqslant \sum_l \frac{\mu(A_i \cap D_l)}{\mu(C_j)} \phi\left(\frac{\mu(A_i \cap D_l)}{\mu(D_l)}\right).$$

Since ϕ is a convex function, the assumption $\bar{C} \leqslant \bar{D}$ implies that

$$\begin{aligned}
&\sum_l \frac{\mu(C_j \cap D_l)}{\mu(C_j)} \phi\left(\frac{\mu(A_i \cap D_l)}{\mu(D_l)}\right) \\
\geqslant &\phi\left(\sum_l \frac{\mu(C_j \cap D_l)}{\mu(C_j)} \frac{\mu(A_i \cap D_l)}{\mu(D_l)}\right) \\
= &\phi\left(\frac{\mu(A_i \cap C_j)}{\mu(C_j)}\right),
\end{aligned}$$

where the last equality follows from the fact that $\mu(C_j \cap D_l) = 0$ or $\mu(D_l)$, and if $\mu(C_j \cap D_l) = \mu(D_l)$, then $\mu(A_i \cap D_l) = \mu(A_i \cap C_j)$. This proves that $H(\bar{A}|\bar{C}) \geqslant H(\bar{A}|\bar{D})$. □

Theorem 8.2.1 *Let (X, Σ, μ) be a probability measure space. If $S : X \to X$ is a measure preserving transformation, then for any finite measurable partition $\bar{A}$ of X,*

$$\lim_{n\to\infty} H\left(S^{-n}(\bar{A})|\bigvee_{i=0}^{n-1} S^{-i}(\bar{A})\right) = \lim_{n\to\infty} \frac{1}{n} H\left(\bigvee_{i=0}^{n-1} S^{-i}(\bar{A})\right).$$

Proof First of all, the assumption that S preserves μ implies that $H(S^{-1}(\bar{A})) = H(\bar{A})$. Using Proposition 8.2.2, we obtain

$$\begin{aligned}
&H\left(S^{-n}(\bar{A})|\bigvee_{i=0}^{n-1} S^{-i}(\bar{A})\right) \\
= &H\left(\bigvee_{i=0}^{n} S^{-i}(\bar{A})\right) - H\left(\bigvee_{i=0}^{n-1} S^{-i}(\bar{A})\right) \\
= &H\left(\bigvee_{i=0}^{n} S^{-i}(\bar{A})\right) - H\left(\bigvee_{i=1}^{n} S^{-i}(\bar{A})\right) \\
= &H\left(\bar{A}|\bigvee_{i=1}^{n} S^{-i}(\bar{A})\right).
\end{aligned} \tag{8.9}$$

By Lemma 8.2.1, the sequence $\left\{H\left(\bar{A}|\bigvee_{i=1}^{n} S^{-i}(\bar{A})\right)\right\}$ of nonnegative numbers is monotonically decreasing, so it has a limit. On the other hand, for $i = 1, 2, \cdots, n-1$, adding up the following $n-1$ equalities obtained from applying Proposition 8.2.2,

$$H\left(S^{-i}(\bar{A})|\bigvee_{j=0}^{i-1} S^{-j}(\bar{A})\right) = H\left(\bigvee_{j=0}^{i} S^{-j}(\bar{A})\right) - H\left(\bigvee_{j=0}^{i-1} S^{-j}(\bar{A})\right),$$

we have

$$
\begin{aligned}
H\left(\bigvee_{i=0}^{n-1} S^{-i}(\bar{A})\right) &= H(\bar{A}) + \sum_{i=1}^{n-1} H\left(S^{-i}(\bar{A}) | \bigvee_{j=0}^{i-1} S^{-j}(\bar{A})\right) \\
&= \sum_{i=0}^{n-1} H\left(S^{-i}(\bar{A}) | \bigvee_{j=0}^{i-1} S^{-j}(\bar{A})\right) \\
&= \sum_{i=0}^{n-1} H\left(\bar{A} | \bigvee_{j=1}^{i} S^{-j}(\bar{A})\right),
\end{aligned}
$$

where the last equality is from (8.9). Therefore,

$$
\frac{1}{n} H\left(\bigvee_{i=0}^{n-1} S^{-i}(\bar{A})\right) = \frac{1}{n} \sum_{i=0}^{n-1} H\left(\bar{A} | \bigvee_{j=1}^{i} S^{-j}(\bar{A})\right).
$$

Using the fact (3.6) and the equality (8.9), we obtain

$$
\begin{aligned}
\lim_{n\to\infty} \frac{1}{n} H\left(\bigvee_{i=0}^{n-1} S^{-i}(\bar{A})\right) &= \lim_{n\to\infty} \frac{1}{n} \sum_{i=0}^{n-1} H\left(\bar{A} | \bigvee_{j=1}^{i} S^{-j}(\bar{A})\right) \\
&= \lim_{n\to\infty} H\left(\bar{A} | \bigvee_{i=1}^{n} S^{-i}(\bar{A})\right) \\
&= \lim_{n\to\infty} H\left(S^{-n}(\bar{A}) | \bigvee_{i=0}^{n-1} S^{-i}(\bar{A})\right).
\end{aligned}
$$

□

Because of Theorem 8.2.1, we have the following definition for measure preserving transformations.

Definition 8.2.4 *Let $\bar{A} = \{A_1, A_2, \cdots, A_k\}$ be a finite measurable partition of X, then the* entropy *of a measure preserving transformation $S : X \to X$ with respect to $\bar{A}$ is defined as*

$$
h_\mu(S, \bar{A}) = \lim_{n\to\infty} \frac{1}{n} H\left(\bigvee_{i=0}^{n-1} S^{-i}(\bar{A})\right).
$$

8.3 Topological Entropy

The concept of the Kolmogorov entropy in the previous section was about the uncertainty with respect to measurable transformations. It can be employed for the uncertainty related to continuous transformations on compact topological spaces since continuous functions are measurable with respect to the Borel algebra. There is, however, one major difficulty to this generalization: in a topological space there is no metric which is similar to the measure of a measure space. Instead we use the means of *finite covering.* In the following we restrict our discussion to only compact metric spaces, although the same idea works the same way for more general compact Hausdorff spaces.

Let X be a compact metric space and let $S : X \to X$ be a continuous transformation. Since X is compact, there is a finite open covering $\bar{A} = \{A_1, A_2, \cdots, A_k\}$ of X. If a subclass of $\bar{A}$ also covers X, it is called a *subcovering.* A subcovering of $\bar{A}$ is said to be *minimal* if there are no other subcoverings of $\bar{A}$ with fewer elements. We denote by $k(\bar{A})$ the number of elements in a minimal subcovering of $\bar{A}$. If each element of a minimal subcovering of $\bar{A}$ is viewed as a basic event with probability $1/k(\bar{A})$, we obtain a finite sample space for which the Shannon entropy is $\ln k(\bar{A})$. This number is called the *entropy* of the open covering $\bar{A}$ and is denoted by $H(\bar{A})$. If $\bar{A}$ is an open covering of X, then $S^{-1}(\bar{A}) = \{S^{-1}(A) : A \in \bar{A}\}$ is also an open covering of X. Furthermore, if $\{A_1, A_2, \cdots, A_{k(\bar{A})}\}$ is a minimal subcovering of $\bar{A}$, then $\{S^{-1}(A_1), \cdots, S^{-1}(A_{k(\bar{A})})\}$ is a subcovering of $S^{-1}(\bar{A})$, but it may not be a minimal subcovering. Hence,

$$k(S^{-1}(\bar{A})) \leqslant k(\bar{A}). \tag{8.10}$$

When $\bar{A}$ and $\bar{B}$ are two open coverings of X, we use $\bar{A} \bigvee \bar{B}$ to denote the open covering $\{A \cap B : A \in \bar{A},\ B \in \bar{B}\}$ of X. If $\{A_1, A_2, \cdots, A_{k(\bar{A})}\}$ is a minimal subcovering of $\bar{A}$ and $\{B_1, B_2, \cdots, B_{k(\bar{B})}\}$ is a minimal subcovering of $\bar{B}$, then $\{A_i \cap B_j : i = 1, 2, \cdots, k(\bar{A}),\ j = 1, 2, \cdots, k(\bar{B})\}$ is a subcovering of $\bar{A} \bigvee \bar{B}$. Thus, $k\left(\bar{A} \bigvee \bar{B}\right) \leqslant k(\bar{A})k(\bar{B})$. It follows that

$$H\left(\bar{A} \bigvee \bar{B}\right) \leqslant H(\bar{A}) + H(\bar{B}). \tag{8.11}$$

We use $k(\bar{A}, S, n)$ to denote the number of elements in a minimal subcovering of the open covering $\bigvee_{i=0}^{n-1} S^{-i}(\bar{A})$ and show that the sequence $\{k(\bar{A}, S, n)/n\}$ converges as n goes to infinity.

Theorem 8.3.1 *Let X be a compact metric space and let $S : X \to X$ be a continuous transformation. Then the limit*

$$\lim_{n \to \infty} \frac{1}{n} \ln k(\bar{A}, S, n)$$

exists for any finite open covering $\bar{A}$ of X.

Proof We need the following result from calculus: if a real number sequence $\{a_n\}$ satisfies the condition that

$$a_{n+p} \leqslant a_n + a_p, \quad \forall\, n,\ p,$$

then $\lim_{n\to\infty} a_n/n$ exists and equals $\inf_n a_n/n$. Denote $a_n = \ln k(\bar{A}, S, n)$. Then from (8.10) and (8.11),

$$\begin{aligned}
a_{n+p} &= H\left(\bigvee_{i=0}^{n+p-1} S^{-i}(\bar{A})\right)\\
&= H\left(\left(\bigvee_{i=0}^{n-1} S^{-i}(\bar{A})\right) \bigvee \left(\bigvee_{i=n}^{n+p-1} S^{-i}(\bar{A})\right)\right)\\
&\leqslant H\left(\bigvee_{i=0}^{n-1} S^{-i}(\bar{A})\right) + H\left(S^{-n}\left(\bigvee_{i=0}^{p-1} S^{-i}(\bar{A})\right)\right)\\
&\leqslant a_n + a_p,
\end{aligned}$$

and so the theorem follows. □

Thus, similar to Definition 8.2.3, we have

Definition 8.3.1 *The* topological entropy *of a continuous transformation S on a compact metric space X with respect to a finite open covering $\bar{A}$ of X is defined as*

$$h_{\text{top}}(S, \bar{A}) = \lim_{n\to\infty} \frac{1}{n} H\left(\bigvee_{i=0}^{n-1} S^{-i}(\bar{A})\right) = \lim_{n\to\infty} \frac{1}{n} \ln k(\bar{A}, S, n).$$

Definition 8.3.2 *The* topological entropy *of a continuous transformation S on a compact metric space X is defined as*

$$h_{\text{top}}(S) = \sup\{h_{\text{top}}(S, \bar{A}) : \bar{A} \textit{ is a finite open covering of } X\}.$$

A compact metric space X becomes a measurable space with the Borel σ-algebra $\mathcal{B}$ generated by the family of all open subsets of X. Given any probability measure μ on $\mathcal{B}$, we obtain a probability measure space $(X, \mathcal{B}, \mu)$ and every continuous transformation $S : X \to X$ is measurable. Thus a Kolmogorov entropy $h_\mu(S)$ is well-defined and depends on μ. On the other hand, the topological entropy $h_{\text{top}}(S)$ is independent of μ. A natural question is: what is the relation between the topological entropy $h_{\text{top}}(S)$ and the values of Kolmogorov entropy $h_\mu(S)$? The following result of Goodwyn, Dinaburg, and Goodman [123] gave an answer to this question and solved a conjecture by Adler, Konhein, and McAndrew [3].

Theorem 8.3.2 *Let X be a compact metric space and let $S : X \to X$ be a continuous transformation. Then*

$$h_{\text{top}}(S) = \sup\{h_\mu(S) : \mu \textit{ is ergodic with respect to } S\}.$$

8.4 Boltzmann Entropy

The original concept of entropy was first introduced by the German theoretical physicist Rudolf Clausius into thermodynamics in 1865, and it was later used in a different form by the Austrian physicist Ludwig Boltzmann in his pioneering work on the kinetic theory of gases in 1866. In thermodynamics and statistical physics, entropy is a measure of the amount of information required to specify the state of a thermodynamic system, and the entropy of substances is a measure of the disorder of randomness of their molecular movements. The famous second law of thermodynamics claims that in an isolated physical system, the thermodynamic entropy never decreases. In this section we give a mathematical description of the Boltzmann entropy for density functions.

Let (X, Σ, μ) be a probability measure space. A density function $f \in L^1(\mu)$ defines a probability measure μ_f on (X, Σ) via $\mu_f(A) = \int_A f \mathrm{d}\mu,\ \forall\, A \in \Sigma$. This new probability space (X, Σ, μ_f) can be viewed as an infinite sample space. Suggested by the Shannon entropy formula (8.5), we are led to the definition of the Boltzmann entropy in the following way. Recall that the set of all densities is denoted by $\mathcal{D}$. Let a function η on $[0, \infty)$ be defined by

$$\eta(u) = \begin{cases} -u \ln u, & u > 0, \\ 0, & u = 0. \end{cases} \tag{8.12}$$

Since

$$\eta''(u) = -\frac{1}{u} < 0, \ \ \forall\, u > 0,$$

η is a strictly concave function on $[0, \infty)$.

Definition 8.4.1 *Let $f \in L^1(\mu)$ be nonnegative. If $\eta \circ f \in L^1(\mu)$, then the* Boltzmann entropy *of f is defined as*

$$H(f) = \int_X \eta(f(x)) \mathrm{d}\mu(x) = -\int_X f(x) \ln f(x) \mathrm{d}\mu(x). \tag{8.13}$$

Remark 8.4.1 If $\mu(X) < \infty$, then the integral (8.13) is well-defined for all $f \geqslant 0$. In fact, since $\eta(u) \geqslant 0$ if and only if $0 \leqslant u \leqslant 1$, the integral $\int_X (\eta \circ f)^+ \mathrm{d}\mu$ is always finite. Therefore, $H(f)$ is either finite or equal to $-\infty$.

Proposition 8.4.1 *The Boltzmann entropy $H(f)$ of f is a concave functional on its domain.*

Proof Since η is a strictly concave function on $[0, \infty)$, for any $u,\ v \geqslant 0$ and $0 \leqslant \alpha \leqslant 1$,

$$\eta(\alpha u + (1 - \alpha)v) \geqslant \alpha\eta(u) + (1 - \alpha)\eta(v).$$

Applying the above inequality to $f,\ g$ in the domain of H gives

$$\eta(\alpha f(x) + (1-\alpha)g(x)) \geqslant \alpha\eta(f(x)) + (1-\alpha)\eta(g(x)), \quad \forall\, x \in X.$$

After integration it follows that

$$H(\alpha f + (1-\alpha)g) \geqslant \alpha H(f) + (1-\alpha)H(g). \qquad \square$$

The concavity of the scalar function η defined by (8.12) leads to the so-called *Gibbs inequality*

$$u - u\ln u \leqslant v - u\ln v, \quad \forall\, u, v \geqslant 0,$$

which implies that, for all $f,\ g \in \mathcal{D}$ such that $\eta\circ f$ and $f\ln g$ are both integrable,

$$\int_X f(x)\ln f(x)\,\mathrm{d}\mu(x) \geqslant \int_X f(x)\ln g(x)\,\mathrm{d}\mu(x), \quad \forall\, f,\ g \in \mathcal{D}. \tag{8.14}$$

Moreover [16], for all $f,\ g \in \mathcal{D}$,

$$\int_X f(x)\ln\frac{f(x)}{g(x)}\,\mathrm{d}\mu(x) \geqslant \frac{1}{2}\left(\int_X |f(x) - g(x)|\,\mathrm{d}\mu(x)\right)^2, \tag{8.15}$$

and the equality sign holds in (8.15) if and only if $f = g$.

Remember that the Shannon entropy $H(p_1, p_2, \cdots, p_n)$ achieves its maximal value $\ln n$ when $p_1 = \cdots = p_n = 1/n$. The Boltzmann entropy has a similar property.

Proposition 8.4.2 *Suppose that* $\mu(X) < \infty$. *Then the constant density* $f^*(x) \equiv 1/\mu(X)$ *satisfies*

$$H(f^*) = \ln\mu(X) = \max\{H(f) : f \in \mathcal{D}\}.$$

Proof Obviously, $f^* \in \mathcal{D}$. Given any $f \in \mathcal{D}$, by (8.14),

$$\begin{aligned} H(f) &= -\int_X f(x)\ln f(x)\,\mathrm{d}\mu(x) \leqslant -\int_X f(x)\ln\frac{1}{\mu(X)}\mathrm{d}\mu(x) \\ &= \ln\mu(X)\int_X f(x)\mathrm{d}\mu(x) = \ln\mu(X) = H(f^*). \end{aligned} \qquad \square$$

Let $\{g_1, g_2, \cdots, g_k\}$ be a finite set of functions in $L^\infty(\mu)$. Consider the following *maximum entropy problem*

$$\max\left\{H(f) : f \in \mathcal{D},\ \int_X f(x)g_i(x)\,\mathrm{d}\mu(x) = b_i,\ 1 \leqslant i \leqslant k\right\}, \tag{8.16}$$

where $b_1, b_2, \cdots, b_k$ are the given real constants. The Gibbs inequality above provides a simple way for a solution to the special constrained optimization problem as the following proposition shows.

Proposition 8.4.3 *Suppose that $a_1, a_2, \cdots, a_k$ are real numbers such that the function*

$$f^*(x) = \frac{\exp\left(\sum_{i=1}^{k} a_i g_i(x)\right)}{\int_X \exp\left(\sum_{i=1}^{k} a_i g_i(x)\right) \mathrm{d}\mu(x)}, \quad \forall\, x \in X$$

satisfies the constraints in (8.16), *that is,*

$$\frac{\int_X g_i(x) \exp\left(\sum_{j=1}^{k} a_j g_j(x)\right) \mathrm{d}\mu(x)}{\int_X \exp\left(\sum_{j=1}^{k} a_j g_j(x)\right) \mathrm{d}\mu(x)} = b_i, \quad i = 1, 2, \cdots, k.$$

Then f^* solves the maximum entropy problem (8.16).

Proof For simplicity, set

$$Z = \int_X \exp\left(\sum_{i=1}^{k} a_i g_i(x)\right) \mathrm{d}\mu(x).$$

Then

$$H(f^*) = \ln Z + \sum_{i=1}^{k} a_i b_i.$$

Given $f \in \mathcal{D}$ that satisfies the constraints of (8.16). Then from (8.14) we have

$$\begin{aligned} H(f) &\leqslant - \int_X f(x) \ln f^*(x) \, \mathrm{d}\mu(x) \\ &= - \int_X f(x) \left[- \ln Z - \sum_{i=1}^{k} a_i g_i(x) \right] \mathrm{d}\mu(x) \\ &= \ln Z + \sum_{i=1}^{k} a_i \int_X f(x) g_i(x) \, \mathrm{d}\mu(x) \\ &= \ln Z + \sum_{i=1}^{k} a_i b_i = H(f^*). \end{aligned}$$

□

Remark 8.4.2 In particular, when $k = 1$, and $g(x)$ is viewed as the energy of the system, $f^*(x) = Z^{-1} \exp(-ag(x))$ is exactly *the Gibbs canonical distribution function*, $Z = \int_X \exp(-ag(x)) \mathrm{d}\mu(x)$ is the corresponding *partition function*, and the maximal entropy $H(f^*) = \ln Z + ab$ is just the well-known *thermodynamical entropy*.

Remark 8.4.3 Proposition 8.4.3 provides a mathematical justification of the *maximum entropy method* for moment problems and related density estimation problems; see, e.g., [99] for more details about this area.

We give two consequences of Proposition 8.4.3 to end this section.

Corollary 8.4.1 *Let $X = \mathbb{R}$ and $\mu = m$. Denote*

$$\bar{\mathcal{D}} = \left\{ f \in \mathcal{D} : \int_{-\infty}^{\infty} x f(x)\mathrm{d}x = 0, \int_{-\infty}^{\infty} x^2 f(x)\mathrm{d}x = 1 \right\}$$

and $f^*(x) = \exp(-x^2/2)/\sqrt{2\pi}$. Then $f^* \in \bar{\mathcal{D}}$ and

$$H(f^*) = \ln \sqrt{2\pi} + \frac{1}{2} = \max\{H(f) : f \in \bar{\mathcal{D}}\}.$$

Corollary 8.4.2 *Let $X = [0, \infty)$, let $\lambda > 0$ be a constant, and let $\mu = m$. Denote*

$$\hat{\mathcal{D}} = \left\{ f \in \mathcal{D} : \int_{0}^{\infty} x f(x)\mathrm{d}x = \frac{1}{\lambda} \right\}$$

and $f^*(x) = \lambda \exp(-\lambda x)$. Then $f^* \in \hat{\mathcal{D}}$ and

$$H(f^*) = 1 - \ln \lambda = \max\{H(f) : f \in \hat{\mathcal{D}}\}.$$

8.5 Boltzmann Entropy and Frobenius-Perron Operators

In this last section we want to see how the Boltzmann entropy of a density changes when a Frobenius-Perron operator is applied. First we examine the behavior of the entropy sequence $\{H(P^n f)\}$. For achieving this goal, we need the following *Jensen inequality.*

Lemma 8.5.1 *Let (X, Σ, μ) be a finite measure space and let $P : L^1 \to L^1$ be a Markov operator such that $P1 = 1$. If $\xi(t)$ is a concave function on $[0, \infty)$, then for every $f \in L^1$ such that $f \geqslant 0$,*

$$\xi \circ (Pf) \geqslant P(\xi \circ f) \quad \textit{whenever } P(\xi \circ f) \textit{ exists.} \tag{8.17}$$

Proof We only prove the inequality for the particular case that $X = \{1, 2, \cdots, n\}$ and μ is the counting measure on the σ-algebra of all subsets of X. In this case P can be viewed as a linear operator $\boldsymbol{P} : \mathbb{R}^n \to \mathbb{R}^n$ of the form

$$(\boldsymbol{P}\boldsymbol{x})_i = \sum_{j=1}^{n} p_{ij} x_j, \quad i = 1, 2, \cdots, n,$$

where $\boldsymbol{x} = (x_1, x_2, \cdots, x_n)^{\mathrm{T}} \in \mathbb{R}^n$, $p_{ij} \geqslant 0$ for all i, $j = 1, 2, \cdots, n$ and $\sum\limits_{j=1}^{n} p_{ij} = 1$ for all $i = 1, 2, \cdots, n$. Since ξ is a concave function,

$$\xi((\boldsymbol{P}\boldsymbol{x})_i) \geqslant \sum_{j=1}^{n} p_{ij}\xi(x_j) = (\boldsymbol{P}(\xi \circ \boldsymbol{x}))_i, \ \forall\, i.$$

That is, $\xi \circ (\boldsymbol{P}\boldsymbol{x}) \geqslant \boldsymbol{P}(\xi \circ \boldsymbol{x})$. In the infinite dimensional case the proof is much more difficult. □

Remark 8.5.1 Jensen's inequality is still valid for any positive operator $P : L^p \to L^p$ with $1 \leqslant p \leqslant \infty$ such that $P1 = 1$.

Theorem 8.5.1 *Let (X, Σ, μ) be a finite measure space, let $S : X \to X$ be a measure preserving transformation, and let $P : L^1 \to L^1$ be the corresponding Frobenius-Perron operator. Then*

$$H(Pf) \geqslant H(f) \tag{8.18}$$

for all $f \in L^1$ such that $f \geqslant 0$. If, in addition, S is invertible, then

$$H(Pf) = H(f). \tag{8.19}$$

Proof Taking $\xi = \eta$ in Jensen's inequality (8.17) and integrating its both sides over X gives, since P preserves the integral, that

$$\begin{aligned} H(Pf) &= \int_X \eta(Pf(x))\mathrm{d}\mu(x) \geqslant \int_X P\eta(f(x))\mathrm{d}\mu(x) \\ &= \int_X \eta(f(x))\mathrm{d}\mu(x) = H(f). \end{aligned}$$

This gives (8.18). If S is also invertible, then $P_S f(x) = f(S^{-1}(x))$ and $P_{S^{-1}} f(x) = f(S(x))$. Thus, $(P_S)^{-1} = P_{S^{-1}}$, and it follows that

$$H(f) = H((P_S)^{-1} P_S f) \geqslant H(P_S f) \geqslant H(f).$$

This proves (8.19). □

Remark 8.5.2 By Proposition 8.4.2 and Theorem 8.5.1, for a measure preserving transformation on a finite measure space, the corresponding entropy sequence $\{H(P^n f)\}$ always converges as $n \to \infty$, and the limit is upper bounded by, but not necessarily equal to, the number $H_{\max} = \ln \mu(X)$. However, for an invertible measure preserving transformation, the entropy sequence $\{H(P^n f)\}$ is a constant sequence.

For exact transformations (so they cannot be invertible), the limit of $\{H(P^n f)\}$ equals $H_{\max}$, as the following result shows.

Theorem 8.5.2 *Let (X, Σ, μ) be a probability measure space, let $S : X \to X$ be an exact measure preserving transformation, and let P be the Frobenius-Perron operator associated with S. Then*

$$\lim_{n\to\infty} H(P^n f) = 0$$

for all $f \in \mathcal{D}$ such that $H(f) > -\infty$.

Proof We only prove the case that f is bounded, and the proof for the general situation is referred to Theorem 9.3.2 of [82]. Suppose that $f(x) \leqslant c$, $\forall\, x \in X$ for some constant $c \geqslant 1$. Then

$$0 \leqslant P^n f \leqslant P^n c = cP^n 1 = c.$$

Let $A_n = \{x : 1 \leqslant P^n f(x) \leqslant c\}$. Then, since $\eta(t) \leqslant 0$ for $t \geqslant 1$,

$$0 \geqslant H(P^n f) \geqslant \int_{A_n} \eta(P^n f(x))\mathrm{d}\mu(x). \tag{8.20}$$

Since $\eta(1) = 0$, by the mean value theorem, we have

$$\begin{aligned}\left|\int_{A_n} \eta(P^n f(x))\mathrm{d}\mu(x)\right| &= \int_{A_n} |\eta(P^n f(x)) - \eta(1)|\mathrm{d}\mu(x)\\ &\leqslant K\int_{A_n} |P^n f(x) - 1|\mathrm{d}\mu(x)\\ &\leqslant K\|P^n f - 1\|,\end{aligned}$$

where $K = \sup\limits_{1\leqslant t\leqslant c} |\eta'(t)|$. Since S is exact, $\|P^n f - 1\| \to 0$ as $n \to \infty$ for all $f \in \mathcal{D}$ by Theorem 4.4.2 (iii). Thus,

$$\lim_{n\to\infty} \int_{A_n} \eta(P^n f(x))\mathrm{d}\mu(x) = 0,$$

which implies via (8.20) that $\{H(P^n f)\}$ converges to zero. □

Remark 8.5.3 In the more general case that S is only nonsingular, the sequence $\{H(P^n f)\}$ may not be an increasing one. For example, for the logistic model $S(x) = 4x(1-x)$, if $f = 1$, then

$$Pf(x) = \frac{1}{2\sqrt{1-x}},$$

and so

$$H(Pf) = -\int_0^1 \frac{1}{2\sqrt{1-x}} \ln \frac{1}{2\sqrt{1-x}}\mathrm{d}x = \ln 2 - 1.$$

Thus, $H(Pf) < H(f) = 0$. □

Because of Remark 8.5.3, it is necessary to introduce the concept of conditional entropy for Frobenius-Perron operators associated with nonsingular transformations.

Definition 8.5.1 *Let* $f,\ g \in \mathcal{D}$ *such that* supp $f \subset$ supp g. *Then the* conditional entropy *of* f *with respect to* g *is defined as*

$$H(f|g) = \int_X g(x)\eta\left[\frac{f(x)}{g(x)}\right] \mathrm{d}\mu(x) = -\int_X f(x)\ln\frac{f(x)}{g(x)}\mathrm{d}\mu(x).$$

Remark 8.5.4 Since g is a density and η is bounded from above, $H(f|g)$ is either finite or $-\infty$. Because of (8.15), the conditional entropy $H(f|g)$ measures the deviation of f from g, and there is always the inequality

$$H(f|g) \leqslant -\frac{1}{2}\|f-g\|^2 \tag{8.21}$$

with equality if and only if $f = g$, and in this case $H(f|g) = 0$.

Since supp $f \subset$ supp g implies supp $Pf \subset$ supp Pg for $f,\ g \in \mathcal{D}$ by Proposition 4.2.1, when $H(f|g)$ is well-defined, $H(Pf|Pg)$ is also well-defined.

Theorem 8.5.3 *Let P be the Frobenius-Perron operator associated with a non-singular transformation S on a measure space X. Then*

$$H(Pf|Pg) \geqslant H(f|g).$$

Proof We only prove the simpler case that supp $g = X$ and supp $Pg = X$, and the function f/g is bounded on X. Define a linear operator $R: L^\infty \to L^\infty$ by

$$Rh = \frac{P(gh)}{Pg}, \quad \forall\, h \in L^\infty.$$

Then R is a positive operator on L^∞ and satisfies $R1 = 1$. By Jensen's inequality (see Remark 8.5.1),

$$\eta \circ (Rh) \geqslant R(\eta \circ h). \tag{8.22}$$

Let $h = f/g$. Then the left-hand side of (8.22) becomes

$$\eta \circ (Rh) = -\frac{Pf}{Pg}\ln\frac{Pf}{Pg}$$

and the right-hand side of (8.22) is

$$R(\eta \circ h) = \frac{1}{Pg}P[(\eta \circ h)g] = -\frac{1}{Pg}P\left(f\ln\frac{f}{g}\right).$$

Hence inequality (8.22) can be written as

$$-Pf\ln\frac{Pf}{Pg} \geqslant -P\left(f\ln\frac{f}{g}\right).$$

Integrating the above inequality over X and noting that P preserves the integral, we obtain that

$$H(Pf|Pg) \geqslant -\int_X P\left[f(x)\ln\frac{f(x)}{g(x)}\right]\mathrm{d}\mu(x)$$
$$= -\int_X f(x)\ln\frac{f(x)}{g(x)}\mathrm{d}\mu(x) = H(f|g). \qquad \square$$

Remark 8.5.5 Theorem 8.5.3 implies that if g is a stationary density of P, then

$$H(Pf|g) \geqslant H(f|g).$$

Thus, the conditional entropy sequence $\{H(P^n f|g)\}$ with respect to a stationary density of P is always monotonically increasing and bounded above by zero, and so it always converges as $n \to \infty$, but the limit is not necessarily 0.

Exercises

8.1 Let (X, Σ, μ) be a probability measure space, and let $\bar{A}$, $\bar{A}'$, $\bar{B}$ be finite measurable partitions of X. Show that
(i) $\bar{A} \leqslant \bar{A}'$ implies $H(\bar{A}) \leqslant H(\bar{A}')$;
(ii) $H(\bar{A} \bigvee \bar{A}' \bigvee \bar{B}) + H(\bar{B}) \leqslant H(\bar{A} \bigvee \bar{B}) + H(\bar{A}' \bigvee \bar{B})$.

8.2 Let (X, Σ, μ) be a probability measure space. Suppose that $\bar{A}$, $\bar{B}$, $\bar{A}'$, $\bar{B}'$ are finite measurable partitions of X. Show that the conditional Shannon entropy given by Definition 8.2.1 has the following properties:
(i) $H(\bar{A} \bigvee \bar{B}|\bar{B}) = H(\bar{A}|\bar{B})$;
(ii) $H(\bar{A}|\bar{B}) = 0$ if and only if $\bar{A} \leqslant \bar{B}$;
(iii) $H(\bar{A} \bigvee \bar{B}) = H(\bar{B}) + H(\bar{A}|\bar{B})$;
(iv) $\bar{A} \leqslant \bar{A}'$ implies that $H(\bar{A}|\bar{B}) \leqslant H(\bar{A}'|\bar{B})$;
(v) $H(\bar{A} \bigvee \bar{A}'|\bar{B}) \leqslant H(\bar{A}|\bar{B}) + H(\bar{A}'|\bar{B})$;
(vi) $\bar{B} \leqslant \bar{B}'$ implies that $H(\bar{A}|\bar{B}') \leqslant H(\bar{A}|\bar{B})$;
(vii) $H(\bar{A} \bigvee \bar{B})|\bar{B}') = H(\bar{B}|\bar{B}') + H(\bar{A}|\bar{B} \bigvee \bar{B}')$.

8.3 Let (X, Σ, μ) be a probability measure space, and let Y be the set of all finite measurable partitions of X with finite entropy. For $\bar{A}, \bar{B} \in Y$ we set $d(\bar{A}, \bar{B}) = H(\bar{A}|\bar{B}) + H(\bar{B}|\bar{A})$. Show that $(Y, d(\cdot,\cdot))$ is a metric space.

8.4 Show that the Boltzmann entropy functional $H : \{f \geqslant 0 : f \in L^1(0,1)\} \to [-\infty, \infty)$ is proper, upper semi-continuous, and concave, and H is strictly concave on its domain that consists of all functions $f \geqslant 0$ with $H(f) > -\infty$ [9].

8.5 Show that for any real number a, the upper level set $\{f \geqslant 0 : H(f) \geqslant \alpha\}$ is weakly compact in $L^1(0,1)$ [9].

8.6 Write a program to study the relation of $\|f - g\|$ and $H(f|g)$ numerically for f, $g \in \mathcal{D} \cap L^1(0,1)$. Compare your numerical results with inequality (8.15).

8.7 Let (X, Σ, μ) be a measure space. Prove that for every two sequences $\{f_n\}$, $\{g_n\}$ in $\mathcal{D}$, the limit $H(f_n|g_n) \to 0$ implies $\|f_n - g_n\| \to 0$. Is the converse of the above statement also true? Exercise 8.6 above may be helpful in guessing the proper answer [95].

The following exercises are related to the *maximum entropy method* [31] and the *minimal energy method* [12] for solving Frobenius-Perron operator equations.

8.8 Show that if $f \in L^1(0,1)$ satisfies the equalities

$$\int_0^1 x^n f(x)\mathrm{d}x = 0, \quad n = 0, 1, \cdots,$$

then $f = 0$.

8.9 Let $P : L^1(0,1) \to L^1(0,1)$ be a Markov operator and let f^* be a density. Show that f^* is a stationary density of P if and only if

$$\int_0^1 f^*(x)(p_n(x) - P^* p_n(x))\mathrm{d}x = 0, \quad n = 1, 2, \cdots,$$

where $P^* : L^\infty(0,1) \to L^\infty(0,1)$ is the dual operator of P and $p_n(x) = x^n$ for each n.

8.10 Let P above be the Frobenius-Perron operator associated with a nonsingular transformation $S : [0,1] \to [0,1]$. Use Exercise 8.9 to give an equivalent condition for f^* to be a stationary density of P.

8.11 Suppose that $S : [0,1] \to [0,1]$ is a nonsingular transformation. Let k be a positive integer and consider the following maximum entropy problem

$$\max\left\{H(f) : f \in \mathcal{D}, \int_0^1 \left[x^i - S(x)^i\right] f(x)\mathrm{d}x = 0, \ 1 \leqslant i \leqslant k\right\}.$$

Use Proposition 8.4.3 to show that the density function

$$f_k(x) = \frac{\exp\left(\sum\limits_{i=1}^k a_i[x^i - S(x)^i]\right)}{\displaystyle\int_0^1 \exp\left(\sum\limits_{i=1}^k a_i[x^i - S(x)^i]\right)\mathrm{d}x}$$

is a solution of the maximum entropy problem if the constants $a_1, a_2, \cdots, a_k$ satisfy the following k equations

$$\int_0^1 \left[x^i - S(x)^i\right] \exp\left(\sum_{j=1}^k a_j[x^j - S(x)^j]\right) \mathrm{d}x = 0, \ i = 1, 2, \cdots, k.$$

8.12 A functional $\Phi : X \to [-\infty, \infty)$ on a normed vector space X is said to be *Kadec* if whenever $x_n \to x$ weakly in X and $\Phi(x_n) \to \Phi(x) < \infty$ then $x_n \to x$ in norm. Show that the Boltzmann entropy functional H is Kadec [9].

8.13 Show that the *energy functional* $V : L^2(0,1) \to (-\infty, 0]$ defined by

$$V(f) = -\frac{1}{2}\|f\|_2^2 = -\frac{1}{2}\int_0^1 |f(x)|^2 \mathrm{d}x$$

is strictly convex on $L^2(0,1)$ with weakly compact upper level sets, and is Kadec [12].

8.14 Show that the *constrained energy functional* $V^+ : L^2(0,1) \to [-\infty, \infty)$ defined by $V^+(f) = -\int_0^1 v_+(f(x))\mathrm{d}x$ is Kadec, where

$$v_+(t) = \begin{cases} \frac{1}{2}t^2, & \text{if } t \geqslant 0, \\ +\infty, & \text{if } t < 0, \end{cases}$$

is strictly convex on the cone of nonnegative integrable functions with weakly compact upper level sets, and is Kadec [12].

8.15 Suppose that the Frobenius-Perron operator P has a unique stationary density f^* that satisfies the condition

$$H(f^*) = -\int_0^1 f^*(x) \ln f^*(x)\mathrm{d}x > -\infty.$$

Let $\{f_k\}$ be the sequence of maximum entropy solutions to the maximum entropy problem in Exercise 8.11. Show that $\lim_{k\to\infty} f_k = f^*$ both weakly and in norm [31].

8.16 Under the same assumptions about P as above, show that the same conclusion of Exercise 8.15 is true when f_k are obtained if the objective function of the maximum entropy problem in Exercise 8.11 is replaced with either the energy functional V in Exercise 8.13 or the constrained energy functional in Exercise 8.14 [12].

8.17 Prove inequality (8.21).

Chapter 9
Applications of Invariant Measures

JiuDing, Department of Mathematics, The University of Southern Mississippi in America.
Aihui Zhou, Academy of Mathematics and Systems Science, Chinese Academy of Sciences.

Abstract Some applications of absolutely continuous invariant measures are illustrated in this last chapter. First we briefly give an application to the estimation of the speed of decay of correlations. Then we present Li-Yorke's work on the application of stationary densities of small variation to random number generation. Two modern applications in the last decade are sketched in the final two sections. One is about the transfer operator in molecular conformation dynamics and computational drug design, and the other is the direct sequence code division multiple access in the third generation wireless communications.
Keywords Decay of correlations, pseudo-random number generation, conformation dynamics, drug design, Hamitonian dynamics, transfer operator, Foias operator, almost invariant set, direct sequence code division multiple access (DSCDMS), spreading sequence.

The mathematical theory and numerical computation of absolutely continuous invariant measures have applications in various fields of sciences and engineering. In this last chapter we briefly give several representative applications of invariant measures to the statistical study of deterministic systems. In the first section we give an application of the classic Ionesou-Tulcea and Marinescu theorem to the estimation of the speed of decay of correlations. Section 9.2, which is from the reference [89] and the lecture notes [87], demonstrates how a stationary density of small variation can be used for the random number generation. In the last two sections we present the latest applications of the theory and methods of absolutely continuous invariant measures. One is in the emerging field of computational molecular dynamics and the drug design, and the other is for the third generation of wireless communications.

9.1 Decay of Correlations

Let (X, Σ, σ) be a probability measure space, and let $S : X \to X$ be a nonsingular transformation. Given two suitable functions f and g from a function space, which are often referred to as the *observables* in statistical physics, the differences

$$\int_X f(g \circ S^n)\mathrm{d}\mu - \int_X f\mathrm{d}\mu \int_X g\mathrm{d}\mu$$

are called the *correlation functions* of the observables f and g. If S preserves μ and is mixing, then from Theorem 4.4.2 (ii), the correlation functions decay to zero as n goes to infinity. Our question is: what is the *rate of decay* of the correlation functions? This rate measures the speed with which the dynamical system determined by S and μ becomes independent of initial conditions.

The problem of decay of correlations is very important in ergodic theory, in particular in the theory of *positive transfer operators* including Frobenius-Perron operators [4, 14], and in many branches of physical and engineering sciences [7]. In this section we apply the Ionescu-Tulcea and Marinescu theorem (Theorem 2.5.3) to estimate the rate of decay for the class of Lasota-Yorke interval mappings. A comprehensive study of decay of correlations is contained in the monograph [4].

Definition 9.1.1 *Let μ be an invariant probability measure for a nonsingular transformation $S : X \to X$ and let n be a positive integer. For any $f \in L^1(\mu)$ and $g \in L^\infty(\mu)$ the quantity*

$$\mathrm{Cor}(f, g, n) = \left| \int_X f(g \circ S^n) \mathrm{d}\mu - \int_X f \mathrm{d}\mu \int_X g \mathrm{d}\mu \right|$$

is referred to as the nth correlation coefficient.

We need the following spectral decomposition result for Frobenius-Perron operators associated with Lasota-Yorke interval mappings (see the paragraph in Section 7.1 containing (7.2)).

Theorem 9.1.1 *Let $S : [0,1] \to [0,1]$ satisfy the conditions of Theorem* 5.2.1. *Suppose further that* $\inf_{x\in[0,1]\setminus\{a_1,\cdots,a_{r-1}\}} |S'(x)| > 2$. *Then the corresponding Frobenius-Perron operator $P : L^1(0,1) \to L^1(0,1)$ is quasi-compact when it is restricted to $BV(0,1)$. In particular as a bounded linear operator from $BV(0,1)$ into itself,*

(i) *P has only finitely many eigenvalues $\lambda_1, \lambda_2, \cdots, \lambda_t$ of modulus* 1 *which are the roots of unity;*

(ii) *the eigenspace E_i associated with λ_i is a finite dimensional vector subspace of $BV(0,1)$ for $i = 1, 2, \cdots, t$;*

(iii) *P has the following spectral decomposition:*

$$P = \sum_{i=1}^{t} \lambda_i \Phi_i + R, \tag{9.1}$$

where each Φ_i is a projection onto E_i such that $\|\Phi_i\| = 1$, $\Phi_i \Phi_j = 0$, $i \neq j$, and $R : BV(0,1) \to BV(0,1)$ can be extended to a bounded linear operator from L^1 into L^1 and from L^1 into $BV(0,1)$ such that

$$\sup_{n\geqslant 0} \|R^n\| < \infty, \; R(BV(0,1)) \subset BV(0,1), \; R\Phi_i = \Phi_i R = 0, \; \forall\, i,$$

and

$$\|R^n\|_{BV} \leqslant Mq^n, \ \forall\, n = 1, 2, \cdots$$

with two constants $0 < q < 1$ *and* $M > 0$.

Proof Let $T = P$ and $V = BV(0,1)$ in Theorem 2.5.4. Then, as in the proof of Theorem 5.2.1,

$$\bigvee_0^1 Pf \leqslant \alpha \bigvee_0^1 f + \beta \|f\|, \ \forall\, f \in BV(0,1),$$

where $\alpha = 2/\inf_{x\in[0,1]\setminus\{a_1,\cdots,a_{r-1}\}} |S'(x)|$ and β are two positive constants. The assumption that $\inf_{x\in[0,1]\setminus\{a_1,\cdots,a_{r-1}\}} |S'(x)| > 2$ implies that $0 < \alpha < 1$, so condition (iii) of Theorem 2.5.4 is satisfied with $k = 1$. All the other conditions of that theorem are consequences of Theorem 2.3.1 and properties of the Frobenius-Perron operator. So the conclusion follows from Theorem 2.5.4. □

Remark 9.1.1 Theorem 9.1.1 is still valid for the weakened assumption $\inf_{x\in[0,1]\setminus\{a_1,\cdots,a_{r-1}\}} |S'(x)| > 1$; see Theorem 7.2.1 of [14] in which it was further proved that

(i) P has finitely many stationary densities $f_1, f_2, \cdots, f_k$ such that $f_i \in BV(0,1)$ and S is ergodic with respect to the invariant probability measure μ_i such that $\mathrm{d}\mu_i/\mathrm{d}m = f_i$ for $i = 1, 2, \cdots, k$.

(ii) For each $1 \leqslant i \leqslant k$, supp μ_i is the union of a finite collection of disjoint sets $A_1^i, A_2^i, \cdots, A_{k_i}^i$ such that for each $j = 1, 2, \cdots, k_i$,

$$P(f_i \cdot \chi_{A_j^i}) = f_i \cdot \chi_{A_{j+1}^i}, \ \ (k_i + 1 \equiv 1),$$

and $S^{k_i} : A_j^i \to A_j^i$ is exact with respect to the invariant measure μ_{ij} defined by $\mu_{ij}(B) = \int_B f_i \mathrm{d}m$ for all measurable subsets B of A_j^i.

Theorem 9.1.2 *Under the same conditions of Theorem* 9.1.1, *if* f^* *is the unique stationary density of* P *and* S *is mixing with respect to the corresponding invariant measure* μ, *then there exists a positive constant* M *and a constant* $0 < q < 1$ *such that for any* $f \in BV(0,1)$ *and all positive integers* n,

$$\left\| P^n f - \left(\int_0^1 f(x)\mathrm{d}x \right) f^* \right\|_{BV} \leqslant Mq^n \|f\|_{BV}. \tag{9.2}$$

Proof It can be shown that under the conditions of the theorem, $t = 1$ in (9.1), and so 1 is the only peripheral spectral point of P. Since f^* is the unique stationary density of P, we see that dim $E_1 = 1$. Thus, from Theorem 9.1.1, we have

$$P^n f = \left(\int_0^1 f(x)\mathrm{d}x \right) f^* + R^n f,$$

where $\|R^n\|_{BV} \leqslant Mq^n$ for some constants $0 < q < 1$ and $M > 0$. This implies (9.2) immediately. □

Theorem 9.1.3 *Under the same conditions of Theorem 9.1.2, for any $f \in BV(0,1)$ and $g \in L^\infty(0,1)$,*

$$\mathrm{Cor}(f,g,n) \leqslant Mq^n \|ff^*\|_{BV}\|g\|_\infty$$

for all positive integers n, where the constants M and q are as in Theorem 9.1.2.

Proof Let $c = \int_0^1 f(x)\mathrm{d}x$. Then the nth correlation coefficient

$$\begin{aligned}\mathrm{Cor}(f,g,n) &= \left|\int_0^1 f(g\circ S^n)\mathrm{d}\mu - c\int_0^1 g\mathrm{d}\mu\right| \\ &= \left|\int_0^1 f(g\circ S^n)f^*\mathrm{d}m - c\int_0^1 gf^*\mathrm{d}m\right| \\ &= \left|\int_0^1 P^n(ff^*)g\mathrm{d}m - \int_0^1 cf^*g\mathrm{d}m\right| \\ &\leqslant \|P^n(ff^*) - cf^*\|\ \|g\|_\infty \leqslant Mq^n\|ff^*\|_{BV}\|g\|_\infty.\end{aligned}$$

□

9.2 Random Number Generation

Random number generators are widely used in many statistical methods for problems that are hard to solve with conventional techniques. For example, generating random numbers is one of the main tools in various Monte Carlo methods. There have been many different ways to generate random numbers. A classic method was proposed by von Neumann in 1946: take the square of the previous random number with 6 digits and extract its middle 6 digits to produce the next random number. This is equivalent to generating the numbers by iterating the mapping $S : [0,1] \to [0,1]$ defined by

$$S(x) = 10^3x^2 \pmod 1.$$

As another example, Hewlett-Packard Company has suggested that the mapping

$$S(x) = (x+\pi)^5 \pmod 1$$

is a generator of uniformly distributed pseudo-random numbers.

One of the most important features of random number generation is randomness or near randomness of the generated number sequences. This means that the corresponding random variable of a good random number generator should have almost uniform distribution. For a pseudo-random number generator given by a deterministic mapping S, a measure of its goodness toward the randomness of generated sequences is how close the stationary density of S is to a constant function, or how small the variation of such a stationary density is.

In this section we shall show that under certain conditions on the mapping S, the variation of the stationary density of the associated Frobenius-Perron operator is small. We are interested in a class of mappings of the form

$$S(x) = p(x) \pmod 1, \tag{9.3}$$

where $p : [0,1] \to [0,\infty)$. The above two concrete mappings are just examples of the general case which was also studied by Rényi in 1957 [110] and by Rochlin in 1964 [111] with some number-theoretic arguments. They considered two classes of differentiable and strictly increasing functions $p : [0,1] \to [0,\infty)$ satisfying the condition that $\inf_{x\in[0,1]} p'(x) > 1$, $p(0) = 0$, and $p(1)$ is a positive integer, and the other is the so-called *Rényi transformation* $p(x) = rx$ where $r > 1$ is a constant. In particular, when r is a positive integer, the Rényi transformation is the r-adic transformation.

In general we have the following uniqueness theorem for mappings $S(x) = p(x) \pmod 1$, which implies ergodicity of S.

Theorem 9.2.1 *Let p be a C^2-function on $[0,1]$ which satisfies the inequality $\inf_{x\in[0,1]} |p'(x)| > 1$, and let S be the corresponding mapping as defined by* (9.3). *Then there exists a unique stationary density of the Frobenius-Perron operator P associated with S.*

Proof The existence of a stationary density is guaranteed by the Lasota-Yorke theorem in Section 5.2. To prove the uniqueness of the stationary density, suppose that there are two densities $f_1, f_2 \in \mathcal{D}$ such that $Pf_1 = f_1$ and $Pf_2 = f_2$. Then $S(\text{supp } f_1) \subset \text{supp } f_1$ and $S(\text{supp } f_2) \subset \text{supp } f_2$ by Proposition 4.2.1. By the Li-Yorke theorem in [89], $m(\text{supp } f_1 \cap \text{supp } f_2) = 0$ and each supp f_i is a union of the intervals containing at least one discontinuity x_i of S in its interior. Let

$$x_1 \in [a_1, b_1] \subset \text{supp } f_1 \quad \text{and} \quad x_2 \in [a_2, b_2] \subset \text{supp } f_2.$$

Then $\lim_{x\to x_i^-} S(x) = 1$ and $\lim_{x\to x_i^+} S(x) = 0$ since S is discontinuous at x_i. Thus $S([a_1,b_1]) \cap S([a_2,b_2]) \subset \text{supp } f_1 \cap \text{supp } f_2$. Since $S([a_1,b_1]) \cap S([a_2,b_2])$ contains a nonempty open interval, this contradicts the fact that $m(\text{supp } f_1 \cap \text{supp } f_2) = 0$. □

Let $p : [0,1] \to [0,\infty)$ be a C^2-function and define

$$\lambda \equiv \inf_{x\in[0,1]} |p'(x)|, \quad \eta \equiv \sup_{x\in[0,1]} \frac{|p''(x)|}{|p'(x)|}.$$

The next theorem gives an upper bound to the variation of the unique stationary density in Theorem 9.2.1. First we need a lemma.

Lemma 9.2.1 *Let f be a function of bounded variation on $[a,b]$. Then*

$$|f(x)| \leqslant \bigvee_a^b f + \frac{1}{b-a}\int_a^b |f(x)|\mathrm{d}x, \quad \forall\, x \in [a,b]. \tag{9.4}$$

Proof Since $(b-a)^{-1}\displaystyle\int_a^b |f(x)|\mathrm{d}x$ is the average value of $|f|$ over $[a,b]$, there exists $y \in [a,b]$ such that

$$|f(y)| \leqslant \frac{1}{b-a}\int_a^b |f(x)|\mathrm{d}x.$$

From $|f(x)| \leqslant |f(x)-f(y)| + |f(y)|$ and $|f(x)-f(y)| \leqslant \bigvee_a^b f$, we obtain (9.4). □

Theorem 9.2.2 *Under the same assumptions of Theorem 9.2.1, let f^* be the unique stationary density of the Frobenius-Perron operator associated with S which is defined by (9.3). If in addition $\lambda > 3$, then*

$$\bigvee_0^1 f^* \leqslant \frac{\eta+2}{\lambda-3}. \tag{9.5}$$

Proof Without loss of generality, we assume that p is strictly increasing on $[0,1]$. Let $0 = a_0 < a_1 < \cdots < a_r = 1$ be the partition of $[0,1]$ such that S is C^2 and $S' > 0$ on every $[a_{i-1}, a_i]$ for $1 \leqslant i \leqslant r$. Writing $S_i = S|_{[a_{i-1},a_i]}$ and $g_i = S_i^{-1}$ for each i. Given any $f \in \mathcal{D} \cap BV(0,1)$. As in the proof of Theorem 5.2.1, we have (5.7), i.e.,

$$Pf(x) = \sum_{i=1}^r \sigma_i(x) f(g_i(x)) \chi_{I_i}(x), \tag{9.6}$$

where $\sigma_i(x) = |g_i'(x)|$ and $I_i = S([a_{i-1}, a_i]) = [S(a_{i-1}^+), S(a_i^-)]$ for all i. It is obvious from the definition of S that for $i = 2, 3, \cdots, r-1$, $I_i = [0,1]$, so $\chi_{I_i}(x) \equiv 1$, and hence,

$$\bigvee_0^1 \sigma_i (f\circ g_i)\chi_{I_i} = \bigvee_0^1 \sigma_i (f \circ g_i).$$

Since $I_1 = [S(0), 1]$ and $I_r = [0, S(1)]$, we have

$$\bigvee_0^1 \sigma_1 (f \circ g_1)\chi_{I_1} = \bigvee_{S(0)}^1 \sigma_1 (f\circ g_1) + \frac{f(0)}{S'(0)}$$

and

$$\bigvee_0^1 \sigma_r(f\circ g_r)\chi_{I_r} = \bigvee_0^{S(1)} \sigma_r(f\circ g_r) + \frac{f(1)}{S'(1)}.$$

Combining the above three equalities with (9.6) gives

$$\begin{aligned}\bigvee_0^1 Pf &\leqslant \sum_{i=1}^{r} \bigvee_{S(a_{i-1}^+)}^{S(a_i^-)} \sigma_i(f\circ g_i) + \frac{f(0)}{S'(0)} + \frac{f(1)}{S'(1)}\\ &\leqslant \sum_{i=1}^{r} \bigvee_{S(a_{i-1}^+)}^{S(a_i^-)} \sigma_i(f\circ g_i) + \frac{f(0)+f(1)}{\lambda}\end{aligned}$$

since $\lambda = \inf_{x\in[0,1]\setminus\{a_1,\cdots,a_{r-1}\}} S'(x)$. As in the proof of Theorem 5.2.1, Proposition 2.3.1 (v) implies that

$$\begin{aligned}\bigvee_{S(a_{i-1}^+)}^{S(a_i^-)} \sigma_i(f\circ g_i) &\leqslant \sup_{x\in I_i}\sigma_i(x) \bigvee_{S(a_{i-1}^+)}^{S(a_i^-)} f\circ g_i + \int_{S(a_{i-1}^+)}^{S(a_i^-)} |\sigma_i'|(f\circ g_i)\mathrm{d}m\\ &\leqslant \frac{1}{\lambda}\bigvee_{a_{i-1}}^{a_i} f + \frac{\eta}{\lambda}\int_{S(a_{i-1}^+)}^{S(a_i^-)} \sigma_i(f\circ g_i)\mathrm{d}m\\ &= \frac{1}{\lambda}\bigvee_{a_{i-1}}^{a_i} f + \frac{\eta}{\lambda}\int_{a_{i-1}}^{a_i} f\mathrm{d}m, \quad \forall\, i=1,2,\cdots,r.\end{aligned}$$

On the other hand, Lemma 9.2.1 implies that

$$f(0)+f(1) \leqslant 2\left(1+\bigvee_0^1 f\right).$$

It follows from such inequalities that

$$\begin{aligned}\bigvee_0^1 Pf &\leqslant \frac{1}{\lambda}\sum_{i=1}^{r}\bigvee_{a_{i-1}}^{a_i} f + \frac{\eta}{\lambda}\sum_{i=1}^{r}\int_{a_{i-1}}^{a_i} f\mathrm{d}m + \frac{2\left(1+\bigvee_0^1 f\right)}{\lambda}\\ &= \frac{1}{\lambda}\bigvee_0^1 f + \frac{\eta}{\lambda}\int_0^1 f\mathrm{d}m + \frac{2\left(1+\bigvee_0^1 f\right)}{\lambda}.\end{aligned}$$

Letting $f = f^*$ in the above inequality, we obtain

$$\bigvee_0^1 f^* = \bigvee_0^1 Pf^* = \frac{1}{\lambda}\bigvee_0^1 f^* + \frac{\eta}{\lambda}\|f^*\| + \frac{2\left(1+\bigvee_0^1 f^*\right)}{\lambda},$$

and thus,

$$\left(1-\frac{3}{\lambda}\right)\bigvee_0^1 f^* \leqslant \frac{\eta+2}{\lambda},$$

which implies (9.5). □

Corollary 9.2.1 *Under the same assumptions as in Theorem* 9.2.2,

$$1-\frac{\eta+2}{\lambda-3} \leqslant f^*(x) \leqslant 1+\frac{\eta+2}{\lambda-3}, \quad \forall\, x \in [0,1]. \tag{9.7}$$

Proof Since $f^* \geqslant 0$ and $\|f^*\| = 1$, By Lemma 9.2.1,

$$|1-f^*(x)| \leqslant \bigvee_0^1 f^* \leqslant \frac{\eta+2}{\lambda-3}.$$

So we have (9.7). □

Corollary 9.2.1 indicates that if the positive number $(\eta+2)/(\lambda-3)$ is very small, the stationary density f^* is almost the constant 1 function, hence the asymptotic behavior of the orbits of S is almost uniformly distributed on $[0,1]$, which is exactly the desired property for a random number generator.

Remark 9.2.1 For Hewlett-Packard Company's mapping $S(x) = (x+\pi)^5$ (mod 1) used in certain HP calculators to simulate a random number generation, $\lambda = 5\pi^4 \approx 487.05$ and $\eta \approx 1.27$, so

$$\frac{\eta+2}{\lambda-3} \approx 0.00676.$$

Therefore this mapping S gives a pretty good pseudo-random number generator. On the other hand, von Neumann's suggestion was not very good since in this case $\eta = +\infty$ due to the fact that $S'(0) = 0$. It can be observed that, since $S(x) < x$ if $0 < x < 10^{-k}$, if one number in the sequence $\{S^n(x)\}$ falls into the interval $(0, 10^{-k})$, the remaining numbers will converge to 0. We may, however, modify von Neumann's procedure with $p(x) = 10^k(1+x)^2$ instead of $p(x) = 10^k x^2$. Then, for $x \in [0,1]$,

$$|p'(x)| \geqslant 2 \cdot 10^k, \qquad \left|\frac{p''(x)}{p'(x)}\right| = \frac{10^k \cdot 2}{10^k \cdot 2(1+x)} \leqslant 1.$$

Hence, even for $k = 3$, we already obtain a rather good result

$$\frac{\eta+2}{\lambda-3} \approx 0.0015.$$

9.3 Conformational Dynamics of Bio-molecules

In recent years, computational molecular dynamics has attracted considerable attention in more effective drug designs for prion diseases such as the mad cow disease and viral diseases like the HIV or the SARS. One indication of such a trend in computational dynamical systems is the plenary lecture entitled "Molecular Conformation Dynamics and Computational Drug Design" by Deuflhard at the fifth International Congress on Industrial and Applied Mathematics (ICIAM) held in Sydney in 2003 [25] and the one delivered by Schütte at the sixth ICIAM in Berlin four years later. In this section we give a brief introduction to this applied area which is of scientific and economic significance, following the presentation of the pioneering work [115] and some related works. For more details we recommend the references [25, 115, 116].

The molecular dynamics is based on the Hamiltonian mechanics. Assume that the dynamics of the molecular system with k atoms under consideration is governed by the *Hamiltonian function*

$$H(\boldsymbol{q},\boldsymbol{p}) = \frac{1}{2}\boldsymbol{p}^{\mathrm{T}}\boldsymbol{M}^{-1}\boldsymbol{p} + V(\boldsymbol{q}),$$

where the first term is the *kinetic energy* that depends on the vector $p = (\boldsymbol{p}_1, \boldsymbol{p}_2, \cdots, \boldsymbol{p}_k)^{\mathrm{T}} \in \mathbb{R}^N (N = 3k)$ of *generalized momenta* via the diagonal mass matrix M, and the second term is the *potential energy* that only depends on the vector $\boldsymbol{q} = (\boldsymbol{q}_1, \boldsymbol{q}_2, \cdots, \boldsymbol{q}_k)^{\mathrm{T}} \in \mathbb{R}^N$ of *generalized positions.* The Hamiltonian function value $H(\boldsymbol{q},\boldsymbol{p})$ denotes the *internal energy* of the system in the state $\boldsymbol{x} = (\boldsymbol{q},\boldsymbol{p}) \in \mathbb{R}^N \times \mathbb{R}^N$, and the resulting *Hamiltonian differential equations* of the motion are

$$\dot{\boldsymbol{q}} = \frac{\partial H}{\partial p} = \boldsymbol{M}^{-1}\boldsymbol{p}, \quad \dot{\boldsymbol{p}} = -\frac{\partial H}{\partial \boldsymbol{q}} = -\nabla V(\boldsymbol{q}). \tag{9.8}$$

With any given initial value $\boldsymbol{x}(0) = \boldsymbol{x}_0 = (\boldsymbol{q}_0, \boldsymbol{p}_0) \in \mathbb{R}^N \times \mathbb{R}^N$, the unique solution, which is called the *Hamiltonian flow*, of the Hamiltonian initial value problem (9.8) can be written in terms of the flow $\boldsymbol{\Phi}^t$ as

$$\boldsymbol{x}(t) = (\boldsymbol{q}(t), \boldsymbol{p}(t)) = \boldsymbol{\Phi}^t x_0. \tag{9.9}$$

Hamiltonian flows $\boldsymbol{\Phi}^t$ have several important conservation and invariance properties: the *conservation of energy* $H(\boldsymbol{\Phi}^t(\boldsymbol{x})) \equiv H(\boldsymbol{x})$, the *conservation of volume* $\det \boldsymbol{J}_{\boldsymbol{\Phi}^t}(x) \equiv 1$, where $\boldsymbol{J}_{\boldsymbol{\Phi}^t}$ is the Jacobian matrix of the flow $\boldsymbol{\Phi}^t$, and the *reversibility* of $\boldsymbol{\Phi}^t : R \circ \boldsymbol{\Phi}^{-t} \circ R = \boldsymbol{\Phi}^t$, where the *momentum reversion* R is defined by $R(\boldsymbol{q},\boldsymbol{p}) = (\boldsymbol{q}, -\boldsymbol{p})$.

It is well-known that the numerical integration of the Hamiltonian differential equation of the molecular system to approximate the individual solution curves (called *trajectories*) $\boldsymbol{x}(t)$ is ill-conditioned when the time t is relatively large, due to the fact that the trajectories of the Hamiltonian initial value problem to be computed are asymptotically chaotic. The difficulty in the bio-molecular design is related to the situation that real times of pharmaceutical interest are in the region of milliseconds to minutes, whereas simulation times are presently in the range of nanoseconds (1 nanosecond = 10^{-9} second) with time steps of less than 5 femtoseconds (1 femtosecond = 10^{-15} second). Consequently, the numerical simulation can only give the information about short-term solutions because of the limitation of available computational facilities. On the other hand, the drug design industry has the desire of long-term prediction.

Because of the above observation, A different approach, called *conformational dynamics*, has been proposed to model long-term dynamics in the investigation of molecular systems, based on the ideas from ergodic theory and statistical physics. The new approach introduces the concept of *transfer operators* that incorporate the stochastic elements of the dynamics. These operators are Markov operators which arrive from the Frobenius-Perron operator associated with the Hamiltonian flow $\boldsymbol{\Phi}^t$. In the following we present the basic idea and methodology in this emerging area.

Let $X = \Omega \times \mathbb{R}^N$ be the *phase space* of the Hamiltonian flow (9.9) in which Ω is the *position space* of the generalized position vector $\boldsymbol{q}$. For each time moment t, let $P_t \equiv P_{\boldsymbol{\Phi}^t}$ be the Frobenius-Perron operator associated with the diffeomorphism $\boldsymbol{\Phi}^t : X \to X$. Since the Lebesgue measure m on X is invariant under $\boldsymbol{\Phi}^t$ from the volume conservation property of the Hamiltonian flow, Proposition 4.2.2 implies that

$$P_t f(\boldsymbol{x}) = f(\boldsymbol{\Phi}^{-t}(\boldsymbol{x}))\frac{\mathrm{d}(m \circ \boldsymbol{\Phi}^{-t})}{\mathrm{d}m}(x) = f(\boldsymbol{\Phi}^{-t}(\boldsymbol{x})), \quad f \in L^1.$$

The Koopman operator U_t corresponding to $\boldsymbol{\Phi}^t$ has the expression

$$U_t g(\boldsymbol{x}) = g(\boldsymbol{\Phi}^t(\boldsymbol{x})), \quad g \in L^\infty.$$

Because of the energy and volume conservation properties of the Hamiltonian flow, the composition of any scalar function with the Hamiltonian function H is an invariant function of both the Frobenius-Perron operator P_t and the Koopman operator U_t.

If the distribution of the initial states of the molecular system is given by the probability density $f_0(\boldsymbol{x})$, then

$$f(\boldsymbol{x}, t) \equiv (P_t f_0)(\boldsymbol{x}) = f_0(\boldsymbol{\Phi}^{-t}(\boldsymbol{x}))$$

is the time-dependent probability density transported along the trajectory $\boldsymbol{\Phi}^t(x)$ of the system. By far most experiments of the molecular dynamics are performed using *equilibrium ensembles*, i.e., ensembles which are described by stationary densities f of the continuous time group of the Frobenius-Perron operators P_t. This means that f satisfies the equality $f(\boldsymbol{x}) = f(\boldsymbol{\Phi}^{-t}(\boldsymbol{x}))$ for all t and $\boldsymbol{x} \in X$. The energy conservation property of $\boldsymbol{\Phi}^t$ implies that for *arbitrary* smooth functions $F : \mathbb{R} \to \mathbb{R}^+$ with the condition $\int_X F(H(\boldsymbol{x}))\mathrm{d}\boldsymbol{x} = 1$, the associated densities $f = F \circ H$ are invariant, i.e.,

$$f = f \circ \boldsymbol{\Phi}^{-t}.$$

Moreover, the reversibility property of $\boldsymbol{\Phi}^t$ implies that the above "energy prepared" densities are also *p-symmetric*, i.e.,

$$f = f \circ R. \tag{9.10}$$

Most experiments on molecular systems are performed under the equilibrium conditions of constant temperature, particle number, and volume. The corresponding stationary density is the *canonical density* associated with the Hamiltonian H:

$$f_{\mathrm{can}}(\boldsymbol{q}, \boldsymbol{p}) = \frac{1}{Z} \exp\left(-\eta\left(\frac{\boldsymbol{p}^{\mathrm{T}}\boldsymbol{M}^{-1}\boldsymbol{p}}{2} + V(\boldsymbol{q})\right)\right), \tag{9.11}$$

where

$$Z = \int_{\mathbb{R}^N}\int_{\Omega} \exp\left(-\eta\left(\frac{\boldsymbol{p}^{\mathrm{T}}\boldsymbol{M}^{-1}\boldsymbol{p}}{2} + V(\boldsymbol{q})\right)\right) \mathrm{d}\boldsymbol{q}\mathrm{d}\boldsymbol{p}$$

is the normalization factor and $\eta = 1/(k_B\mathcal{T})$, with $\mathcal{T}$ the system's temperature and k_B the *Boltzmann constant.* This density can be factorized as

$$f_{\mathrm{can}}(\boldsymbol{q}, \boldsymbol{p}) = \mathcal{P}(\boldsymbol{p})\mathcal{Q}(\boldsymbol{q}),\ Z = Z_{\boldsymbol{p}}Z_{\boldsymbol{q}},\ \int_{\mathbb{R}^N} \mathcal{P}(\boldsymbol{p})\mathrm{d}\boldsymbol{p} = \int_{\Omega} \mathcal{Q}(\boldsymbol{q})\mathrm{d}\boldsymbol{q} = 1,$$

where

$$\mathcal{P}(\boldsymbol{p}) = \frac{1}{Z_{\boldsymbol{p}}} \exp\left(-\frac{\eta}{2}\boldsymbol{p}^{\mathrm{T}}\boldsymbol{M}^{-1}\boldsymbol{p}\right),\quad \mathcal{Q}(\boldsymbol{q}) = \frac{1}{Z_{\boldsymbol{q}}} \exp\left(-\eta V(\boldsymbol{q})\right).$$

Assume that an arbitrary stationary density f_0 is given. We would like to define a transition probability from one region, $B \subset X$, of the phase space to another one, $C \subset X$, after a time span τ. The new ensemble of all such systems with states $\boldsymbol{x} \in B$ selected from the ensemble f_0 at $t = 0$ has the density

$$f_B(\boldsymbol{x}) = \left(\int_B f_0(\boldsymbol{x})\mathrm{d}\boldsymbol{x}\right)^{-1} \chi_B(\boldsymbol{x}) f_0(\boldsymbol{x}).$$

Since all systems evolve due to $\boldsymbol{\Phi}^t$, after time τ, the relative frequency of systems in the ensemble f_B with states in C equals

$$\int_X \chi_C(\boldsymbol{\Phi}^\tau(\boldsymbol{x})) f_B(\boldsymbol{x})\mathrm{d}\boldsymbol{x}.$$

Thus, the *transition probability* should be defined as

$$w(B, C, \tau) = \frac{\displaystyle\int_B \chi_C(\boldsymbol{\Phi}^\tau(\boldsymbol{x})) f_0(\boldsymbol{x}) \mathrm{d}\boldsymbol{x}}{\displaystyle\int_B f_0(\boldsymbol{x}) \mathrm{d}\boldsymbol{x}}. \tag{9.12}$$

From (9.12), a subset $B \subset X$ is *invariant* under the flow $\boldsymbol{\Phi}^t$ if and only if $w(B, B, t) = 1$ for all $t \in \mathbb{R}$, which is *independent* of the choice of the stationary density f_0. In the conformational dynamics of the molecular system, we are interested in *almost invariant* subsets B in the sense that $w(B, B, \tau)$ is sufficiently close to 1. Therefore,

$$B \subset X \text{ is almost invariant} \quad \Leftrightarrow \quad w(B, B, \tau) \approx 1. \tag{9.13}$$

Remark 9.3.1 The concept of "almost invariant sets" was first introduced by Dellnitz and Junge [24] to the computational problem of dynamical systems.

Any conformation in the conformational dynamics of bio-molecules is an almost invariant set of the ensemble in the sense of (9.13). But the main aspects of the chemical intuition indicate that the phrase "conformation" does not refer to the momentum information. So *spatial* subsets of Ω will be used in the definition of transfer operators below. The transition probability between $B \subset \Omega$ and $C \subset \Omega$ is defined as the transition probability between the associated phase space fibers (or cylinders) $X(B) = \{(\boldsymbol{q}, \boldsymbol{p}) \in X : \boldsymbol{q} \in B\}$ and $X(C) = \{(\boldsymbol{q}, \boldsymbol{p}) \in X : \boldsymbol{q} \in C\}$, i.e.,

$$w(B, C, \tau) = w(X(B), X(C), \tau). \tag{9.14}$$

Naturally, $B \subset \Omega$ is said to be *almost invariant* if $w(B, B, \tau) \approx 1$.

Now we introduce the concept of the spatial transfer operator which incorporates the Frobenius-Perron operator with a stationary density $f_0 \in L^1(X)$ that satisfies condition (9.10) such that the reduced density

$$F(\boldsymbol{q}) = \int_{\mathbb{R}^N} f_0(\boldsymbol{q}, \boldsymbol{p}) \mathrm{d}\boldsymbol{p}$$

is positive, finite, and smooth on Ω. Then, the *spatial transfer operator* $T \equiv T(\tau)$ is defined by

$$Tu(\boldsymbol{q}) = \frac{1}{F(\boldsymbol{q})} \int_{\mathbb{R}^N} u(\Pi\, \boldsymbol{\Phi}^{-\tau}(\boldsymbol{q}, \boldsymbol{p})) f_0(\boldsymbol{q}, \boldsymbol{p}) \mathrm{d}\boldsymbol{p}, \tag{9.15}$$

where $u = u(\boldsymbol{q})$ is a function on Ω and Π is the projection $\Pi(\boldsymbol{q}, \boldsymbol{p}) = \boldsymbol{q}$ onto the position space. T can be interpreted as a suitable *weighted average* of the Frobenius-Perron operator over the momenta in each cross section $X(\boldsymbol{q}) = \{\boldsymbol{q}\} \times \mathbb{R}^N$ with the weights given by f_0.

In the special case that $f_0 = f_{\text{can}}$ defined by (9.11), $F(\boldsymbol{q}) = \mathcal{Q}(\boldsymbol{q})$, and so (9.15) becomes

$$Tu(\boldsymbol{q}) = \int_{\mathbb{R}^N} u(\Pi\, \boldsymbol{\Phi}^{-\tau}(\boldsymbol{q}, \boldsymbol{p}))\mathcal{P}(\boldsymbol{p})\mathrm{d}\boldsymbol{p}. \tag{9.16}$$

Hence, in this case, the spacial transfer operator T describes the momentum weighted fluctuations inside the canonical ensemble with respect to the time scale τ.

We shall prove that, with $r = 1$ or 2, T is a bounded linear operator on the weighted L^r-space

$$L_F^r(\Omega) = \left\{u : \int_\Omega |u(\boldsymbol{q})|^r F(\boldsymbol{q})\mathrm{d}\boldsymbol{q} < \infty\right\}.$$

As usual, the inner product on the Hilbert space $L_F^2(\Omega)$ is defined as

$$\langle u, v\rangle_F = \int_\Omega u(\boldsymbol{q})v(\boldsymbol{q})F(\boldsymbol{q})\mathrm{d}\boldsymbol{q}$$

and it induces the L^2-norm $\|u\|_{2,F} = \sqrt{\langle u, u\rangle_F}$. The L^1-norm on $L_F^1(\Omega)$ is $\|u\|_{1,F} = \displaystyle\int_\Omega |u(\boldsymbol{q})|F(\boldsymbol{q})\mathrm{d}\boldsymbol{q}$.

Proposition 9.3.1 *$T : L_F^1(\Omega) \to L_F^1(\Omega)$ is a Markov operator.*

Proof Let $u \in L_F^1(\Omega)$ be nonnegative. Then, via definition (9.15),

$$\begin{aligned}
\|Tu\|_{1,F} &= \int_\Omega \frac{1}{F(\boldsymbol{q})}\left|\int_{\mathbb{R}^d} u(\Pi\, \boldsymbol{\Phi}^{-\tau}(\boldsymbol{q}, \boldsymbol{p}))f_0(\boldsymbol{q}, \boldsymbol{p})\mathrm{d}\boldsymbol{p}\right| F(\boldsymbol{q})\mathrm{d}\boldsymbol{q} \\
&= \int_\Omega \int_{\mathbb{R}^N} u(\Pi\, \boldsymbol{\Phi}^{-\tau}(\boldsymbol{q}, \boldsymbol{p}))f_0(\boldsymbol{q}, \boldsymbol{p})\mathrm{d}\boldsymbol{p}\mathrm{d}\boldsymbol{q} \\
&= \int_X u(\Pi\, \boldsymbol{\Phi}^{-\tau}(\boldsymbol{x}))f_0(\boldsymbol{x})\mathrm{d}\boldsymbol{x} \\
&= \int_X u(\Pi \boldsymbol{x})f_0(\boldsymbol{x})\mathrm{d}\boldsymbol{x} = \int_\Omega u(\boldsymbol{q}) \int_{\mathbb{R}^N} f_0(\boldsymbol{q}, \boldsymbol{p})\mathrm{d}\boldsymbol{p}\mathrm{d}\boldsymbol{q} \\
&= \int_\Omega u(\boldsymbol{q})F(\boldsymbol{q})\mathrm{d}\boldsymbol{q} = \|u\|_{1,F},
\end{aligned}$$

where the fourth equality is from the invariance of f_0 with respect to the substitution $\boldsymbol{x} \to \boldsymbol{\Phi}^\tau(\boldsymbol{x})$ and the volume conservation property of the flow. Thus, T is a Markov operator. □

Proposition 9.3.2 *$T : L_F^2(\Omega) \to L_F^2(\Omega)$ is well-defined and satisfies that $\|Tu\|_{2,F} \leqslant \|u\|_{2,F}$ for all $u \in L_F^2(\Omega)$.*

Proof For any $u \in L^2_F(\Omega)$,

$$\|Tu\|^2_{2,F} = \int_\Omega \frac{1}{F(\boldsymbol{q})^2}\left(\int_{\mathbb{R}^N} u(\Pi\,\boldsymbol{\Phi}^{-\tau}(\boldsymbol{q},\boldsymbol{p}))f_0(\boldsymbol{q},\boldsymbol{p})\mathrm{d}\boldsymbol{p}\right)^2 F(\boldsymbol{q})\mathrm{d}\boldsymbol{q}$$
$$= \int_\Omega \frac{1}{F(\boldsymbol{q})}\left(\int_{\mathbb{R}^N} u(\Pi\,\boldsymbol{\Phi}^{-\tau}(\boldsymbol{q},\boldsymbol{p}))f_0(\boldsymbol{q},\boldsymbol{p})\mathrm{d}\boldsymbol{p}\right)^2 \mathrm{d}\boldsymbol{q}. \tag{9.17}$$

For $\boldsymbol{q} \in \Omega$ let $u_{\boldsymbol{q}}(\boldsymbol{p}) = u(\Pi\,\boldsymbol{\Phi}^{-\tau}(\boldsymbol{q},\boldsymbol{p}))$. Then, similar to the proof of Proposition 9.3.1, we have

$$\int_\Omega\int_{\mathbb{R}^N} [u_{\boldsymbol{q}}(\boldsymbol{p})]^2 f_0(\boldsymbol{q},\boldsymbol{p})\mathrm{d}\boldsymbol{p}\mathrm{d}\boldsymbol{q} = \|u\|^2_{2,F} < \infty. \tag{9.18}$$

The Cauchy-Schwarz inequality yields

$$\left(\int_{\mathbb{R}^N} u(\Pi\,\boldsymbol{\Phi}^{-\tau}(\boldsymbol{q},\boldsymbol{p}))f_0(\boldsymbol{q},\boldsymbol{p})\mathrm{d}\boldsymbol{p}\right)^2$$
$$= \left(\int_{\mathbb{R}^N} u_{\boldsymbol{q}}(\boldsymbol{p})f_0(\boldsymbol{q},\boldsymbol{p})\mathrm{d}\boldsymbol{p}\right)^2$$
$$\leqslant \int_{\mathbb{R}^N} [u_{\boldsymbol{q}}(\boldsymbol{p})]^2 f_0(\boldsymbol{q},\boldsymbol{p})\mathrm{d}\boldsymbol{p} \cdot \int_{\mathbb{R}^N} f_0(\boldsymbol{q},\boldsymbol{p})\mathrm{d}\boldsymbol{p}$$
$$= \int_{\mathbb{R}^N} [u_{\boldsymbol{q}}(\boldsymbol{p})]^2 f_0(\boldsymbol{q},\boldsymbol{p})\mathrm{d}\boldsymbol{p} \cdot F(\boldsymbol{q}).$$

Combining the above inequality with (9.17) and (9.18) gives

$$\|Tu\|^2_{2,F} \leqslant \int_\Omega\int_{\mathbb{R}^N} |u_{\boldsymbol{q}}(\boldsymbol{p})|^2 f_0(\boldsymbol{q},\boldsymbol{p})\mathrm{d}\boldsymbol{p}\mathrm{d}\boldsymbol{q} = \|u\|^2_{2,F}.$$ □

We can use the transfer operator $T : L^2_F(\Omega) \to L^2_F(\Omega)$ to express the transition probability (9.14) as follows. Since the characteristic functions χ_B and χ_C belong to $L^2_F(\Omega)$, we have

$$\langle T\chi_B, \chi_C\rangle_F = \int_\Omega \frac{1}{F(\boldsymbol{q})}\int_{\mathbb{R}^N} \chi_B(\Pi\,\boldsymbol{\Phi}^{-\tau}(\boldsymbol{q},\boldsymbol{p}))f_0(\boldsymbol{q},\boldsymbol{p})\mathrm{d}\boldsymbol{p}\chi_C(\boldsymbol{q})F(\boldsymbol{q})\mathrm{d}\boldsymbol{q}$$
$$= \int_X \chi_B(\Pi\,\boldsymbol{\Phi}^{-\tau}(\boldsymbol{x}))f_0(\boldsymbol{x})\chi_{X(C)}(\boldsymbol{x})\mathrm{d}\boldsymbol{x}$$
$$= \int_X \chi_{X(B)}(\boldsymbol{x})\chi_{X(C)}(\boldsymbol{\Phi}^{\tau}(\boldsymbol{x}))f_0(\boldsymbol{x})\mathrm{d}\boldsymbol{x}, \tag{9.19}$$

where the last equality is from the transformation $\boldsymbol{x} = \boldsymbol{\Phi}^{\tau}(\boldsymbol{y})$ together with the invariance of f_0 and the volume conservation property of the flow. Since

$$\langle \chi_B, \chi_C\rangle_F = \int_B F(\boldsymbol{q})\mathrm{d}\boldsymbol{q} = \int_{X(B)} f_0(\boldsymbol{x})\mathrm{d}\boldsymbol{x},$$

we finally have

$$w(B,C,\tau)=\frac{\displaystyle\int_{X(B)}\chi_{X(C)}(\boldsymbol{\Phi}^{\tau}(\boldsymbol{x}))f_0(\boldsymbol{x})\mathrm{d}\boldsymbol{x}}{\displaystyle\int_{X(B)}f_0(\boldsymbol{x})\mathrm{d}\boldsymbol{x}}=\frac{\langle T\chi_B,\chi_C\rangle_F}{\langle\chi_B,\chi_C\rangle_F}.$$

Proposition 9.3.3 $T: L^2_F(\Omega)\to L^2_F(\Omega)$ *is* self-adjoint. *Hence, its spectrum* $\sigma(T)\subset[-1,1]$.

Proof Given measurable subsets B and C of Ω, (9.19) and the reversibility property of $\boldsymbol{\Phi}^t$ implies that

$$\langle T\chi_B,\chi_C\rangle_F=\int_X\chi_{X(B)}(\boldsymbol{x})\chi_{X(C)}(R(\boldsymbol{\Phi}^{-\tau}(R(\boldsymbol{x})))f_0(\boldsymbol{x})\mathrm{d}\boldsymbol{x}.$$

Since f_0 is chosen to satisfy (9.10) and the sets $X(B)$ and $X(C)$ include all possible momenta, the transformation $\boldsymbol{y}=R(\boldsymbol{x})$ yields

$$\begin{aligned}\langle T\chi_B,\chi_C\rangle_F&=\int_X\chi_{X(B)}(\boldsymbol{y})\chi_{X(C)}(R(\boldsymbol{\Phi}^{-\tau}(\boldsymbol{y})))f_0(\boldsymbol{y})\mathrm{d}\boldsymbol{y}\\&=\int_X\chi_{X(B)}(\boldsymbol{\Phi}^{\tau}(\boldsymbol{x}))\chi_{X(C)}(R(\boldsymbol{x}))f_0(\boldsymbol{x})\mathrm{d}\boldsymbol{x}\\&=\int_X\chi_{X(C)}(\boldsymbol{x})\chi_{X(B)}(\boldsymbol{\Phi}^{\tau}(\boldsymbol{x}))f_0(\boldsymbol{x})\mathrm{d}\boldsymbol{x}=\langle\chi_B,T\chi_C\rangle_F,\end{aligned}$$

where the second equality is from the volume conservation property of the flow. Since the simple functions are dense in $L^2_F(\Omega)$, we get $\langle Tu,v\rangle_F=\langle u,Tv\rangle_F$ for all $u,\ v\in L^2_F(\Omega)$. □

We give a concrete example with $f_0=f_{\text{can}}$ to have a taste of the transfer operator in the simplest case.

Example 9.3.1 Consider the one-dimensional harmonic oscillator with $H(q,p)=(q^2+p^2)/2$ and $\Omega=\mathbb{R}$. Then $\mathcal{P}_\cdot(p)=\sqrt{\beta/(w\pi)}\exp\left(-\frac{\eta}{2}p^2\right)$. From (9.16) we have

$$Tu(q)=\int_{-\infty}^{\infty}u((\cos\tau)q-(\sin\tau)p)\sqrt{\frac{\eta}{w\pi}}\exp\left(-\frac{\eta}{2}p^2\right)\mathrm{d}p.$$

We distinguish two different cases. First assume that $\tau=2j\pi$ for some integer $j>0$. Then $\cos\tau=1$ and $\sin\tau=0$, so

$$Tu(q)=\int_{-\infty}^{\infty}u(q)\sqrt{\frac{\eta}{w\pi}}\exp\left(-\frac{\eta}{2}p^2\right)\mathrm{d}p=u(q)$$

for all $u\in L^2_Q$. In other words, T is the identity operator on L^2_Q, hence $\sigma(T)=\{1\}$ and every $B\subset\mathbb{R}$ is invariant. Similarly, if $\tau=(2j+1)\pi$ for some nonnegative integer j, then T is the identity operator on the subspace of L^2_Q consisting of all the even functions.

For all the other time scales τ, we have $|\cos\tau| < 1$. Suppose that $u \in L^2_Q$ is a smooth eigenvector of T associated with eigenvalue λ. Differentiating $Tu(q) = \lambda u(q)$ with respect to q yields

$$(Tu)'(q) = \cos\tau(Tu')(q) = \lambda u'(q).$$

Thus, $\lambda/\cos\tau$ is also an eigenvalue of T with eigenvector u'. Since $T1 = 1$, by induction we can find a sequence of eigenvectors which are polynomials $u_n \in L^2_Q$ such that

$$u_1'(q) \equiv 1, \quad u_n'(q) \equiv u_{n-1}(q), \quad \forall\, n \geqslant 1.$$

With an additional condition that such polynomials are pairwise orthogonal under $\langle\cdot,\cdot\rangle_Q$, they are uniquely determined as

$$u_1(q) = q, \; u_2(q) = q^2 - \frac{1}{\eta^2}, \; u_3(q) = q^3 - \frac{3}{\eta^2}q, \cdots$$

with the corresponding eigenvalues $\lambda_n = \cos^n\tau$ for all $n \geqslant 1$. This means that, for $\tau \neq j\pi$ with any positive integer j, T has a purely discrete spectrum with 0 as its limit point.

Remark 9.3.2 In general, under some mild conditions it has been shown [115] that the transfer operator $T : L^2_F(\Omega) \to L^2_F(\Omega)$ is quasi-compact.

We give a stochastic interpretation of the transfer operator in terms of Markov chains. Stochastic perturbations of dynamical systems have been extensively studied in, e.g., [75] and [100]; see also some related chapters of [82]. For the sake of simplicity we only consider the case of the canonical ensemble $f_0(\boldsymbol{q},\boldsymbol{p}) = f_{\text{can}}(\boldsymbol{q},\boldsymbol{p}) = \mathcal{P}(\boldsymbol{p})\mathcal{Q}(\boldsymbol{q})$. We shall need the concept of Foias operators on measures whose definition is taken from [82].

Definition 9.3.1 *Given two positive integers i and j, let $\boldsymbol{S} : W \times Y \subset \mathbb{R}^i \times \mathbb{R}^j \to W$ such that $\boldsymbol{S}(\cdot,\boldsymbol{y})$ is continuous for each $\boldsymbol{y} \in Y$ and $\boldsymbol{S}(\boldsymbol{x},\cdot)$ is Borel measurable for each $\boldsymbol{x} \in W$. The dynamical system with random perturbations*

$$\boldsymbol{x}_{n+1} = \boldsymbol{S}(\boldsymbol{x}_n,\boldsymbol{\xi}_n), \quad n = 0, 1, \cdots, \tag{9.20}$$

where $\boldsymbol{\xi}_n$ are independent random vectors such that the measure

$$\nu(B) = \text{prob}(\boldsymbol{\xi}_n \in B), \quad \forall\, B \in \mathcal{B}(Y)$$

is the same for all n and the random vectors $\boldsymbol{x}_0, \boldsymbol{\xi}_0, \boldsymbol{\xi}_1, \cdots$ are independent, is called a regular stochastic dynamical system. *The linear operator $P : \mathcal{M}_{\text{fin}} \to \mathcal{M}_{\text{fin}}$ defined by*

$$P\mu(A) = \int_W \int_Y \chi_A(\boldsymbol{S}(\boldsymbol{x},\boldsymbol{y}))\mathrm{d}\nu(\boldsymbol{y})\mathrm{d}\mu(\boldsymbol{x}), \quad \forall\, \mu \in \mathcal{M}_{\text{fin}},\ A \in \mathcal{B}(W),$$

where $\mathcal{M}_{\text{fin}}$ *is the set of all finite measures on W, is called the* Foias operator *corresponding to the dynamical system* (9.20).

For a given initial position $\boldsymbol{q}_0 \in \Omega$ we define the following regular stochastic dynamical system

$$\boldsymbol{q}_{n+1} = \Pi\,\boldsymbol{\Phi}^{\tau}(\boldsymbol{q}_n, \boldsymbol{p}_n)),\ \boldsymbol{p}_n \text{ is } \mathcal{P}\text{-distributed},\ n = 0, 1, \cdots, \tag{9.21}$$

which gives rise to a sequence of probability measures μ_k derived from finding the probability of the position vector $\boldsymbol{q}_k$ in a subset $B \in \mathcal{B}(\Omega)$, i.e.,

$$\mu_k(B) = \operatorname{prob}(\boldsymbol{q}_k \in B),\ \ \forall\, B \in \mathcal{B}(\Omega).$$

Actually the sequence $\{\mu_k\}$ is also given by the iterates of the following Foias operator P defined by

$$P\mu(B) = \int_{\Omega}\int_{\mathbb{R}^N} \chi_B(\Pi\,\boldsymbol{\Phi}^{\tau}(\boldsymbol{q},\boldsymbol{p}))\mathcal{P}(\boldsymbol{p})\mathrm{d}\boldsymbol{p}\mathrm{d}\mu(\boldsymbol{q}),\ \ \mu \in \mathcal{M}_{\text{fin}}. \tag{9.22}$$

That is, $\mu_k = P^k\mu_0$ if $\mu_0 \in \mathcal{M}_{\text{fin}}$ is the probability measure according to which the initial random position $\boldsymbol{q}_0$ is distributed.

If we only consider absolutely continuous finite measures $\mu_u \in \mathcal{M}_{\text{fin}}$ with Radon-Nikodym derivatives $u \in L^1_{\mathcal{Q}}$, then (9.22) implies that

$$\begin{aligned} P\mu_u(B) &= \int_X \chi_B(\Pi\,\boldsymbol{\Phi}^{\tau}(\boldsymbol{q},\boldsymbol{p}))u(\Pi\boldsymbol{x})f_{\text{can}}(\boldsymbol{x})\mathrm{d}\boldsymbol{x} \\ &= \int_{X(B)} u(\Pi\,\boldsymbol{\Phi}^{-\tau}(\boldsymbol{x}))f_{\text{can}}(\boldsymbol{x})\mathrm{d}\boldsymbol{x} \\ &= \int_B\int_{\mathbb{R}^N} u(\Pi\,\boldsymbol{\Phi}^{-\tau}(\boldsymbol{q},\boldsymbol{p}))\mathcal{P}(\boldsymbol{p})\mathrm{d}p\mathcal{Q}(\boldsymbol{q})\mathrm{d}\boldsymbol{q}. \end{aligned}$$

Therefore, the density of the measure $P\mu_u$ has the expression

$$Tu(\boldsymbol{q}) = \int_{\mathbb{R}^N} u(\Pi\,\boldsymbol{\Phi}^{-\tau}(\boldsymbol{q},\boldsymbol{p}))\mathcal{P}(\boldsymbol{p})\mathrm{d}\boldsymbol{p},$$

which is exactly the same as (9.16) given by the transfer operator T. Hence, the transfer operator $T : L^1_{\mathcal{Q}}(\Omega) \to L^1_{\mathcal{Q}}(\Omega)$ is equivalent to the Foias operator $P : \mathcal{M}_{\text{fin}} \to \mathcal{M}_{\text{fin}}$ restricted to the finite measures with Radon-Nikodym derivatives in $L^1_{\mathcal{Q}}$, and the stochastic dynamical system (9.21) is a realization of the Markov chain induced by the Markov operator T, which combines a short term deterministic model, characterized by the flow $\boldsymbol{\Phi}^{\tau}$, with a statistical model, characterized by the $\mathcal{P}$-distribution, the momentum part of the canonical distribution. Consequently, if the initial position $\boldsymbol{q}_0$ of (9.21) is distributed according to the probability density $u \in L^1_{\mathcal{Q}}(\Omega)$, the probability density $u_k \in L^1_{\mathcal{Q}}(\Omega)$ of finding $\boldsymbol{q}_k = \boldsymbol{q}$ is given by $u_k(\boldsymbol{q}) = T^k u(\boldsymbol{q})$.

A dynamical system is said to have a *metastable decomposition*, if its state space can be decomposed into a finite number of disjoint sets such that the probability of exit from each of these sets is extremely small. In the context of the molecular dynamics discussed here, the metastability is related to the dominant eigenvectors of the transfer operator. More specifically, metastable decompositions can be detected via the discrete eigenvalues of the transfer operator $T(\tau)$ close to its spectral radius 1 which is the maximal eigenvalue of $T(\tau)$. They can be identified from the structure of the corresponding eigenfunctions with the help of the concept of almost invariant sets.

Since the metastability of a set B may be measured by $w(B, B, \tau)$ in (9.12), for any finite decomposition $\{B_1, \cdots, B_k\}$ of the state space into k disjoint subsets B_i, the metastability of the decomposition can be defined by

$$w_k(\tau) = \sum_{i=1}^{k} w(B_i, B_i, \tau).$$

The possibility of identifying an almost optimal metastable decomposition of the state space is seen from the following result [67].

Theorem 9.3.1 *Let $T : L^2_Q \to L^2_Q$ be a reversible transfer operator whose* essential spectral radius *is less than* 1 *and* 1 *is its simple eigenvalue. Then T is self-adjoint and its spectrum has the form*

$$\sigma(T) \subset [a, b] \cup \{\lambda_k\} \cup \cdots \cup \{\lambda_2\} \cup \{1\},$$

where $-1 < a \leqslant b < \lambda_k \leqslant \cdots \leqslant \lambda_1 = 1$. Such λ_is are isolated eigenvalues of finite multiplicity that are counted according to the multiplicity with the corresponding eigenfunctions $v_k, v_{k-1}, \cdots, v_1$ normalized to $\|v_i\|_{L^2_Q} = 1$. Let Q be the orthogonal projection of L^2_Q onto span $\{\chi_{A_1}, \cdots, \chi_{A_k}\}$, *where each $A_i =$* supp v_i. *Then the following bounds hold:*

$$1 + \sum_{i=2}^{k} c_i\lambda_i + c \leqslant w_k(\tau) \leqslant 1 + \sum_{i=2}^{k} \lambda_i,$$

where $c_i = \|Qv_i\|^2_{L^2_Q} \leqslant 1$, $i = 1, 2, \cdots, k$, and $c = |a| \prod_{i=2}^{k} (1 - c_i) < 1$.

From the above theorem, if the transfer operator T can be well approximated by finite dimensional operators, a Perron cluster analysis (see [25, 26]) can be used to find the number, the life times, and the decay pattern of the metastable conformations.

Since the definition of the transfer operator is based on the Hamiltonian flow at time τ, one can numerically solve the initial value problem of the Hamiltonian equations of motion (9.8) with time step size $h = \tau/s$ to yield a discrete flow $\boldsymbol{\Psi}^h$ such that $\boldsymbol{\Phi}^\tau \boldsymbol{x}_0$ is approximated by $\boldsymbol{x}_s$ via the iteration

$$\boldsymbol{x}_{j+1} = \boldsymbol{\Psi}^h \boldsymbol{x}_j, \ j = 0, 1, \cdots, s-1.$$

Then the chain

$$\boldsymbol{q}_{n+1} = \Pi(\boldsymbol{\Psi}^h)^s(\boldsymbol{q}_n, \boldsymbol{p}_n), \ \ \boldsymbol{p}_n \text{ is } \mathcal{P}\text{-distributed}, \ \ n = 0, 1, \cdots$$

approximates the Markov chain (9.21) after using the Metropolis acceptance procedure to yield a hybrid Monte Carlo (HMC) chain which contains good approximations of sub-trajectories of the Hamiltonian system.

Now the idea of Ulam's method can be applied to approximate the transfer operator T in solving the eigenvalue problem $Tu = \lambda u$. In the context of the $L^2_{\mathcal{Q}}$ space here, Ulam's method is the same as the orthogonal projection method. Given a finite partition of the position space Ω in terms of the N-dimensional rectangles $\{B_1, B_2, \cdots, B_n\}$, let $\chi_i = \chi_{B_i}$ and $V_n = \text{span}\{\chi_1, \chi_2, \cdots, \chi_n\}$. Let $\Pi_n : L^2_{\mathcal{Q}} \to V_n$ be the operator that projects each $u \in L^2_{\mathcal{Q}}$ orthogonally onto V_n. Then

$$\Pi_n u = \sum_{i=1}^n \frac{1}{\langle \chi_i, \chi_i \rangle_{\mathcal{Q}}} \langle \chi_i, u \rangle_{\mathcal{Q}} \cdot \chi_i = \sum_{i=1}^n \frac{1}{\displaystyle\int_{B_i} \mathcal{Q}(\boldsymbol{q}) \mathrm{d}\boldsymbol{q}} \langle \chi_i, u \rangle_{\mathcal{Q}} \cdot \chi_i.$$

The resulting discretized eigenvalue problem is $\Pi_n T u_n = \lambda u_n$. If $u_n = \sum\limits_{i=1}^n v_i \chi_i$, then the resulting matrix eigenvalue problem is

$$\sum_{j=1}^n \frac{\langle T\chi_i, \chi_j \rangle_{\mathcal{Q}}}{\langle \chi_i, \chi_i \rangle_{\mathcal{Q}}} v_j = \lambda v_i, \ \ \forall\, i = 1, 2, \cdots, n.$$

Thus, the (i, j)-entry of the $n \times n$ stochastic matrix $\boldsymbol{T}_n$ is

$$t_{ij} = \frac{\langle T\chi_i, \chi_j \rangle_{\mathcal{Q}}}{\langle \chi_i, \chi_i \rangle_{\mathcal{Q}}} = w(B_i, B_j, \tau). \tag{9.23}$$

For the numerical evaluation of the t_{ij}'s in (9.23), using a realization $\{\boldsymbol{q}_1, \boldsymbol{q}_2, \cdots, \boldsymbol{q}_l\}$ of the HMC chain from above, the (i, j)-entry of the transition matrix $\boldsymbol{T}_n$ can be computed by the formula

$$T_{ij} = \frac{\#\{\boldsymbol{q}_{n+1} \in B_j, \ \boldsymbol{q}_n \in B_i\}}{\#\{\boldsymbol{q}_n \in B_i\}}, \ \ i, j = 1, 2, \cdots, n.$$

More details on the computational algorithm is referred to [25, 26, 115] and the references therein.

9.4 DS-CDMA in Wireless Communications

With the increasing demand of wireless communications around the world, the research on the *third generation* wireless communication technologies has been extensive in electrical engineering. Unlike the first generation with the time division multiple access (TDMC) and the second generation with the frequency division multiple access (FDMC), the third generation wireless communication technique uses the *direct sequence code division multiple access* (DS-CDMA) in which the statistical properties of the so-called *spreading sequences* generated by chaotic transformations play an important role. We give a short introduction to this technique; more details can be found in [117].

We consider a simplified baseband equivalent of an asynchronous DS-CDMA system model as an example. Suppose that there are s users of the system with the ith user's information signal given by $S^i(t) = \sum_{k=-\infty}^{\infty} S_k^i g_T(t - kT)$, where $S_k^i = \pm 1$ are the information symbols and g_T is the rectangular pulse function which is 1 on $[0, T]$ and 0 elsewhere. The spreading signal depends on sequences $x^i = \{x_k^i\}$ of finite length l (called the spreading factor) from the alphabet set X, which are mapped into the set Y of all the dth complex roots of 1 by a quantization function $Q : X \to Y$. The combined signal $Q^i(t) = \sum_{k=-\infty}^{\infty} Q(x_k^i) g_{T/l}(t - kT/l)$ is then multiplied by $S^i(t)$ and transmitted along the (equivalent) channel together with the spread-spectrum signals from the other users.

At the receiver of the useful jth user, the global signal is multiplied by $\overline{Q}^j(t)$ and fed into an integrate-and-dump stage, where $\overline{Q}^j$ is the complex conjugate function of Q^j defined by $\overline{Q}^j(t) = \overline{Q^j(t)}$. In theory, the performance of a DS-CDMA system tends to be optimal if the codes for different users are almost *orthogonal*, i.e., the spreading sequences are generated with almost vanishing cross correlations. Nevertheless, for the asynchronous communication environment in which a CDMA link is from the mobile transmitters to a fixed base station in a cellular system, the sequence design problem becomes much more complicated than the determination of a set of sequences with vanishing cross correlations. As signals of different links are not synchronized and the spreading sequences are not perfectly uncorrelated, orthogonality cannot be achieved and contributions from undesired signals appear at the output of the integrate-and-dump as the co-channel interference c_k^j to S_k^j for each k.

A commonly adopted approach to the co-channel interference estimation is to consider each c_k^j as a Gaussian random variable whose variance can be estimated from the cross-correlation characteristics of the spreading sequences. If we define the *partial correlation function* Γ_τ between the ith and the jth spreading sequences $x^i = \{x_k^i\}$ and $x^j = \{x_k^j\}$ as

$$\Gamma_\tau(x^i, x^j) = \begin{cases} \sum_{k=0}^{l-\tau-1} Q(x_k^i)\overline{Q}(x_{k+\tau}^j), & 0 \leqslant \tau < l, \\ \overline{\Gamma}_\tau(x^j, x^i), & -l < \tau < 0, \\ 0, & |\tau| \geqslant l, \end{cases}$$

then the *bit-error probability* P_{err} of the communication link can be expressed as

$$P_{\text{err}} \approx \frac{1}{2}\text{erfc}\left(\left[\frac{1}{(s-1)R}\right]^{\frac{1}{2}}\right), \tag{9.24}$$

where $\text{erfc}(t) = 2\pi^{-1/2}\int_t^\infty \exp(-\xi^2)\text{d}\xi$ and the quantity

$$R = \frac{1}{3l|\Gamma_0(x^j, x^j)|^2} \sum_{\tau=-l+1}^{l-1} E_{x^i,x^j}^{i \neq j}[2|\Gamma_\tau(x^i, x^j)|^2 + \text{Re}\left(\Gamma_\tau(x^i, x^j)\overline{\Gamma}_{\tau+1}(x^i, x^j)\right)] \tag{9.25}$$

assumes, in terms of the signal power, the significance of an expected *signal-to-interference ratio* (SIR) per interfering user. This is the expected degradation in system performance when a new user is added.

The global SIR R is the *performance merit figure* that one intends to optimize. This quantity depends on the spreading sequences. Traditionally they are chosen to be random-like such as the Gold or maximum-length sequences. It has been shown in recent years that suitable chaotic transformations can be used to deliver spreading sequences which are better than the random-like methods in terms of the statistical merit figure R. It follows that an optimal performance can be achieved if one is able to design spreading sequences characterized by the desired statistical features. Thus, it is the statistical study of chaotic deterministic systems that is essential in the performance analysis of wireless communications.

The property of decay of correlations for chaotic transformations plays a key role in the statistical analysis of the spreading sequence performance. The expression (9.24) indicates that the bit-error probability P_{err} is decreased when the global SIR R is decreased for a fixed constant s, the number of the users. So the purpose of performance improvements is for the minimization of the quantity R. From its expression (9.25), R can be evaluated by means of the expectations of the partial correlation functions Γ_τ and their products as random variables with respect to the collection of all the spreading sequences. Since the spreading sequences x^i are generated by chaotic transformations starting from randomly chosen initial points, and any initial point x_0^i corresponds to the resulting iteration sequence $x^i = \{x_k^i\}$, expectations of $\Gamma_\tau(x^i, x^j)$ and their products taken over all the spreading sequences x^i and x^j can be translated into expectations over randomly chosen initial points x_0^i and x_0^j which determine the sequences themselves. Thus, the expression (9.25) for R can be written as

$$\begin{aligned}
R= & \frac{1}{3l|\Gamma_0(x^j,x^j)|^2} \sum_{\tau=-l+1}^{l-1} E^{i\neq j}_{x_0^i,x_0^j} \left[2|\Gamma_\tau(x^i,x^j)|^2\right. \\
& \left. + \operatorname{Re}\left(\Gamma_\tau(x^i,x^j)\overline{\Gamma}_{\tau+1}(x^i,x^j)\right)\right] \\
= & \frac{1}{3l^3}\left\{2E^{i\neq j}_{x_0^i,x_0^j}\left[|\Gamma_0(x^i,x^j)|^2\right]\right. \\
& +2\sum_{\tau=1}^{l-1} E^{i\neq j}_{x_0^i,x_0^j}\left[|\Gamma_\tau(x^i,x^j)|^2+|\Gamma_\tau(x^j,x^i)|^2\right] \\
& +\operatorname{Re}E^{i\neq j}_{x_0^i,x_0^j}\left[\Gamma_0(x^i,x^j)\overline{\Gamma}_1(x^i,x^j)\right] \\
& +\operatorname{Re}\sum_{\tau=1}^{l-1} E^{i\neq j}_{x_0^i,x_0^j}\left[\Gamma_\tau(x^i,x^j)\overline{\Gamma}_{\tau+1}(x^i,x^j)\right. \\
& \left.\left. + \Gamma_{\tau-1}(x^j,x^i)\overline{\Gamma}_\tau(x^j,x^i)\ \right]\right\} \\
= & \frac{1}{3l^3}\left\{2E^{i\neq j}_{x_0^i,x_0^j}\left[|\Gamma_0(x^i,x^j)|^2\right]+4\sum_{\tau=1}^{l-1}E^{i\neq j}_{x_0^i,x_0^j}\left[|\Gamma_\tau(x^i,x^j)|^2\right]\right. \\
& +\operatorname{Re}E^{i\neq j}_{x_0^i,x_0^j}\left[\Gamma_0(x^i,x^j)\overline{\Gamma}_1(x^i,x^j)\right] \\
& +\operatorname{Re}\sum_{\tau=1}^{l-1} E^{i\neq j}_{x_0^i,x_0^j}\left[\Gamma_\tau(x^i,x^j)\overline{\Gamma}_{\tau+1}(x^i,x^j)\right. \\
& \left.\left. + \Gamma_{\tau-1}(x^i,x^j)\overline{\Gamma}_\tau(x^i,x^j)\ \right]\right\},
\end{aligned} \tag{9.26}$$

where E is the expectation and the random variables x_0^i and x_0^j have the density f_0.

For the spreading sequences generated by the chaotic mapping $S:[0,1]\to[0,1]$, we have $x_k^i=S^k(x_0^i)$ and $x_{k+\tau}^j=S^{k+\tau}(x_0^j)$. So

$$|\Gamma_0(x^j,x^j)|^2=\left(\sum_{k=0}^{l-1}|Q(S^k(x_0^j))|^2\right)^2,$$

$$\begin{aligned}
& E^{i\neq j}_{x_0^i,x_0^j}\left[|\Gamma_\tau(x^i,x^j)|^2\right] \\
& = \sum_{k=0}^{l-\tau-1}\sum_{n=0}^{l-\tau-1} E^{i\neq j}_{x_0^i,x_0^j}\left[Q(S^k(x_0^i))\overline{Q}(S^{k+\tau}(x_0^j))\overline{Q}(S^n(x_0^i))Q(S^{n+\tau}(x_0^j))\right] \\
& = \sum_{k=0}^{l-\tau-1}\sum_{n=0}^{l-\tau-1}\int_0^1\int_0^1 Q(S^k(x_0^i))\overline{Q}(S^{k+\tau}(x_0^j))\overline{Q}(S^n(x_0^i))
\end{aligned}$$

$$\times Q(S^{n+\tau}(x_0^j))f_0(x_0^i)f_0(x_0^j)\mathrm{d}x_0^i\mathrm{d}x_0^j$$
$$= \sum_{k=0}^{l-\tau-1}\sum_{n=0}^{l-\tau-1}\int_0^1 Q(S^k(x))\overline{Q}(S^n(x))f_0(x)\mathrm{d}x$$
$$\times \int_0^1 \overline{Q}(S^{k+\tau}(x))Q(S^{n+\tau}(x))f_0(x)\mathrm{d}x$$

for $0 < \tau < l$, and for $0 \leqslant \tau < l-1$,

$$E_{x_0^i,x_0^j}^{i\neq j}\left[\Gamma_\tau(x^i,x^j)\overline{\Gamma}_{\tau+1}(x^i,x^j)\right]$$
$$= \sum_{k=0}^{l-\tau-1}\sum_{n=0}^{l-\tau-2} E_{x_0^i,x_0^j}^{i\neq j}\left[Q(S^k(x_0^i))\overline{Q}(S^{k+\tau}(x_0^j))\overline{Q}(S^n(x_0^i))Q(S^{n+\tau+1}(x_0^j))\right]$$
$$= \sum_{k=0}^{l-\tau-1}\sum_{n=0}^{l-\tau-2}\int_0^1\int_0^1 Q(S^k(x_0^i))\overline{Q}(S^{k+\tau}(x_0^j))\overline{Q}(S^n(x_0^i))$$
$$\times Q(S^{n+\tau+1}(x_0^j))f_0(x_0^i)f_0(x_0^j)\mathrm{d}x_0^i\mathrm{d}x_0^j$$
$$= \sum_{k=0}^{l-\tau-1}\sum_{n=0}^{l-\tau-2}\int_0^1 Q(S^k(x))\overline{Q}(S^n(x))f_0(x)\mathrm{d}x$$
$$\times \int_0^1 \overline{Q}(S^{k+\tau}(x))Q(S^{n+\tau+1}(x))f_0(x)\mathrm{d}x.$$

Thus, it follows from (9.26) that the evaluation of R involves integrals of the type

$$\int_0^1 (f\circ S^k)(g\circ S^n)f_0\mathrm{d}m = \int_0^1 fP^k\left[(g\circ S^n)f_0\right]\mathrm{d}m, \tag{9.27}$$

which is closely related to the rate of decay of correlations for the chaotic mapping S.

As an illustration how to proceed along this direction, we let $k = 0$ in (9.27). The discussion in Section 9.1 tells us that if S is piecewise C^2 with $\inf_{x\in[0,1]\setminus\{a_1,\cdots,a_{r-1}\}}|S'(x)| > 1$ (i.e., S is a Lasota-Yorke interval mapping [88]), and if the invariant probability measure μ^* with density f^* is mixing with respect to S, then it is also *exact*. That is, $\lim_{n\to\infty}\mu^*(S^n(A)) = 1$ for any measurable subset A of $[0,1]$ with $\mu^*(A) > 0$ [14]. Furthermore, there is a constant M such that for any function $f \in BV(0,1)$,

$$\left\|P^n f - \left(\int_0^1 f(x)\mathrm{d}x\right)f^*\right\|_{BV} \leqslant M\|f\|_{BV}r_{\mathrm{mix}}^n,\ \ \forall\, n, \tag{9.28}$$

where $r_{\text{mix}} \in (0,1)$ is the *rate of mixing.* (9.28) implies the following inequality

$$\left|\int_0^1 f(x)g(S^n(x))f^*(x)\mathrm{d}x - \int_0^1 f(x)f^*(x)\mathrm{d}x \cdot \int_0^1 g(x)f^*(x)\mathrm{d}x\right|$$
$$\leqslant C\|f\|_\infty\|g\|_\infty r_{\text{mix}}^n$$

for the correlation coefficient

$$\left|\int_0^1 f(x)g(S^n(x))f^*(x)\mathrm{d}x - \int_0^1 f(x)f^*(x)\mathrm{d}x \cdot \int_0^1 g(x)f^*(x)\mathrm{d}x\right|,$$

where C is a constant independent of the choice of the functions f and g. In particular, if $\int_0^1 f(x)f^*(x)\mathrm{d}x = 0$ or $\int_0^1 g(x)f^*(x)\mathrm{d}x = 0$, then

$$\left|\int_0^1 f(x)g(S^n(x))f^*(x)\mathrm{d}x\right| \leqslant C\|f\|_\infty\|g\|_\infty r_{\text{mix}}^n, \quad \forall\, n.$$

The above analysis implies that the performance of the spreading sequences depends on the correlation property of the underlying mapping S, which is determined by the rate of mixing $r_{\text{mix}} \in (0,1)$. We refer to [117] for more details.

Exercises

9.1 Let μ be a unique absolutely continuous invariant measure of $S: [0,1] \to [0,1]$ with the density f^*. Let $P: L^1(0,1) \to L^1(0,1)$ be the corresponding Frobenius-Perron operator with respect to the Lebesgue measure m. Define an operator P_μ on $L^1(\mu)$ by

$$P_\mu f = \frac{P(f \cdot f^*)}{f^*}.$$

Show that P_μ is the Frobenius-Perron operator induced by S from $L^1(\mu)$ into itself (see Exercise 4.12).

9.2 Let $X = [0,1]$ and let μ be a unique absolutely continuous invariant measure of $S: [0,1] \to [0,1]$. Express the correlation coefficient Cor (f,g,n) in Definition 9.1.1 in terms of the Frobenius-Perron operator with respect to the Lebesgue measure m and the invariant measure μ, respectively.

9.3 Show that if $f \in L^1(0,1)$ is of bounded variation and $m(\text{supp } f) > 0$, then the interior of supp f is nonempty.

9.4 Show that if $f \in L^1(0,1)$ is of bounded variation, then the support of f can be expressed as

$$\text{supp } f = \bigcup_{n=1}^{s} I_n \ \text{ a.e.}, \quad 1 \leqslant s \leqslant \infty,$$

where I_n are mutually disjoint open intervals.

9.5 Let f and g be two distinct stationary densities of the Frobenius-Perron operator associated with a nonsingular transformation $S : [0,1] \to [0,1]$ that satisfies the conditions of Theorem 5.2.1. Show that there exist two stationary densities f^* and g^* of P such that $\operatorname{supp} f^* \cap \operatorname{supp} g^* = \varnothing$.

9.6 Show that if f is a stationary density of the Frobenius-Perron operator associated with S and if $S^{-1}(A) = A$, then $f \cdot \chi_A$ is a fixed point of P.

9.7 Let $f : [a,b] \to \mathbb{R}$ be of bounded variation. Let x, $y \in [a,b]$ and $x < y$. Show that

$$|f(x)| + |f(y)| \leqslant \bigvee_0^1 f + \frac{2}{y-x} \int_x^y |f(t)| \mathrm{d}t.$$

Bibliography

[1] Adams R A. Sobolev Spaces. New York: Academic Press, 1975

[2] Adl-Zarabi K. Absolutely continuous invariant measures for piecewise expanding C^2 transformations on $\mathbb{R}^n$ on domains with cusps on the boundaries. Ergod Th & Dynam Sys, 1996, 16: 1–18

[3] Adler R, Konhein A, McAndrew M. Topological entropy. Trans Amer Math Soc. 1965, 114: 309–319

[4] Baladi V. Positive Transfer Operators and Decay of Correlations. World Scientific, 200

[5] Baladi V, Young L S. On the spectra of randomly perturbed expanding maps. Comm Math Phys, 1993, 156(2): 355–385

[6] Barnsley M. Fractals Everywhere. Boston: Academic Press, 1988

[7] Beck C, Schlögl F. Thermodynamics of Chaotic Systems. Cambridge: Cambridge University Press, 1993

[8] Billingsley P. Convergence of Probability Measures. Wiley, 1968

[9] Borwein J M, Lewis A S. On the convergence of moment problems. Trans Amer Math Soc, 1991, 325(1): 249–271

[10] Bose C. On the existence and approximation of invariant densities for nonsingular transformations on $\mathbb{R}^d$. J Approx Th, 1994, 79(2): 260–270

[11] Bose C, Murray R. The exact rate of approximation in Ulam's method. Disc Cont Dynam Sys, 2001, 7(1): 219–235

[12] Bose C, Murray R. Minimum 'energy' approximations of invariant measures for nonsingular transformations. Discrete Contin Dyn Syst, 2006, 14(3): 597–615

[13] Bowen R. Equilibrium States and the Ergodic Theory of Anosov Diffeomorphisms. Springer-Verlag, 1975

[14] Boyarsky A, Góra P. Laws of Chaos: Invariant Measures and Dynamical Systems in One Dimension. Birkhäuse, 1997

[15] Boyarsky A, Góra P, Lou Y S. Constructive approximations to the invariant densities of higher-dimensional transformations. Construc Approx, 1994, 10: 1–13

[16] Boyarsky A, Lou Y S. Approximating measures invariant under higher-dimentional chaotic transformation. J Approx Th, 1991, 65: 231–244

[17] Chatelin F. Spectral Approximation of Linear Operators. Academic Press, 1983

[18] Chen G, Ueta T. Chaos in Circuits and Systems. World Scientific, 2002

[19] Chiu C, Du Q, Li T Y. Error estimates of the Markov finite approximation of the Frobenius-Perron operator. Nonlin Anal, Th Meth Appl, 1992, 19(4): 291–308

[20] Ciarlet P G. The Finite Element Method for Elliptic Problems. North-Holland, 1978

[21] Clément P. Approximation by finite element functions using local regularization. RAIRO Anal Numér, 1975, 9: 77–84

[22] Collet P, Eckmann J P. Iterated Maps on the Interval as Dynamical Systems. Boston: Birkhäuser, 1980

[23] de Melo W and Van Strien S One-Dimensial Dynamics. Berlin: Springer-Verlag, 1993

[24] Dellnitz M, Junge O. On the approximation of complicated dynamical behavior. SIAM J Numer Anal, 1999, 36(2): 491–515

[25] Deuflhard P, Schütte Ch. Molecular conformation dynamics and computational drug design. Applied Mathematics Entering the 21st Century, Invited Talks from ICIAM 2003 Congress, Sydney, 2004, 91–119

[26] Deuflhard P, Weber M. Robust Perron cluster analysis in conformation dynamics. Linear Algb Appl, 2005, 398: 161–184

[27] Devaney R L. An Introduction to Chaotic Dynamical Systems. Red wood: Addison-Wesley, 1989

[28] Ding J. Finite Approximations of a Class of Frobenius-Perron Operators. Ph.D. Dissertation, Michigan State University, USA, 1990

[29] Ding J. Computing invariant measures of piecewise convex transformations. J Stat Phys, 1996, 83(3/4): 623–635

[30] Ding J. Decomposition theorems for Koopman operators. Nonlin Anal, Th Meth Appl, 1997, 28(6): 1011–1018

[31] Ding J. A maximum entropy method for solving Frobenius-Perron operator equations. Appl Math Comput, 1998, 93: 155–168

[32] Ding J, Du Q, Li T Y. High order approximation of the Frobenius-Perron operator, Appl Math Comput, 1993, 53: 151–171

[33] Ding J, Du Q, Li T Y. The spectral analysis of Frobenius-Perron operators. J Math Analy Appl, 1994, 184(2): 285–301

[34] Ding J, Hitt L, Zhang X. Markov chains and dynamic geometry of polygons. Linear Algb Appl, 2003, 367: 255–270

[35] Ding J, Hornor W E. A new approach to the Frobenius-Perron operator. J Math Analy Appl, 1994, 187(3): 1047–1058

[36] Ding J, Hunt F. Error estimates for quasi-compact Markov operators. Nonlinear Anal, 2000, 42(1): 85–95

[37] Ding J, Jin C, and Zhou A. A posteriori error estimates for Markov approximations of Frobenius-Perron operators. Nonlinear Anal, 2007, 67: 763–772

[38] Ding J, Li T Y. Markov finite approximation of Frobenius-Perron operator. Nonlinear Anal, 1991, 17(8): 759–772

[39] Ding J, Li T Y. Entropy - an introduction. In: Shan-Tao Liao et al. ed. Dynamical Systems. Nankai Series Pure Applied Math. Theore. Phys. World Scientific, 1993, 4: 26–53

[40] Ding J, Li T Y. Projection solutions of the Frobenius-Perron operator equations. Inter J Math Sci, 1993, 16(3): 465–484

[41] Ding J, Li T Y, A convergence rate analysis for Markov finite approximations to a class of Frobenius-Perron operators. Nonlinear Anal, Th, 1998, 31(5/6): 765–777

[42] Ding J, Li T Y, Zhou A. Finite approximations of Markov operators. J Compt Appl Math, 2002, 147: 137–152

[43] Ding J, Mao D, Zhou A. A Monte Carlo approach to piecewise linear Markov approximations of Markov operators. Monte Carlo Methods Appl, 2003, 9(4): 295–306

[44] Ding J, Mead L. Maximum entropy approximation for Lyapunov exponents of chaotic maps. J Math Phys, 2002, 43(5): 2518–2522

[45] Ding J, Rhee N. Approximations of Frobenius-Perron operators via interpolation. Nonlinear Anal, 2004, 57(5/6): 831–842

[46] Ding J, Wang Z. Parallel computation of invariant measures. Ann Oper Res, 2001, 103: 283–290

[47] Ding J, Zhou A. The projection method for computing multi-dimensional absolutely continuous invariant measures. J Stat Phys, 1994, 77(3/4): 899–908

[48] Ding J, Zhou A. Piecewise linear Markov approximations of Frobenius-Perron operators associated with multi-dimensional transformations. Nonlinear Anal, 1995, 25(4): 399–408

[49] Ding J, Zhou A. Finite approximations of Frobenius-Perron operators. A solution of Ulam's conjecture to multi-dimensional transformations. Physica D, 1996, 92: 61–68

[50] Ding J, Zhou A. Invariant measures and their computation. Adv. Math. (China), 1998, 27(4): 309–323

[51] Ding J, Zhou A. Computing fixed points of quasi-compact Frobenius-Perron operators. Nonlinear. Anal. 1999, 39(2): 231–239

[52] Ding J, Zhou A. Absolutely continuous invariant measures for piecewise C^2 and expanding mappings in higher dimensions. Disc Cont. Dynam Sys, 2000, 6(2): 451–458

[53] Ding J, Zhou A. On the spectrum of Frobenius-Perron operators. J Math Anal Appl, 2000, 250: 610–620

[54] Ding J, Zhou A. Constructive approximations of Markov operators. J Stat Phys, 2001, 105(5/6): 863–878

[55] Ding J, Zhou A. Structure preserving finite element approximations of Markov operators. Nonlinearity, 2002, 15: 923–936

[56] Dugundji J. Topology. Boston: Allyn and Bacon, 1966

[57] Dunford N, Schwartz L. Linear Operators, Part I. Amsterdam: Interscience, 1957

[58] Foguel S F. The Ergodic Theory of Markov Processes. Van Nostrand Reinhold, 1969

[59] Froyland G. Finite approximations of Sinai-Bowen-Ruelle measures for Anosov systems in two dimensions. Random Comp Dynam, 1995, 3(4): 251–264

[60] Froyland G. Approximating physical invariant measures of mixing dynamical systems in higher dimensions. Nonlinear Anal, 1988, 32(7): 831–860

[61] Gilbert D, Trudinger N S. Elliptic Partial Differential Equations of Second Order. Berlin: Springer-Verlag, 1983

[62] Girault V, Raviart P A. Finite Element Method for Navier-Stokes Equations-Theory and Algorithms. Berlin-New York: Springer-Verlag, 1986

[63] Giusti E. Minimal Surfaces and Functions of Bounded Variation. Basel: Birkhäuser, 1984

[64] Góra P, Boyarsky A. Absolutely continuous invariant measures for piecewise expanding C^2 transformations in $\mathbb{R}^n$. Israel J Math, 1989, 67(3): 272–286

[65] Hitt L, Zhang X. Dynamic geometry of polygons. Elemente der Math, 2001, 56(1): 21–37

[66] Hofbauer F, Keller G. Ergodic properties of invariant measures for piecewise monotonic transformations. Math Z, 1982, 180: 119–140

[67] Huisinga W, Schmidt B. Metastability and dominant eigenvalues of transfer operators. Lecture Notes in Computational Science and Engineering. Berlin Springer-Verlag, 2005, 167–182

[68] Hunt F Y. A Monte Carlo approach to the approximation of invariant measures. Random Comp Dynam, 1994, 2(1): 735–749

[69] Hunt F Y, Miller W M. On the approximation of invariant measures. J Stat Phys, 1992, 66(1/2): 535–548

[70] Jabłoński M. On invariant measures for piecewise C^2-transformations of the n-dimensional cube. Ann. Polon. Math., 1983, XLIII: 185–195

[71] Jenkinson O, Pollocott M. Entropy, exponents and invariant measures for hyperbolic systems: dependence and computation. Modern Dynamical Systems and Application. Cambridge University Press, 2004, 365–384

[72] Keane M, Murrary R, Young L S. Computing invariant measures for expanding circle maps. Nonlinearity, 1998, 11(1): 27–46

[73] Keller G. Probabilités-ergodicité et mesures invariantes pour les transformations dilatantes par morceaux d'une région bornée du plan. C R Acad Sci Paris, Série I Math, 1979, 289: 625–627

[74] Keller G. Stochastic stability in some chaotic dynamical systems. Mona Math, 1982, 94: 313–333

[75] Kifer Y. Random Perturbations of Dynamical Systems. Boston: Birkhäuse, 1988

[76] Kohda T, Murao K. Piecewise polynomial Galerkin approximation to invariant densities of one-dimensional difference equations. Electro Communi Japan, 1982, 65A(6): 1–11

[77] Komornik J, Lasota A. Asymptotic decomposition of Markov operators. Bull Polon Acad Sci Math, 1987, 35: 321–327

[78] Koopman B O. Hamiltonian systems and transformations in Hilbert spaces. Proc Nat Acad Sci USA, 1931, 17: 315–318

[79] Krengel U. Ergodic Theorems. Berlin: Walter de Gruyter, 1985

[80] Krzyzewski K, Szlenk W. On invariant measures for expanding differentiable mappings. Studia Math, 1969, 33: 83–92

[81] Lasota A, Li T Y, Yorke J A. Asymptotic periodicity of the iterates of Markov operators. Trans Amer Math Soc, 1984, 286: 751–764

[82] Lasota A, Mackey M. Chaos, Fractals, and Noise: Stochastic Aspects of Dynamics. New York: Springer-Verlag, 1994

[83] Lasota A, Yorke J A. On the existence of invariant measures for piecewise monotonic transformations. Trans Amer Math Soc, 1973, 186: 481–488

[84] Lasota A, Yorke J A. Exact dynamical systems and the Frobenius-Perron operator. Trans Amer Math Soc, 1982, 273: 375–384

[85] Lax P. The ergodic character of sequence of pedal triangles. Amer Math Monthly, 1990, 97: 377–381

[86] Li T Y. Finite approximation for the Frobenius-Perron operator, a solution to Ulam's conjecture. J Approx Th, 1976, 17: 177–186

[87] Li T Y. Ergodic Theory on $[0, 1]$. Lecture Notes at Kyoto University, Japan, 1987

[88] Li T Y, Yorke J A. Period three implies chaos. Amer Math Monthly, 1975, 82: 985–992

[89] Li T Y, Yorke J A. Ergodic maps on $[0, 1]$ and nonlinear pseudorandom number generators. Nonlin Anal, 1978, 2(2): 473–481

[90] Li T Y, Yorke J A. Ergodic transformations from an interval into itself. Trans Amer Math Soc, 1978, 235: 183–192

[91] Lin M. Mixing for Markov operators. Z Wahrscheinlichkeitstheorie Verw. Gebiete, 1971, 19: 231–242

[92] Liverani C. Decay of correlations for piecewise expanding maps. J Stat Phys, 1995, 78(3/4): 1111–1129

[93] Liverani C. Rigorous numerical investigation of the statistical properties of piecewise expanding maps. A feasibility study. Nonlinearity, 2001, 14: 463–490

[94] Lorenz E N. Deterministic nonperiodic flow. J Atmos Sci. 1963, 20: 130–141

[95] Loskot K, Rudnicki R. Relative entropy and stability of stochastic semigroups. Ann Polon Math, 1991, 53: 139–145

[96] Losson J, Mackey M. Evolution of probability densities in stochastic coupled map lattices. Phys Rev E, 1995, 52(2): 1403–1417

[97] Mané R. Ergodic Theory of Differentiable Dynamical Systems. Berlin: Springer-Verlag, 1987

[98] May R M. Simple mathematical models with very complicated dynamics. Nature, 1976, 261: 459–467

[99] Mead L R, Papanicolaou N. Maximum entropy in the problem of moments. J Math Phys, 1984, 25: 2404–2417

[100] Meyn S P, Tweedie R L. Markov Chains and Stochastic Stability. London Springer-Verlag, 1993

[101] Miller W M. Stability and approximation of invariant measures for a class of nonexpanding transformations. Nonlinear. Anal 1994, 23(8): 1013–1025

[102] Miller W M. Discrete approximation of invariant measures for multidimensional maps. DIMACS Ser Disc Math Theoret Compt Sci, 1997, 34: 29–46

[103] Minc H. Nonnegative Matrices. New York: Wiley, 1988

[104] Misiurewicz M. Absolutely continuous invariant measures for certain maps of an interval. Publ Math IHES, 1981, 53: 17–51

[105] Jr. Morrey C B. Multiple Integrals in the Calculus of Variation. New York: Springer-Verlag, 1966

[106] Murray R. Discrete Approximation of Invariant Densities. Ph.D. thesis, Cambridge: Cambridge University, 1997

[107] Murray R. Approximation error for invariant density calculations. Disc Cont Dynam Sys. 1998, 4(3): 535–557

[108] Murray R. Existence, mixing and approximation of invariant densities for expanding maps on $\mathbb{R}^r$. Nonlinear Anal, 2001, 45(1): 37–72

[109] Natanson I P. Theory of Functions of a Real Variable. New York: F Ungar Pub Co., 1955

[110] Rényi A. Representation for real numbers and their ergodic properties. Acta Math Acad Sci Hungar, 1957, 8: 477–493

[111] Rohlin V A. Exact endomorphisms of Lebesgue spaces. Amer Math Soc Tranl, 1964, 2: 1–36

[112] Roydon H L. Real Analysis. New York: Macmillan, 1968

[113] Ruelle D. Statistical mechanics of a one-dimensional lattice gas. Commun Math Phys, 1968, 9: 267–278

[114] Schaefer H H. Banach Lattices and Positive Operators. New York-Heldelberg: Springer-Verlag, 1974

[115] Schütte C. Conformational Dynamics: Modeling, Theory, Algorithm, and Application to Biomoleculaes. Habilitation Thesis, Berlin: Freie Universität, 1999

[116] Schütte C. Huisinga W, Deuflhard P. Transfer operator approach to conformational dynamics in biomolecular systems, Ergodic Theory, Analysis, and Efficient Simulation of Dynamical Systems. Berlin: Springer-Verlag, 2001, 191–223

[117] Setti G, Mazzini G, Rovatti R, Callegari S. Statistical modeling of discrete-time chaotic processes-Basic finite-dimensional tools and applications. Proc IEEE, 2002, 90: 662–690

[118] Shannon C. A mathematical theory of communication. Bell Syst Tech J, 1948, 27: 379–423

[119] Stark J, Bressloff P. Iterative function systems and their applications. New York: Oxford University Press, 1993

[120] Ulam S M. A Collection of Mathematical Problems, vol. 8. New York-London: Interscience, 1960

[121] Ulam S M, von Neumann J. On combination of stochastic and deterministic processes. Bull Amer Math Soc, 1947, 53: 1120

[122] Wyk M A, Steeb W H. Chaos in Electronics, vol 2. Dordreeht: Kluwer Academic Publishers, 1977

[123] Walters P. An Introduction to Ergodic Theory. New York-Berlin: Springer-Verlag, 1982

[124] Wong S. Some metric properties of piecewise monotonic mappings of the unit interval. Trans Amer Math Soc, 1978, 246: 493–500

[125] Xu J, Zikatanov L. A monotone finite element scheme for convection-diffusion equations. Math Comput, 1999, 68: 1429–1446

[126] Ziemer W P. Weakly Differentiable Functions. Springer-Verlag, 1989

Index

GPSR Compliance

The European Union's (EU) General Product Safety Regulation (GPSR) is a set of rules that requires consumer products to be safe and our obligations to ensure this.

If you have any concerns about our products, you can contact us on ProductSafety@springernature.com

In case Publisher is established outside the EU, the EU authorized representative is:

Springer Nature Customer Service Center GmbH
Europaplatz 3
69115 Heidelberg, Germany

Batch number: 10584448

Printed by Printforce, the Netherlands